THINKING THROUGH STATISTICS

Thinking Through Statistics

John Levi Martin

The University of Chicago Press   Chicago and London

The University of Chicago Press, Chicago 60637
The University of Chicago Press, Ltd., London

Published 2018
Printed in the United States of America

27 26 25 24 23 22 21 20 19 18   1 2 3 4 5

ISBN-13: 978-0-226-56746-4 (cloth)
ISBN-13: 978-0-226-56763-1 (paper)
ISBN-13: 978-0-226-56777-8 (e-book)
DOI: https://doi.org/10.7208/chicago/9780226567778.001.0001

Library of Congress Cataloguing-in-Publication Data
Names: Martin, John Levi, 1964– author.
Title: Thinking through statistics / John Levi Martin.
Description: Chicago ; London : The University of Chicago Press, 2018. | Includes bibliographical references and index.
Identifiers: LCCN 2017053885 | ISBN 9780226567464 (cloth : alk. paper) | ISBN 9780226567631 (pbk. : alk. paper) | ISBN 9780226567778 (e-book)
Subjects: LCSH: Statistics—Methodology. | Social sciences—Methodology.
Classification: LCC HA29 .M135 2018 | DDC 001.4/22—dc23
LC record available at https://lccn.loc.gov/2017053885

♾ This paper meets the requirements of ANSI/NISO Z39.48-1992 (Permanence of Paper).

Contents

Preface

Why another book on statistics, and from a non-statistician?

The answer is because—and I mean this in a good way—if you are a practicing sociologist, statisticians are not your friends.

They're nice enough people, and statistics is a wonderful field, but the problems that *they* solve are not *your* problems. And in fact, we'll see over the course of this book, that a lot of the time, the solutions they propose to *their* problems are going to make *your* problems *worse*.

Why? Simple. Statisticians have the job of determining the problems that come from the tasks of estimation and inference to a population, and solving them if possible. They can guarantee you the best answers, *if* you already know the *right* model.

But in sociology (and in most of its kin), we never already know the right model—we never have full knowledge about what is going on out there in the world. I'd love it if some day we did, and all we need to do was to pin down our numerical estimates. But we use statistical data analysis to try to tell us which model to believe in. That leads to different problems, and those are the problems I'm going to deal with in this book.

Whom This Book Is For

You are a social scientist—a sociologist, a political scientist, a public health researcher, an applied economist—and you want to learn from formal data analysis. You've taken at least one statistics class, and you are comfortable with the idea of multiple regression. You can read an equation, but don't necessarily deal with matrices a lot. You might also be interested in these new cool things you've seen and heard about, like network analysis or spatial statistics. But you aren't necessarily looking to quantify a causal effect of one variable on another such that you can predict an intervention. There are plenty of books to help you with that task. The problem is that most of our actual problems can't be squeezed into forms of that task, and trying doesn't help us learn anything.

If you are a good methodologist, there are going to be a number of times when I start patiently walking down a road, setting up a demonstration, and you're going to know where we're going. You're going to be impatient, and think, "but that's *obvious*! Everyone knows that!" But trust me, they don't. Even if you once taught it to your own students in class, there's a good chance that it's no longer with them.

A Note on Sources

I'm going to tell true stories here. That is, I'm going to give examples of prominent work done by people in my field. And sometimes I'm going to say that the work is wrong. Not wrong as in "I would do it differently" or wrong as in "That's not my perspective" or wrong as in "There are better classes of assumptions to make." I mean wrong. If *I'm* wrong, I'm going to look like an idiot. And chances are good I'll be wrong *somewhere* here. This sort of unreserved critique—one of us messed up—is not something that we feel good about in sociology.

I used to share that feeling. But my thinking changed as I saw good work being rejected by reviewers who didn't understand it, and good people being unemployed. I realized this sort of conflict aversion leads our discipline to reward hasty, bad work. This really hit home when I was teaching Bosk's *Forgive and Remember* (discussed in *Thinking Through Methods*; hence, *TTM*). Every year, I grew increasingly outraged at the parts where the high-ranking doctors disclose that if they learned that a colleague was incompetent and harming patients, they would . . . do absolutely nothing. "Why if I were they," I would think . . . and then, one year, it hit me. I *was* they. I knew of other sociologists who were publishing erroneous claims and I said nothing, because to do so seemed "mean" or "uncool" or "hierarchical" (if they were younger or less well placed).

That attitude, which I once had, confuses professional ethics with personal prudence. Let me be more blunt: it is more than prudence. Academics are among the most cowardly people with whom I have ever had to interact. If there is anyone who isn't a coward, she will be torn down by the others as being "aggressive." Ugh.

We also have a disciplinary habit, I have found, of using the idea of "complexity" to ignore evidence. I really hope that we do continue to oppose the degradation of scientific debate to 255 character newspeak "critweets" ("Martin book doubleplusungood!"). But we can use our faith that things are "complex" to abrogate our responsibilities. Faced with a strong critique, it's easy for us to decide to leave everything as it is, by thinking to ourselves, "well, I'm sure it's more complicated than that." But sometimes the truth *isn't* a little bit here and a little bit there. You may need to think about some of this yourself. And you might need to take a stand.

At a certain point, there's a fork in the road. Either you believe in what you are doing, that the social sciences deserve support because they're a real field of serious study, or they're just a joke, or a form of entertainment, or a sinecure. If you go with the former—and I do—then it isn't okay to let your field give the seal of approval to things that are wrong. We all need to work together to understand that we aren't trying to tear individuals down, we're trying to build our field *up*. You aren't a bad person for finding and publicizing errors in others' work, and just because you've made an error doesn't mean you can't do many other great things, before and afterwards.

Of course, I completely understand that by doing this, I'm really asking people to go over *my* work with a fine-toothed comb. Since I've started this, I've been good at saving intermediate data files and all my programs. If people find embarrassing and incompetent errors, then I suppose at least I can say that this supports my general point, if not my own career—our practice is one that allows for bad work to drive out good.

One last thing: one of the fastest ways of improving social research is for leading journals to refuse to publish any papers that do not make the data public. That doesn't mean the *whole* data set, but it does mean everything needed to replicate the analyses. It's time.

A Note on Notation

I try to be consistent across chapters with notation; thus while I tend to rely on the conventions of my source, I may adopt for clarity. In general, a random variable is denoted in italics (note that a "random" variable simply means one that can take on any value in a distribution, not that it is inherently stochastic), but so is a constant, which is usually lower case. When I have an independent variable, a dependent variable, and a control, these will be x, y, and z respectively. I denote vectors as bold lower case type, matrices in bold upper case type, though where the nature of these as random variables (and not compositions of elements) is emphasized, they may instead be set as italics. Where I am discussing sets and elements of sets, both are in italics, with elements lower case and sets upper case. Where it does not cause confusion, I may use $\mathbf{X}$ to represent a large set of independent variables (and not the full data matrix).

When I am discussing the construction of data sets with error, I will use ε to denote this error term, and the distribution of this error will be described either as $N(m, sd^2)$, where this indicates a normal distribution with mean *m* and standard deviation *sd*, or *U*(*min, max*), a uniform distribution running between *min* and *max*. For simplicity, I generally notate coefficients as *b*, not distinguishing between sample and population parameters. Model intercepts (constants) are denoted *c*.

For most data structures, the index *i* will refer to the individual (out of *N* observations), *j* to a context (out of *J* contexts), though *j* may also refer to an alter (of *J* alters) for dyadic data, and *k* to the variable (out of *K* total). Where a data matrix has a number of columns not equal to the number of variables, I will use *M* to describe the number of columns. Thus a conventional data matrix will be given as $\mathbf{X}_{N \times M}$ (since some columns may be, say, interactions of variables). Finally, arbitrary actors are A, B, and C. I also italicize important terms upon first use.

Usage

Two issues of usage: First, I believe that the word *data* is making the shift from a plural (the plural of *datum*) to an "uncountable" noun (like *rice*). This is of course a bit ironic, because it's for reasons of counting that we talk about data in the first place. But it's not all the way there, and when the employment of the word emphasizes plurality and differentiation, I treat it as plural, and when not, I treat it as singular, so that it sounds harmonious to contemporary ears. Similarly, when I use "statistics" to mean data, I interpret it as plural, and when I mean the accumulated wisdom of the field of investigation, I interpret it as singular.

Second, I found that I often use the word "defensible" to mean "doing this is or was okay." That is, I'm not saying "correct" or "valid." The reason is that I think that we have to accept that there are going to be many places where reasonable people differ . . . where you can make very different analytic choices. A *defensible* approach is one that, if we all were to gather around, and really look closely at the sequence of choices that it involved, we wouldn't find something that was flagrantly incompetent. It might not be the way you would do it, but you can see that there is a logic to it . . . and no weak spots.

That might seem like a remarkably low bar. It isn't. An analytic approach usually involves dozens of such decisions, and an approach is only as strong as its weakest link. Much of our work involves making one link—the one we're all used to arguing about—very strong . . . and allowing others to be hilariously weak. If all our work rose to the standards of "defensible," we'd be sitting pretty.[1]

1. And "defensibility" is basically the same thing that John Dewey meant by "warranted assertibility" . . . which was as close to "true" as he thought we came!

Finally, I give a number in each example for the R code that was used to generate it—thus R 3.1 means the code for the first example in chapter 3. You can find the code on this website: www.press.uchicago.edu/sites/martin. My code is ugly (I learned to program in an already outdated FORTRAN) but it is clear enough for anyone to follow. Feel free to improve it and I'll post yours.

Plan of the Whole

The first chapter sketches our problem—and why existing statistics books are unlikely to help us. We have to figure out how to learn from data when there is no "true" or "best" model to fit, and when our parameters don't actually correspond to any real world processes. I propose a way of thinking about the role that statistical analysis can play in social science, one which fits what we actually do a lot better than most contemporary theories. We are trying to learn from data—estimating parameters is a means to that end, and not an end in itself. In chapter 2, I emphasize that before we do any computational work, we need to understand the contours of our data. I show some of the most common issues that can arise when we don't.

In chapter 3, I quickly go over the idea of causality so as to highlight conventional ways of thinking about selectivity—the ways in which observational data *fails* the causal model—that I think are very useful whether or not one is taking a causal approach. That's because our real problem has to do with important but unobserved variables. Thinking in terms of selectivity can help us identify likely problems to our analyses. But most social scientists aren't going to work hard isolating a causal estimate; instead they're going to rely on "control" strategies and in chapter 4, I discuss how we can do those better or worse.

I then go on to discuss somewhat more complicated data structures. In chapter 5, I start with the issue of *variance*, emphasizing that we need to have a sense for where the important variance is in our data. With this we turn to "nested" data, that with variance at different levels, and see how we can make sure that our analyses match up with our theoretical claims. In chapter 6, I turn to the problems that arise when we try to compare units that have some different "risk" of producing an observation. The most important case here involves aggregates (like cities) that produce some count variables (like "number of churches"). We have been adopting rules-of-thumb that are as likely to introduce spurious findings as to control for them.

I then examine data structures in which our observations have a common embedding that can contain information about unobserved predictors. These embeddings are time, geographical space (these two discussed in chapter 7), and social space (or social networks, discussed in chapter 8). I show that it is very easy to produce false findings by assuming that we have "neutralized" the effect of these embeddings (like, "we did a fixed effects model"), when in fact, our cases still have correlations due to unobserved predictors associated with these embeddings.

In chapter 9, I then turn to three examples of "too good to be true" analytic approaches—those that tend to generate too many false findings, because they have lower (or no) bars for rejection of claims. I deal with latent class and mixture models, simulations, and Qualitative Comparative Analysis. A conclusion draws the threads together and deals with issues of ethics.

Acknowledgments

First, I'd like to thank Ken Frank and Tom Dietz: as readers of this manuscript they pushed me in a number of directions; it's been great to be in dialogue with them. Another anonymous reviewer also made important corrections and suggestions. At the press, I've been fortunate to work with Kyle Adam Wagner, Levi Stahl, Mary Corrado, Joan Davies, Matt Avery and, of course, Douglas Mitchell; I am grateful for their contributions and tolerant good humor. Joe Martin made the maps of New Jersey in chapter 7; thanks, Joe! Finally, I sent the portions of the manuscript, often critical, that used real world examples to the authors of those examples. In many cases they pointed out problems with my analysis: usually mistaken assumptions about what they had done, exaggerations, or other errors. I am deeply grateful for their corrections, and have learned a great deal from them. Matt Salganik read chapter 8, and while I doubt he agrees with everything said, he was an outstanding interlocutor.

I'd also like to acknowledge a set of a number of brilliant and highly ethical methodologists, whom I've been fortunate enough to be influenced by, and sometimes to study with. First I want to note that it was Mike Hout who got me excited about statistics, when I considered myself one of the anti-statistical types. He showed me—and continues to show me—that if you want to learn about the social world then, by gum, numbers often work really well. Second, it was Leo Goodman's work that really has been my guiding way of thinking about statistics; I've been truly blessed to have him as a teacher and a colleague.

Third, there is Ron Breiger. Breiger has really pioneered the approach to data analysis that I'd like to put forward here. If you know Breiger's work in mathematical sociology, especially with the equally astounding Philippa Pattison, you know that this is a mind capable of the most abstruse complexity when theoretically necessary. But what he did was to try to find the *basis* of our techniques, so that we could understand what we know, and what we can know, in any particular case. And he has tied this to a deep understanding of the nature of regularity in the universe. I've been privileged to see some of his work—sometimes sketched on a napkin—before it came out, and that's helped me adapt my thoughts while I still could.

Fourth, there is one of my best teachers, Adam Slez. Before I met Adam, in my mind, I was the somewhat idealistic believer in rigor, forcing students to shake their findings, try alternate approaches, and so on, though every now and then I'd secretly start to wonder whether it was all worth it. Now if I ever have a doubt about what is the right thing to do, I ask myself, "What would Adam do?" The idea of disappointing him by cutting corners is a greater motivation than anything else.

Fifth, there is Herbert Hyman. I was fortunate enough, as an undergraduate, to participate in a data analysis seminar with Hyman. I didn't appreciate it at the time. It was only when I read his 1954 book on interviewing as I was finishing up *Thinking Through Methods* that I was struck by the fact that his thinking therein was far ahead of our current understanding of the social psychology of interviewing. I wish I could tell him how impressed I am. May his memory be for a blessing.

Sixth, I want to recognize someone whose influence has been very important for me, though I never knew the man. This is Otis Dudley Duncan. Now I continue to side with Goodman on everything except insofar as he wielded the sword of statistical inference against the mathematical sociology of Harrison White (in a nearly forgotten debate over models for group size that I'm going to talk about later). And I can't speak to Duncan's character as a person the way I can about Goodman, who has been a model to me as a father as well as a methodologist.

But Duncan had set his sights on a *serious* social science, which meant one that made the achievements of people obsolete, the faster the better. He introduced methods, important and brilliant ones, and when he thought we now had better ones, he told people to drop the ones associated with his name. One of the things I love about Duncan I think for him was actually associated with a bit of depression and resignation—his recognition of the limits of what he had accomplished. His (1984a) *Notes on Social Measurement* is a brilliant self-critique and he was always looking for ways of being better and more rigorous. And it was his enthusiasm for psychometric approaches that had energized Hout and from Hout to me. Let's toast his memory.

Seventh, there is Stanley Lieberson. Lieberson tried to push us to be more serious about what we did, and never to paper over *conceptual* weaknesses with irrelevant mathematics. He called people out on their b******t when he had to, but he also modeled good analytic behavior. This book is intended as a contribution to what he began. *Making It Count* (1985) is still required reading for everyone.

But most of all, Jim Wiley was my mentor and collaborator on a number of exciting projects, many of which we never published. He loves math, and loves good problems—but he also once refused to look at some cool complicated statistics I had developed for him for work on the Young Men's Health Study. He just wanted to see tables, lots of them. "This is public health, John. When we're wrong, people die." That sober lesson stuck with me. As did much else as well. I dedicate this work to Jim Wiley, mentor, collaborator, and friend. He has combined everything that I think is valuable in a scientist who cares about the world, his craft, and those around him. This book tries to codify core principles of learning from data that I first learned from him. If we were all Jim Wileys, what a wonderful world this would be.

* 1 *
Introduction

Map: I'm going to start by identifying a meta-problem: most of what we learn in "statistics" class doesn't solve our actual problems, which have to do with the fact that we don't know what the true model is—not that we don't know how best to fit it. This book can help with that—but first, we need to understand how we can use statistics to learn about the social world. I will draw on pragmatism—and falsificationism—to sketch out what I think is the most plausible justification for statistical practice.

Statistics and the Social Sciences

What Is Wrong with Statistics

Most of statistics is irrelevant for us. What we need are methods to help us adjudicate between substantively different claims about the world. In a very few cases, refining the estimates from one model, or from one class of models, is relevant to that undertaking. In most cases, it isn't. Here's an analogy: there's a lot of criticism of medical science for using up a lot of resources (and a lot of monkeys and rabbits) trying to do something we know it can't do—make us live forever. Why do researchers concentrate their attention on this impossible project, when there are so many more substantively important ones? I don't deny that this might be where the money is, but still, there are all sorts of interesting biochemical questions in how you keep a ninety-nine-year-old millionaire spry. But if you look worldwide, and not only where the "effective demand" is, you note that the major medical problems, in contrast, are simple. They're things like nutrition, exercise, environmental hazards, things we've known about for years. But those things, simple though they are, are *difficult* to solve in practice. It's a lot more fun to concentrate on complex problems for which we can imagine a magic bullet.

So too with statistical work. Almost all of the discipline of statistics is about getting the absolutely best estimates of parameters from *true* models (which I'll call "bestimates"). Statisticians will always admit that they consider their job only this—to figure out how to estimate parameters given that we already know the most important things about the world, namely the model we should be using. (Yes, there is also work on model selection that I'll get to later, and work on diagnostics for having the wrong model that I won't be able to discuss.) Unfortunately, usually, if we knew the right model, we wouldn't bother doing the statistics. The problem that we have isn't getting the bestimates of parameters from true models, it's about not having model results mislead us. Because what we need to do is to propose ideas about the social world, and then have the world be able to tell us that we're wrong . . . and having it do this more often when we *are* wrong than when we aren't.

How do we do this? At a few points in this book, I'll use a metaphor of carpentry. To get truth from data is a craft, and you need to learn your craft. And one part of this is knowing when *not* to get fancy. If you were writing a book on how to make a chair, you wouldn't tell someone to start right in after sawing up pieces of wood with 280 grit, extra fine, sandpaper. You'd tell them to first use a rasp, then 80 grit, then 120, then 180, then 220, and so on. But most of our statistics books are pushing you right to the 280. If you've got your piece in that kind of shape, be my guest. But if you're staring at a pile of lumber, read on.

Many readers will object that it simply isn't true that statisticians always assume that you have *the* right model. In fact, much of the excitement right now involves adopting methods for classes of models, some of which don't even require that the true model be in the set you are examining (Burnham and Anderson 2004: 276). These approaches can be used to select a best model from a set, or to come up with a better estimate of a parameter *across* models, or to get a better estimate of parameter uncertainty given our model uncertainty. In sociology, this is going to be associated with Bayesian statistics, although there are also related information-theoretic approaches. The Bayesian notion starts from the idea that we are thinking about a range of models, and attempting to compare a posteriori to a priori probability distributions—before and after we look at the data.

Like almost everyone else, I've been enthusiastic about this work (take a look at Raftery 1985; Western 1996). But we have to bear in mind that even with these criteria, we are only looking at a teeny fraction of all possible models. (There are some Bayesian statistics that don't require a set of models, but those don't solve the problem I'm discussing here.) When we do model selection or model averaging, we usually have a fixed set of possible variables (closer to the order of 10 than that of 100), and we usually don't even look at all possible combinations of variables. And we usually restrict ourselves to a single family of specifications (link functions and error distributions, in the old GLM [General Linear Models] lingo).

Now I don't in any way mean to lessen the importance of this sort of work. And I think because of the ease of computerization, we're going to see more and more such exhaustive search through families of models. This should, I believe, increasingly be understood as "best practices," and it can be done outside of Bayesian framework to examine the robustness of our methods to other sorts of decisions. (For example, in an awesome paper recently, Frank et al. [2013] compared their preferred model to all possible permutations of all possible collapsings of certain variables to choose the best model.) But it doesn't solve our basic problem, which is not being able to be sure we're somewhere even *close* to the true model.

You might think that even if it doesn't solve our biggest problems, at least it can't *hurt* to have statisticians developing more rigorously defined estimates of model parameters. If we're lucky enough to be close to the true model, then our estimates will be way better, and if they aren't, no harm done. But in fact, it is often—though, happily, not invariably—the case that the approaches that are best for the *perfect* model can be *worse* for the wrong model.

When I was in graduate school, there was a lot of dumping on Ordinary Least Squares (OLS) regression. Almost never was it appropriate, we thought, and so it was, we concluded, the thing that thoughtless people would do and really, the smartest people wouldn't be caught dead within miles from a linear model anyway. We loved to list the assumptions of regression analysis, thereby (we thought) demonstrating how implausible it was to believe the results.

I once had two motorcycles. One was a truly drop dead gorgeous, 850 cc parallel twin Norton Commando, the last of the kick start only big British twins, with separate motor, gearbox, and primary chain, and a roar that was like music. The other was a Honda CB 400 T2—boring, straight ahead, what at the time was jokingly called a UJM—a "Universal Japanese Motorcycle." No character whatsoever.

I knew nearly every inch of that Commando—from stripping it down to replace parts, from poring over exploded parts diagrams to figure out what incredibly weird special wrench might be needed to get at some insignificant part. And my wife never worried about the danger of me having a somewhat antiquated motorcycle when we had young children. The worst that happened was that sometimes I'd scrape my knuckles on a particularly stuck nut. Because it basically stayed in the garage, sheltering a pool of oil on the floor, while I worked on it.

The Honda, on the other hand, was very boring. You just pressed a button, it started, you put it in gear, it went forward, until you got where you were going and turned it off.[1] If I needed to make an impression, I'd fire up the Norton. But if I needed to be somewhere "right now," I'd jump on the Honda. OLS regression is like that UJM. Easy to scorn, hard to appreciate—until you really need something to get done.

Proof by Anecdote

I find motorcycle metaphors pretty convincing. But if you don't, here's a simple example from some actual data, coming from the American National Election Study (ANES) from 1976. Let's say that you were interested in consciousness raising at around this time, and you're wondering whether the parties' different stances on women's issues made some sort of difference in voting behavior, so you look at congressional voting as a simple dichotomy, with 1 = Republican and 0 = Democrat. You're interested in the gender difference primarily, but with the idea that education may also make a difference. So you start out with a normal OLS regression. And we get what is in table 1.1 as model 1 (the R code is R1.1).

Gender isn't significant—there goes *that* theory—but education is. That's a finding! You write up a nice paper for submission, and show it to a statistician friend, very interested that those with more education are more likely to vote Republican. He smiles, and says that makes a lot of sense given that education would make people better able to understand the economic issues at hand (I think *he* is a Republican), but he tells you that you have made a major error. Your dependent variable is a dichotomy, and so you have run the wrong model. You need to instead use a logistic regression. He gives you the manual.

You go back, and re-run it as model 2. You know enough not to make the mistake of thinking that your coefficients from model 1 and model 2 are directly comparable. You note, to your satisfaction, that your basic findings are the same: the gender coefficient is around half its standard error, and the education coefficient around four times *its* standard error. So you add this to your paper, and go show it to an even more sophisticated statistician friend, and he says that your results make a lot of sense (I think he too is a Republican) but that you've made a methodological error. Actually, your cases are not statistically independent. ANES samples congressional districts,[2] and persons in the same congressional

1. In fact, it was so boring that I attached deer antlers to the front for a while, until a helpful highway patrolman told me that if I ever was in an accident, and an antler actually impaled a pedestrian, unlikely though that was, I would certainly get the electric chair, and then go straight to hell.

2. Actually in the 1976 ANES, the sampling units were not congressional districts but

Table 1.1. Proof by Anecdote

	Model 1	*Model 2*	*Model 3*	*Model 4*	*Model 5*
Type of Model	OLS	LOGISTIC	HGLM	OLS	HGLM
GENDER	−.017	−.071	−.011	−.039	−.107
	(.031)	(.130)	(.152)	(.033)	(.164)
EDUCATION	−.083***	−.348***	−.383***	−.029	−.166
	(.018)	(.078)	(.093)	(.022)	(.108)
CONSTANT	.620	.497	.734	.633	.808
RANDOM EFFECTS VARIANCE			1.588		1.552
SECRET				−.081***	−.352**
				(.021)	(.111)
R^2	.021			.028	
AIC		1374.9	1264.5		1112.6

$N = 1088$; Number of districts = 117; *** $p < .001$; ** $p < .01$

district have a non-independent chance of getting in the sample. This is especially weighty because that means that they are voting for the same congressperson. "What do I do?" you ask, befuddled. He says, well, a robust standard error model could help with the non-independence of observations, but the "common congressperson" issue suggests that the best way is to add a random intercept at the district level.

So you take a minicourse on mixed models, and finally, you are able to fit what is in model 3: a hierarchical generalized linear model (HGLM). Your statistician friend (some friend!) was right—your coefficient for education has changed a little bit, and now its standard error is a bit bigger. But your results are all good! Pretty robust! But then you show it to me. It doesn't make sense to me that education would increase Republican vote by making people smarter (strike one) or by helping you understand economic issues (strike two). I tell you that I bet the problem is that educated people tend to be *richer*, not *smarter*. Your problem is a misspecification one, not a statistical one.

I have the data and quickly run an OLS and toss income in the mix

looped across them; in 1978 they were congressional districts. But your statistician friend is a bit foggy on these details. . . .

(model 4). The row marked "SECRET" is the income measure (I didn't want you to guess where this is going—but you probably did anyway). Oh no! Now your education coefficient has been reduced to a *thirteenth* of its original size! It really looks like it's *income*, and *not* education, that predicts voting. Your paper may need to go right in the trash. "Hold on!" you think. "Steady now. None of these numbers are *right*. I need to run a binary logistic HGLM model instead! That might be the ticket and save my finding!" So you do model 5. And it basically tells you the exact same thing.

At this point, you are seriously thinking about murdering your various statistician friends. But it's not their fault. They did *their* jobs. But never send in a statistician to do a sociologist's job. They're only able to help you get bestimates of the *right* parameters. But you don't know what they are. The lesson—I know you get it, but it needs to stick—is that it rarely makes any sense to spend a lot of time worrying about the bells and whistles, being like the fool mentioned by Denis Diderot, who was afraid of pissing in the ocean because he didn't want to contribute to drowning someone. Worry about the omitted variables. That's what's really drowning you.

So moving away from OLS might be important for you, but in most cases, that isn't your problem. Indeed, OLS turns out to be pretty robust to violations of its assumptions. Sure, it doesn't give you the *best* estimates, but it doesn't go bonkers when you have restricted count data, even (often) just 0s and 1s. Further, and more important, it has a close relation to some model-independent characteristics of the data. You can interpret a "slope" coefficient as some sort of estimate of a causal effect, if you really want to . . . or you can see it as a re-scaled partial correlation coefficient. And those descriptive interpretations can come in handy. Most methodologists these days are going to tell you to work closer and closer to a behavioral model. And I'm going to say that's one half of the story. Just like some politicians will say, "Work toward peace, prepare for war," I will say, "Work toward models, but prepare for description." And so I'm going to take a moment to lay out the theory of the use of data that guides the current work. But first, a little terminology.

Models, Measures, and Description

We're often a bit casual in talking about "models," "measures" and so on—statisticians aren't, and I think we should follow them here. A model is a statement about the real world that has testable implications: it can be a set of statements about *independencies*—certain variables can be treated as having no intrinsic association in the population—as in Leo Goodman's loglinear system. Or it can be a statement about mechanisms or processes—either causal pathways, or behavioral patterns.

Models usually have parameters in them. If the model is correct, these parameters *may* have a "real worldly" interpretation. For example, one of them might indicate the probability that persons of a certain type will do a certain type of thing. They might indicate the "elasticity" of an exchange between two types of resource. But they don't always need to have a direct real-world analogue. And our *estimates* of these parameters can be useful even when they aren't really very interpretable. In many cases, we use as a rule-of-thumb the statistical test of whether a parameter is likely non-zero in the population as a way of constraining the stories we tell about data. This approach has come in for a lot of hard knocks recently, perhaps deservedly, but I'll defend it below. For now, the point is that not all parameters have to be real-world interpretable for us to do something with them.

Let's go on and make a distinction between the *estimate* of a real-world parameter (should there be any) and *measures* that (as said in *TTM*) I'll use to refer to a process whereby we interact with the units of measurement (individually and singly, one might say) and walk away with information. Finally, when possible, we'll try to distinguish model parameters (and measurements) from *descriptive statistics*. While this distinction may get fuzzy in a few cases, the key is that descriptions are ways of summarizing information in a set of data that are *model-independent*. No matter what is going on in the world, a mean is a mean.[3] Not to be intentionally confusing, but what a mean *means* stays the same. In contrast, a parameter in a complex model (like a structural equation measurement model) has no meaning if the model is seriously off.

That's the nature of good description. What you do is figure out ways of summarizing your data to simplify it, give you a handle on it, and the best descriptive methods structure the data to help you understand some especially useful aspects, ones that have to do with the nature of your data and the nature of your questions, but still without making use of particular assumptions about the world. When it comes to continuous data expressed in a correlation matrix (which is what OLS among others deals with), conventional factor analysis is a classic descriptive approach.

3. If you're a sophist, you'll immediately try to come up with reasons why this isn't so, and you'll be wrong. If you have a single distribution of a continuous variable, with N observations, there are N moments to that distribution. The mean is the first, the standard deviation the second, and so on. If you use all of them, you recreate the whole distribution. The first two moments adequately characterize *some* distributions; for a wider family, you need the first four, but every distribution can be characterized with all of them. The best that you can do, oh beloved sophist, is to argue that there is a continuum, with some quantities useful for practically any model, and others useful for very few. I'll accept that, and it doesn't undermine my point.

Now when I was in graduate school, factor analysis was the one thing we despised even more than OLS. It was, as Mike Hout called it, "voodoo"—magic, in contrast to a theoretically informed model. That's because something always came out, and, as I'll argue in chapter 9, when something "always comes out" it's usually very bad. But in contrast to other techniques that always give an interpretable result, factor analysis turns out to be pretty robust. Is it perfect? Of course not. Should you accept it as a model of the data? Of course not—because it isn't a *model*. It's a description, a reduction of the data. And chances are, it's going to point out something about the nature of the data you have.

Now as Duncan (1984b, c) emphasized, knowing a correlation matrix doesn't always help you. The pattern of covariation in a set of data, he said, is far from a reliable guide to the structural parameters that should explain the data. But most of our methods are based on the same fundamental mathematical technique of singular value decomposition. This is a way of breaking up a data matrix into row and column spaces. As Breiger and Melamed (2014) have demonstrated, most of our techniques basically take these results and rescale them (like correspondence analysis) or project them (like regression). So our common linear model, rather than being an *alternative* to description, can be understood as a particular *projection* of the description for certain analytic purposes.

Many methodologists think that the best approach is one that precisely translates a set of behavioral assumptions into a model with parameters that quantify the linkages in the model. To them, the fact that OLS is close to description shows how primitive it is. I'm going to argue that we're best off attempting to use our data to eliminate theories using models that are actually as close as possible to description. So it's time I laid out the approach to statistics that guides this work.

What Is, What Are, and What Should Be, Statistics?

It seems that there isn't real agreement as to where the word "statistics" comes from, and it currently is ambiguous. We use it both to mean the raw materials—the numbers—that we analyze, as well as the set of tools that we use to analyze them. It seems pretty certain that the word first referred only to the former; it is also pretty certain that it comes from the root of words for "state," but it shares the ambiguity of that word itself, for it appears that it originally meant "those numbers that tell us the state of the state," that is, the condition of the government (see Pearson 1978 [1921–1933]; Stigler 1986).

But what we think of as statistics as a field of applied mathematics came from work that demonstrated that a finite, indeed, rather small, sample could be used to estimate (1) the population value of some numerical characterization (such as average height); (2) the population variance; (3) the likely error of each of these estimates. That's still the heart of statistics—what we call the "central limit theorem." It isn't the limit that's "central," it's the theorem. It's the basis for what we think of as statistics. This enterprise of statistics, then, is all about inference to populations from samples.

This emphasis on inference continued as statistics began to move away from descriptions (like a mean or a correlation coefficient) to what were increasingly interpreted as models (like a set of slope coefficients). But this opened up a new form of error. There are three ways that a slope coefficient can be wrong. The first is that we've made a mistake of *calculation*. These days, this usually means a programming error. The second is that while our calculation is correct, the value in the sample isn't the same as that in the population, and it's the latter that we're trying to get at. If we had a complete sample, we'd get the right value. This, then, is a mistake of *inference*. The third is that there's nothing wrong with our calculation and inference, but our model is wrong. We have a mistake of *interrogation*—it's not that we have the wrong answer, but that we asked the world the wrong question. While we can make errors of calculation or of inference for descriptive statistics, errors of interrogation only arise when we move from description to models.

Our basic problem is that statistics, as generally taught, has relatively little to say about our problems of interrogation. We can't ask for the right estimates until we know what is going on, or at least, what might be going on—and until we actually have measures of these factors. So how can we proceed? We're often told that we should be using our data to test theories. I'm pretty sure that *that* approach hasn't worked out too well. Instead, I think we should start with a rudimentary theory of social science based on pragmatism (especially that of C. S. Peirce and John Dewey). If you aren't interested, feel free to skip to the next section. But I think we can derive a far more coherent theory of the relation of data to knowledge than we currently have, and one that will better help us with our actual practice, than our current orthodoxy. To do this, you need to imagine that there is something you are interested in, something that you don't know about. That's step one. Step two is that you think about the set of possible plausible explanations (this notion was famously laid out by the great geographer Chamberlain 1965 [1890]).

When we assemble this set of possibilities, you don't allow something that you call "theory" to lead you to ignore hypotheses or interpretations that draw some interest from competent social scientists. That is, if by "theory" you mean "what we already know," like one might talk about "plate tectonic theory," then sure, your work should be theoretically informed. But if by "theory" you mean "my presuppositions" or "my claims," then any analysis that assumes it is a waste of all our time. Don't let anyone cow you into thinking that he has some special reason why he gets to ignore what he wants to ignore (we'll see examples of how this leads to bad practice in chapter 9).

This notion that you start not with *your* theory, but with the competing notions of a community of inquiry, doesn't quite fit conventional ways of thinking, though it is compatible with the neo-Chamberlainism of Anderson (2012). The pragmatist conception, however, differs even more fundamentally from our current vision of statistics.[4] In the conventional philosophy of science—one that lies at the bottom of our frequentist interpretations of most of our statistical practice—you start from scratch every time. You have a model, a theory of reality, and you want to test it. That means you have a billion things that you test all at the same time. It means you aren't just asking, "Does income affect vote?," but simultaneously, "Does this measure income? Is the central limit theorem applicable? Do other people have consciousness? Am I perhaps a brain in a vat?" A weakness in any one of the assumptions that you need to make your practice defensible undermines your conclusion.

In the Bayesian version, you try to fold more of your current understandings and beliefs into how you interpret the evidence the world gives you. The pragmatist idea is somewhat different. The consistent Bayesian will allow you to assign a .9 probability to the model that says that the presence of capitalist relations of production leads to greater psychological health than do socialist relations of production if you *really, really, really, really, really, really, really, really, really, really, really, really*, believe it.[5] The pragmatists, in contrast, understood science as the efforts of a *community*, and argued that no one could unilaterally declare something likely or unlikely.

Further, in the pragmatist conception, rather than start from the ground and build up, we start from where we are, and where we are is with what we, as regular people, already think that we know. Maybe you can't prove that any of this is true knowledge. Maybe we are hanging in the air without a foundation, but, still, *that's* where we are. Whether what is in our heads be philosophically justifiable or not, either way, science can

4. You might wonder why, given this pragmatist grounding, I don't focus on solving *practical* problems. I don't want to get side-tracked, but the simple story is that too often, our problems are due to *other people*. . . .

5. I recognize that there are now some non-subjectivist Bayesians, or so they think (with high subjective confidence). But they still accept the notion that a model *has a probability*. That actually isn't an easy notion to explain unless you are a subjectivist or believe that there are a large set of parallel universes constantly splitting from one another.

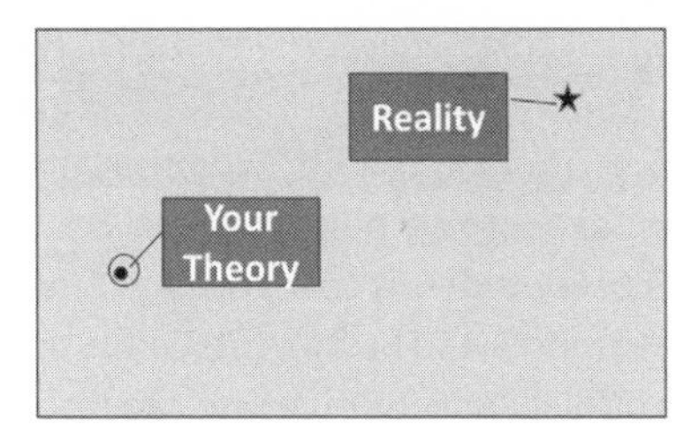

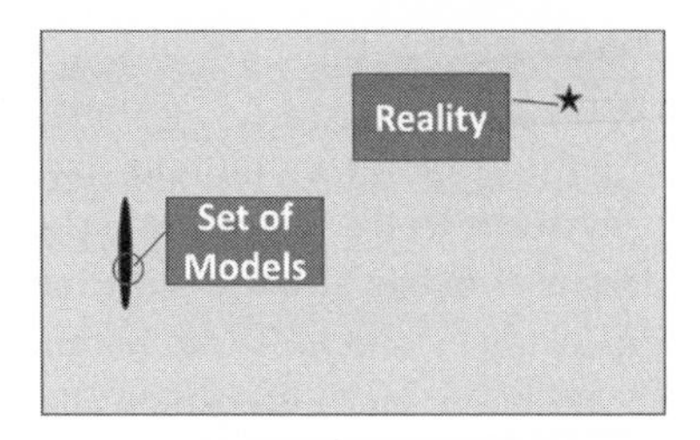

FIGURE 1.1. Conventional and Bayesian Approaches to the Use of Statistics

make it better—if we want it to. Some of our everyday knowledge might be wrong, but if it doesn't matter, let it be. If something matters, and our existing knowledge isn't good enough—that's when we say, "Stand back, I'm going to try science."

That means that if a statistical analysis improves on what we think we know from everyday life, we want to take that evidence seriously . . . even if our quantifications show that there's lots of imprecision. Our goal is to use the power of our data and the rigor of our methods to exert leverage on this process of deciding what interpretation to accept. How should we do that?

I propose that we adopt a strictly *falsificationist* ideal for our practice. This is an old-fashioned idea, and, even worse, associated with a stodgy liberal German philosopher of science, Karl Popper (1959 [1934]). There are lots of good critiques of his theory, and of falsificationism in general. But that isn't determinative—maybe it isn't a good philosophy of science. But it is good practice for *our* field. We should be using statistics not to estimate real world parameters, but to force us away from interpretations of the world that are inconsistent with the body of evidence as it stands.

Of the possible ways of formalizing this connection, falsification is best at working with our current state of statistics. Imagine that the space of all possible explanations of some phenomenon is a plane, like that drawn in figure 1.1, left. Conventional statistics work by us coming up with a single model. We try to estimate the parameters, and the better the estimates (the smaller the radius around the star, impressionistically speaking), the better. But if the truth is far away from where we even are, we don't get much traction on it. We're just tightening a lasso on a steer who's already got clear away.

Now you might respond that conventional methods *are* falsificationist. We're rejecting the null hypothesis, right? Didn't we learn that that's all we can do—falsify models? Well, not quite. Although some loglinear modelers still reject whole models, most of us don't. We might reject one *parameter* in a model, but we don't use that to say, "You know, maybe the *whole idea of regression isn't true* here." You might also say, well, Bayesian methods get us past this "choke chain on nothing" view of statistics. And there's going to be a lot of promise in Bayesian methods, not only in terms of getting better maximum likelihood estimates, but in widening our view of the world. Bayesian methods can take into account a large class of possible models, and use this to choose a best model, or to incorporate our model uncertainty into our parameter estimates. (To be fair, information-theory based methods can do all of this as well, without leaving the Fisherian approach to statistics I'll advocate in chapter 5.)

Bayesian methods, however, also only explore a small fraction of this space. Even in some of the wilder approaches that try to examine multiple link functions as well as selections of variables, they are limited to a relatively small set of variables, and they certainly don't include the "unknown unknowns." (See fig. 1.1, right.) Although many Bayesian approaches can be used in service of the falsificationist approach I'm advocating here, I'm actually not going to discuss them here, for a simple reason: while Bayesianism as a philosophy is pretty old, most specific techniques are too new. You *should* be excited about the increasing ease of doing what are called "fully Bayesian" techniques that allow for taking one portion of our uncertainty—our distribution of belief across a set of models, both a priori and a posteriori—into account when we estimate parameters and their standard errors, as well as approaches that suggest how to trade off between fit and parsimony to choose reasonable models. But not only are none of these techniques going to offer us a silver bullet,[6] most are simply too new for us to have a "feel" for how they work in practice. Given that your tools will *all* be imperfect, it's better to use one whose tendencies and biases you understand than one you don't . . . of course, while you are also trying to develop new ones.

What's most important, then, isn't any particular number that you create, but the logic of your exploration of a set of substantively important alternative models. Rejecting false models is good, but since there are an infinite number of wrong ideas, we might just bounce from one to another. So we want to see whether, as we move away from a bad model to a different one, we are increasing our inferential scope. If so, we have reason to think that we're moving toward better ideas by throwing out the worse ones. To do this, all we really need are methods that have a better-than-

6. We can never take into account our true uncertainty, because the realm of possible models is actually an infinite one, and not restricted to the handful of predictors we happen to be looking at. Our biggest problems aren't about *estimation*, they're about *specification*, and what we need is rarely going to be solved by new statistics—rather, it will be solved by industrious workers applying the techniques they understand to get a good grasp of what's going on in their data.

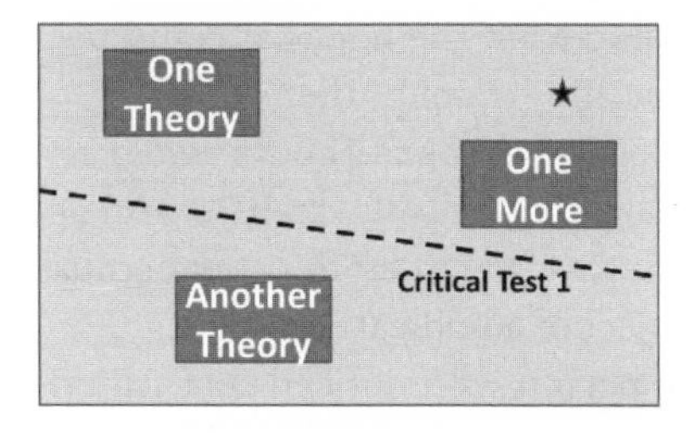

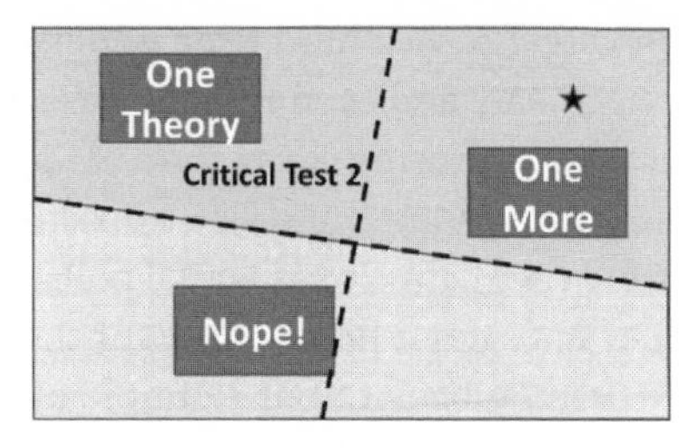

FIGURE 1.2. The Falsificationist Use of Statistics

random chance of leading us to give up on a false theory in favor of a different one that is more likely to be true.

And the cool thing about this is that we can actually say something useful even when there are "unknown unknowns"—relevant factors that affect our findings that we don't know about—so long as we understand something about the *other* possible options. So the way this falsificationist approach works is that we start with a set of possible explanations, and try to ask a question, the answer to which, divides up that space (see fig. 1.2, left). While this is classically called a "critical test," the point isn't that it's a one-shot, never-think-about-it-again answer. It's that it shifts the weight of the evidence from one possibility to another. Once we do that, we're going to want to see if we can further narrow down our options, by asking other relevant questions (see fig. 1.2, right). We might end up unable to narrow things down to more than a large section of this space. But this space has two important characteristics. First, it includes one of the three competing theories. And it includes reality.

This is, it should be emphasized, an *ideal*. We'd like to push one whole side of this space into "nope!" Often we can't, and so what we are instead doing is assigning some sort of credence to various hypotheses—shifting the weight of the evidence back and forth. That sort of endeavor is mathematized in various information criteria, but I'm arguing that we need to be doing something like that as a community of investigators, even if we can't mathematize the answer (for example, when the evidence for one theory comes from ethnographies, and that for another from statistics). There aren't easy answers. But there might be good ones. Let's translate this to practice.

Enter Statistics

What we want, then, is to use statistical practice to help us in this goal. Let's take the example of studying individual political behavior with data on votes. If you regress political vote on income and education, you can, of course, be trying to get a sense of what the causal effect of education is on vote. I have a hard time understanding what the hell that could possibly mean, and even if there was such a thing, I doubt there's any way of getting at it using these sorts of approaches. Many stats books would say, then go do something else. I disagree. Regression results aren't useful because they capture the actual real world causal parameters . . . but because they have descriptive utility for the project of adjudication.

Let's go back to our example of the person who noticed that Americans with more education are more likely to vote Republican than those with less education. Our researcher might have an explanation, namely that those with education are more likely to have an accurate grasp of the complexities of economic policy, and the need for responsible budgeting. As this set of economic policies is (rightly or wrongly) more strongly associated with the Republican party, those with more education are more likely to support the Republicans.

That's a story. Now you might think about education as a "cause" and something that might have "confounders" or all that sort of thing. And you could work very hard at establishing the correct estimate of this causal parameter. But you don't have to. You could stick with this as a comparison—those of more versus less education. You might notice that the logic of this person's claim implies that if you were, say, to stratify your sample by income level, within every income level, those with more education should be more likely to vote Republican. That is, there's no reason why his explanation wouldn't continue to hold. So let's say that you do this—you divide up income into deciles, and you look at the difference in Republican vote between those with more and those with less education *within* each decile. As you might guess from before, this will actually lead the relation to disappear or reverse; those with more education, in any income category, will vote *less* Republican!

We're going to throw out that first guy's story. And we're going to have an understanding of *why* it should be thrown out—in our sample, education is associated with income, and those with higher household income are more likely to vote Republican. Said somewhat differently, the relation between education and vote is very different from the relation between education-conditional-on-income and vote. You don't know much about American politics unless you understand both of these.

So we could split the sample into deciles like I said here, but if we found that in all deciles, the relationship between education and vote was relatively similar, then we might make a single number that expresses this relation, instead of ten different numbers. That's what a regression slope is. It's a way of taking a complex data space, rotating it so that we can ignore trivial dimensions, and then projecting it in such a way that it answers a question about a complex comparison.

We do a regression like this when we are going to *rely upon* the assumption that there is a simple relationship between subcategories formed by the cross-classification of many variables. We do that—and we do it *justifiably*—when some of the following are true: 1) in the past, we've found that sort of simple relationship; 2) we don't really have enough data to do anything else; 3) we don't really care all that much about our findings—we won't be in deep trouble if we're wrong about this. (This might seem shockingly casual, but I think we'll see that as we do serious research, we often have to budget our forces reasonably—we can't chase down every possibility fully.) Of course, if the comparison we're trying to get out of our regression is *too* complex, there's indeed a good chance that our regression won't give us what we want. One assumption of a linear relation is pretty innocuous. Add a second, and an assumption of independence of the two predictors, and it gets a bit shaky. Add a dozen, and you're talking Vegas odds. In general, we want our models to do two things: first, let us see patterns in the data that are relatively robust, and second, rule out some interpretations of those patterns.

Now of course, there *could* be a reason why our first person's theory turns out to still be true. It could be that there's *another* variable that we can use to stratify further that reverses our conclusions. Or it could be there's a really good reason, when we think about it carefully, that the person could still be right, though maybe we can't show it with these data. But the burden of proof has certainly shifted. Until this person comes back with something really strong, *that* story is out. Do we understand the association between education and vote? We don't. But we won't learn anything by just *assuming* that it *causes* vote in some mysterious way, and then trying to *estimate* the strength of this causal pathway.

In sum, statistics aren't there to estimate the parameters from your story. They're there to *eliminate* stories, to falsify some claims. We want to understand the possible explanations for simple descriptive patterns, and then bring data to bear on them to so that we can throw some of those out. And the task of statistics is to make sure that we're more likely to eliminate the interpretations that *should* be eliminated than those that shouldn't. If what we're left with really looks like it is pretty close to the "true" story, then let's break out the champagne.

Let me give an "existence proof" of the plausibility of this approach in the form of a true story. When my first book on *Social Structures* came out, it was given a very thorough, but ultimately negative, review by Neil Gross in *Contemporary Sociology*.[7] How delighted I was, then, to find that Gross—who is not an expert statistician—was writing a book on an obviously im-

7. "'Lack of insight!' I'll lack-of-insight him, I will!" I went around muttering for a week.

possible question, "Why are professors liberal?" All I had to do was hang back, watch him fall flat on his face, and then step daintly over.

Actually, he pulled it off (2013). How? First, he doesn't try to hang everything on a single coefficient in a single model. Instead, he starts with a robust finding—that professors *are* disproportionately liberal—and a small set of well-formulated alternative explanations of this fact: self-selection, discouragement of others, capture, and conversion. He judiciously goes over the different bits of evidence that we have, doing multivariate analyses where he can find something relatively robust, and continually assessing where the balance of the evidence lies. And, most important, while doing this, he attempts to root his investigations in what we can figure out about the concrete processes whereby careers unfold over the lifecourse. He never rests on one clever silver bullet to come up with "the" right estimate of a parameter from a simplified model.

To conclude, we've been a little bit too scared of ambiguity. Because our statistics would be totally defensible *if* the world was such-and-such a way, we've been tempted to *pretend* that it is in fact that way (cf. Gigerenzer 1991).[8] I'm saying that if we are willing to stop pretending, we have a different way of working, one that involves trying to learn from data by pursuing different interpretations and conducting multiple tests. And, sadly, many of us have been taught that this is the *worst* thing we can do!

Overfitting and Learning from Data

When I was in graduate school, we were severely warned against "data mining," which meant trawling through our data, looking for interesting things. The reason for this is that all the statistics we were learning required that one first (without looking at the data) construct a hypothesis, and then test it. If you didn't "call it in the air," the statistics were meaningless. If you *really* needed to look at your data first, you could have *one* peek, but only if you split your data, and peeked at one half, and then tested on the other.

In case you're rusty on this, and because sociologists can be casual in their language, let's review our basic terms. *Fit* is how close our predictions are to the data we used. *Model selection* is about choosing which model to believe. Often the best model is not the best-fitting one. Why?

8. I often hear the response, "But if you can think it, you can model it." Indeed, if you devote five minutes to thinking about something about the social world, you should be able to write down a fitable model for it. But if you spend two hours *really* thinking about it, you should be able to figure out why the model isn't right. And if you spend *eight* hours, you should be able to figure out why you *can't* do a plausible model that doesn't assume almost everything you're supposedly trying to find out.

Because the one that stretches itself to best fit *these* data might be terrible on the *next* set of data. It's trained itself to fight the last war, as it were. Of equally well-fitting models, we tend to prefer those with fewer parameters, not because we arbitrarily prefer simplicity, but because we think it's less likely to have parameters that only fit the particularities of *this* sample. We use the combination of fit and parsimony in the quest of model selection—our decision as to what we think holds in the population from which we've sampled. Fit, then, is about *this sample*, and when we are interested in inferences, we have to control our desire to fit. If we spend too much time looking at our data, we will increase our fit—but decrease our capacity to make inferences, because we choose the wrong model for the population. Hence the rule: don't choose your models based on what the data tell you! It's a basic notion that goes back to Francis Bacon.[9]

That follows logically from the basic axioms of our statistics. But no one I know who went on to be a skilled data analyst did this. Guiltily, we pored over our data, looking for what the hell was going on. Yes, that means we sometimes "over-fit"—that is, we came up with "false positives" where it looked like our theory was true, but it really was just the "good" luck due to the sampling fluctuations in *this* data set. . . . in next year's, say, we wouldn't come up with the same finding.[10] But we over-fit a lot less than you would think. In fact, I think those of us who combed through the data had *fewer* false positives than the rigid "testers." Remember that if your statistics teacher urges you to take a more orthodox approach, that's because in his class of 40, he is fine with the idea that two or three of you will go to press with a *totally* wrong finding just due to sampling error. You do things his way, that's as good as you can expect. You look more closely, and I think you're less likely to be one of those false positives.

Here's why. The classic "tester" has a finding she wants to be true. Let's say that she wants to argue against those who say that "feminism" (in the sense of being against traditional gender roles) is intrinsically a pro-

9. "In forming our axioms from induction, we must examine and try, whether the axiom we derive, be only fitted and calculated for the particular instances from which it is deduced, or whether it be more extensive and general. If it be the latter, we must observe, whether it confirm its own extent and generality by giving surety, as it were, in pointing out new particulars, so that we may neither stop at actual discoveries nor with a careless grasp catch at shadows and abstract forms instead of substances of a determinate nature and as soon as we act thus well authorized hope may with reason be said to beam upon us" (*Novum Organum*, 106; 1901 [1620]: 128). Thus statisticians correctly caution us against overfitting.

10. Now even then, truth be told, there were some suitable ways of determining the statistical significance of a parameter when one conducts multiple tests and so on, and now it's more common for people to carry these out.

woman position. So she takes data from a survey (say the General Social Survey [GSS]) and shows that men and women don't differ significantly in their support for less traditional gender roles for women (and indeed, that's what you'd get, at least circa 1993 when I last did this). But a critic might say, "Maybe this is due to education differences between men and women." Sweating, our researcher adds "education" as a control, crosses her fingers, and . . . ah! No change. The gender parameter is still insignificant. Thank God.

What this approach (theory testing) is doing is preventing our researcher from using the data to learn much. She's so intent on preserving her own theory, that she can't see what she didn't already think of . . . unless it was suggested by someone else. And even then, rather than push us to be increasingly clear about the contours of the processes or patterns we are trying to uncover, this way of prompting research blurs them. The game she plays is that she tries to hang onto her finding, while doing justice to everyone else's idea by tossing in more and more controls. All these controls make it hard to know what the hell we are doing. (We'll return to these issues in chapter 4.)

And what this sort of approach encourages is that we *stop* our investigation at a convenient place. That's not a "conservative" strategy at all! A conservative strategy is one where we take our current leading interpretation (even if it's "nothing is going on") and see whether it implies any *other* testable hypotheses. Our researcher won't stop with just finding an insignificant gender coefficient, because she's trying to paint a picture of what men and women are like. And then she'll think, "Well, if men and women really are no different, if I split the sample by sex, within each group, all the predictors should have the same coefficients." And she'll do that and find that the education coefficient is the same for both men and women. But if she throws in a measure of income, suddenly, men and women *aren't* the same.

Why not? Now our researcher isn't interested in testing, but learning. Hmmm. . . . that measure of income was total *household* income. Maybe it matters *whose* income it is. What if I look only at married people here, and only at women who have jobs, and men whose wives have jobs, and break up the income into two portions, one, her income, and the other his. Guess what jumps out at you? Every dollar that the woman makes predicts *decreased* traditionalism on her part. Every dollar that the man makes predicts *increased* traditionalism on her part.

Don't stop here. Is it the dollars, the percentage she contributes, or just the fact of working, that is the best predictor? Can we do the model separately by year? Are the parameters changing? And so on. When you're done with this, you're more likely to have a true finding, despite the "data mining," than if you tested. Why? Because you follow the trail pointed to by a working hypothesis, and you follow it until the trail dissolves. Everything seems to be coming together, and that implies you should see *X* . . . and there's no *X*. So you have to revise your working hypothesis and go on. That's what we call learning. In other words, we put robustness (the same finding appears when we do things differently) and internal validity (the findings all support one another in interpretation) above that magical * sign that indicates statistical significance.

Further, this approach helps you remember what you have—cases, usually people. Instead of jumping to test an abstract theory, you start thinking concretely. It's a mystery you need to solve, but like a good detective, you want to determine your suspects before you start worrying about their motives. Who is driving my finding? Who has high values on my key variables? Who has low? Is the finding driven more by lows or highs, or both?

Finally, if you're still queasy about this idea of really plumbing your data, because your statistics teacher has drummed into your head that only cheaters look at the data before conducting tests, ask your teacher if *the entire scientific community* should do only one test for each research question. Presumably, she or he will say absolutely not. We should use the results of past research to come up with new hypotheses and test them. Now ask why *you* can't do that on the basis of your *own* past results. Why is *your* p-value wrong when you pursue the logical next step, when the p-value of the person in the office next door isn't?

Science does, just like the squares will teach you, work by constructing hypotheses. But, as I said in *TTM*, this is something you should be doing *a dozen times a week*. Otherwise, you just make the rest of us work through the implications of your claims. And that slows us all down. If the scientific community as a whole can do it, you can try to do it too.

Standards of Proof

There's one last issue about how we compare our explanations, which has to do with our standards of evidence. I think we need to understand that different questions, and different comparisons, require different standards. We usually first think about this when we face the problem that classical statistics is able only to reject a null hypothesis—not to establish your own. The null hypothesis is something boring and dumb, like "everything is completely random" or "this variable doesn't predict that one at all!" or "maybe we just pulled a weird sample." In the rare chance that we test our *own* theory, as opposed to the null, we want to make it easier to reject. That's because it's a bit unfair to imply that you are right simply because you can stump the chump. That doesn't establish which of the infinite number of *other* hypotheses is worth believing in. It's sort of like Albert Einstein claiming he's right, because he's shown that Mortimer Snerd is wrong.[11]

Approaches that look at the weight of evidence across a class of models seem a lot better: they can include a null model, but aren't restricted to this. But they come with their own problems; such equal weight across all models isn't always what we want, because we can't always determine the proper set of possible explanations, and how we populate this set influences our conclusions (I'll return to this shortly). Instead of looking for a general numerical solution to this problem, we need to think about the different types of claims we might be investigating.

Here, for want of a better term, I'm going to borrow a notion from Boltanski and Thénevot (2006 [1991]) and say that what researchers are trying to do with their statistics is to "qualify" something in the world. For example, they may say that employers discriminate against African-Americans. If successful, the admittedly obscure object "employers" now has a new quality—*discriminatory*. Others may attempt to use evidence to prevent or remove this qualification.

Here we can draw upon a different way that we have learned to think about this process of qualification, one coming from our legal system.[12] First, let's consider what I will call the *criminal* understanding of the relation between claim and evidence. This is an asymmetric understanding, where the burden of proof is against the claims-maker. Such asymmetry makes sense in some settings: it makes sense for criminal trials, given the asymmetry between a state and a citizen. The *p*-value test and the 95% confidence interval are our "beyond a reasonable doubt." When is this approach appropriate for social science? Perhaps when there is a single contending qualifier of the world (that is, someone who says, "the world is like this") and a single world to be qualified, where we want desperately to avoid being wrong, but don't particularly mind not being right. And when might that be? Very often, when someone is making a claim to support an *intervention* (e.g., a new policy). The deck *should* be stacked against her—at least, to the extent that her argument is made using statistical reasoning.

11. Mortimer Snerd was a literal dummy, used by ventriloquist Edgar Bergen. The great sociological theorist and grouch Pitrim Sorokin used him as an example of a real idiot. Because Sorokin probably listened to Bergen on the radio, I'm not sure whether he knew that Snerd was made of wood. If not, who was the real dummy?

12. If you would like some pedigree here, C. S. Peirce proposed just this way of thinking about differences between intellectual endeavors and their standards of proof. See "The Logic Notebook," 1985 [1865–1866], *Writings*, vol. 1, p. 337–350; "The Logic of Science; Or, Induction and Hypothesis, Lecture III, Lowell Lectures of 1866," 1985 [1866]: p. 357–504.

In contrast is a *civil* understanding, in which we have a balanced burden, and whichever side is over 50% wins. This arises in law when we have not a state against a citizen, but two citizens arguing over where the fence line should be. It has to be *somewhere*, and it would be unfair to imagine that whoever was the first to complain should face a higher burden of proof. When is this standard of "over 50%" the reasonable one? We might at first suspect that it is necessary in conditions where we are oriented by a need for practice: where we must do something—to do, or to forebear. But in a moment, I'll try to argue that this *isn't* the case; rather, this is most useful when we have a problem that is relatively disconnected from practice, and where the alternatives are few and data scarce.

For example, historians may have narrowed down their theory of the basis of the emerging party system in revolutionary America to two general theories. One is that it has to do with, basically, class relations (though not everyone uses these terms!). Some elites were tied to land, others to commerce; some had an interest in westward expansion, and others didn't. The second is that it was historically specific—whichever faction of elites was able to capture the governorship alienated the elites in different networks, as the former monopolized patronage opportunities. For this question, there is no pressing need to come to closure on our decision. Data is difficult to bring to bear on the question; the participants are long dead, and written statements of motivation are suspect. Each new bit of analysis brought by adherents of one theory can push the property line a bit further in the other direction, as it were.

But why isn't the same thing true if we are deciding, say, whether or not to decriminalize cocaine use? Either we do or we don't (cf. Peirce 1985 [1865–1866], "Logic Notebook," 339). So we should assess the evidence at hand, and whichever way it points, that's the way we should act, right? Unfortunately, this way of thinking about the issue is misleading, and here I'd like to review the identical problem as it defeated a number of early theories of probability (and I want to rely on C. S. Peirce's masterful work here).

Imagine we are hoping to test whether education affects income. We might think, well, either education causes income or it does not. Whichever side has the most evidence is the side I will pick. And (to use Bayesian terminology) you might think that you should therefore assume that your prior estimate of the probability of "no effect" is .5. But rephrase this as "I wonder whether the effect of education on income is precisely zero." Then the a priori chance of a zero effect is vanishingly small, so small that you'd have to put a rather crazy distribution on the priors to avoid it being completely unnecessary to carry out any research at all, because nothing could shake your commitment to a non-zero effect. Peirce pointed out that a number of previous attempts to mathematize probability had foundered on this problem, confusing our ignorance regarding *one particular enunciation of a question*—such as, "will the die turn up a one or not?"—with a defensible assignment of probabilities. We may indeed be completely unsure as to the next roll, argued Peirce, but the *only* stable denominator for the construction of a probability is not the number of alternatives we happen to be thinking about, but *the number of possible states of the world*.[13]

There are more than two options for drug policy. Whenever we take *our own* preferred one, and put it up against all others, we're coming up with an incredibly overblown estimate of its probability, just like a would-be Bayesian who claimed that the prior probability of rolling a 1 versus a non-1 in a conventional die was ½. That said, we can make the inverse error—rather than using our interest in one particular option to push it to the front, as in this example of drug policy, we use our interest to push it off the table completely! How do we do this? We confuse our failure to reject a null hypothesis with our having established it. For example, as we'll see in chapter 3, it's quite true that, as many statistical methodologists now emphasize, observational studies often can't really identify putatively causal effects of environmental factors in the presence of strong selectivity. That fact can be used to deny the claim to make a policy to affect some outcome. But just as the capacity to reject the null hypothesis in one data set doesn't mean that the hypothesis held by the investigator is demonstrated to be true, so the failure to reject doesn't mean it's false. In many cases, the failure to reject the null hypothesis comes not because our results show that the true effect is close to zero, but because they simply tell us that our estimates are extremely imprecise. If the data is saying "this effect could be zero, or it could be very, very strong indeed," it's irresponsible for us to look at the data, and then turn around and tell people, "statistical analysis demonstrated that there was no effect."

How do we balance these two different errors? First, weak statistical evidence shouldn't lead us to throw out our belief in processes that we have other sorts of evidence for—even if that evidence is of the kind that doesn't qualify as social science, it's still evidence (for example, personal experience, hearsay, common sense—all of which we'd like to improve upon, but none of which we can live without).

Second, I would suggest that we will want to rely more and more on

13. For this reason, our conclusions can be changed depending on how we populate our set of possibilities: if, instead of considering three hypotheses, A, B and C, we divide A into two variants, A1 and A2, many methods will now give $A = A1 \cup A2$ more a priori weight!

Bayesian approaches, in which we consider which model among a set to prefer, when our problems are more like *civil* ones, and more on classical approaches when our problems are more like *criminal* ones. And most important, what we want is to move to situations in which civil understandings are appropriate. That means that we, as a scientific community, have narrowed down the set of possible alternatives to a manageable number, and we are doing different types of research to assess the relative weight of the evidence for each. Where we *aren't* in that situation, either because there are far too many possibilities (put differently, we don't really know much) or because we are proposing some very new theory, criminal standards—and therefore classical statistics—are more appropriate. Because this is rapidly becoming a minority position, I'd like to defend it explicitly.

A Mindful Defense of Mindlessness

If you've ever read Evans-Pritchard's (1974 [1956]) book on *Nuer Religion*—and you should if you haven't!—you'll be familiar with a poison oracle. This oracle, like many others, is a way of taking luck and using it to make a difficult binding decision. Many societies do this where there is an important matter of life and death and no one can really be sure what is the right thing to do. In such cases, sometimes using a Magic 8-Ball turns out to be a better way of making the decision than trying to think it through. At least then no one is responsible for making a really bad decision.

This particular oracle involves preparing a special poison in a dose that is as likely as not to kill a fowl. The bird is made to ingest the poison. You ask it a question. It answers by dying or living. Think of this as the equivalent to running a regression and getting a significant or non-significant *p* value. It is a ritual that tells us whom (or, in our version, whose career) we should go off and kill.

There are all sorts of good arguments against using statistical significance as a criterion. Most impressively, the American Statistical Association recently came out with a strongly worded rejection of the use of such tests for the purposes of making claims (see Wasserstein and Lazar 2016). The arguments *against* are pretty good—it describes a practice we don't actually carry out (one-sided tests) to solve a problem we rarely have (rejecting the inference that a null hypothesis holds in the population). The arguments *for* are a lot weaker. If that's so, how can I defend it? Easy. If I had reason to think that the poison oracle has a non-zero correlation with the true answer, and I couldn't think of anything better, I'd say we should use *that*.

It's for this reason, when we are conducting research that needs to be judged according to a *criminal* standard, I think it makes sense to use *p*-values in the old-style. The virtue of the *p*-value is that it is arbitrarily rigid. There are two findings, one with *p*-value = .0499999 and the other with *p*-value = .0500000. Accept the first, reject the second. Why? Because social science is a discipline that does not produce technology (the way physics produces refrigerators), we cannot be trusted to bargain with nature.[14] We need formulae like this, at least for the everyday, run-of-the-mill research that we are going to do. We should *increase* the rigidity and formulaic nature of *this* side of our work—just we should *decrease* the rigidity and formulaic nature of our work when it can be judged according to *civil* standards of proof.

In the civil understanding, we have a set of researchers who know what the relevant possible interpretations are. They use the compiled data in effect to bargain with each other about which to accept. In the criminal understanding, each researcher is out on her own . . . trying to bargain not with nature but only with herself. We have had too much sophistication, special pleading, bargaining, and "nuance." We need more constraint on our thinking, because we have far too many false positives. If you have a finding with a significance of $p = .07$, even if you have a story for why it's still something worth investigating, I'm going to counsel moving on to something more robust. And if you want to bargain with me I'll sit it out with the same glazed smile I use on the Maoists on Telegraph Avenue. Whatever.

Of course, you should put learning from your data above conducting tests. And that means I don't endorse the orthodox view of how to interpret the *p*-values. The statistics behind these are usually for a *single* test you have run on data you have not peeked at. We *never* do that, and you *never* should. When we are exploring, we might use the *p*-value as a rough guide, but there are other ones (raw numbers, percentages, change in slopes, and sub-sample sizes). So in data exploration, you should use the *p*-values the way meteorologists use their models—as one bit of information among others.

However, when we're done, we should have something that we believe in, and that is also significant according to conventional standards. It's quite true that because we haven't done the one-shot, no peeking, test, the *p*-value isn't quite right—our procedures are too liberal. But as if to

14. And this leads to a serious ethical issue about borderline results, which I explore in our conclusion. It also leads to what is called "*p*-hacking," where analysts tweak their models to get that .053 *p*-value down to .049. I think my emphasis on testing the implications of an interpretation is the best counter, but you should also know, it's pretty common for analysts to find that *nothing* they do can get the *p*-value below .053. (Not that I've tried.)

make up for it, we conduct two-tailed tests by convention, though we really have one-sided theories, and that's being overly conservative.[15]

It's worth emphasizing that this defense makes sense only for conventional data analysis—where we have an omnibus data set, collected for general purposes, and accessible to many different researchers for replications on basically even terms (as opposed to the original researchers collecting data that are already impregnated with their own assumptions). I certainly don't defend the use of p-values for experimental work, like that done in social psychology.

To conclude, our goal is not the estimation of parameters. That is a *means* to our goal. Our goal is seeing whether there are some interpretations of the world that we can, for now, dismiss. Does that seem a bit weak to you? Maybe, but knowing your limitations is a big part of science. If you want a more inspiring vision to strive for, the (optional) coda to this chapter describes the notion of mathematical sociology, which I think should be, if not our goal in actual practice, our orienting dream. Our reach must exceed our grasp, else what's a heaven for?

Coda: A Plea for Mathematical Sociology

There are good reasons to imagine that in the current world, the field of statistics would be at death's door, while mathematical sociology would be rising like a phoenix. Statistics as a discipline was oriented around issues of inference from samples, and, in a world where we have millions of observations, many of the issues that classical statistics was best at solving have retreated in importance. In contrast, the penetration of social network analysis by physicists has led to an infusion of much more rigorous mathematics to a number of the questions that historically were central to mathematical sociology.

Yet not only is mathematical sociology *not* reviving, I think it's in danger of completely being forgotten. You, dear reader, *may not even understand what I am talking about*. That is, you won't understand the opposition between the two approaches.

Mathematical sociology is about looking for the mathematics that (might) underlie social processes and structures. It's perhaps a quixotic quest for a true social science. The great breakthroughs here were Levi-

15. There should be *no actual* two-tailed tests, which makes around as much sense as Newton saying that his theory is right if objects either accelerate constantly *downwards* or constantly *upwards*. Pathological research can be identified by tests of a set of coefficients that are allowed to go in *either* direction, and get a story no matter which way. But using the statistics as if we *were* conducting two-sided tests has turned out to be a good practice.

Strauss's adoption of structural models, and then their mathematization by Andre Weil and Harrison White (as well as other attempts by people like George A. Lundberg and of course George K. Zipf!); the later work of White and collaborators like Scott Boorman and Ronald Breiger, and their collaborators Philippa Pattison and her student Carter Butts. Also important is the work by Paul Lazarsfeld on static models and his student James Coleman on processes.

I'll give one image for mathematical sociology—it's a lot like early crystallography. Before the age of the electron microscope, chemists tried to understand something about the structure of molecules by growing them as crystals. Water forms hexagonal crystals (that's why snowflakes look like they do), salt forms rectangular prisms, and so on. To make such crystals with recognizable forms, you generally need a pure solution of the molecule in question. If someone were to criticize you because you didn't have a good "random" sample of naturally occurring salt water, say (some from the Atlantic, some from the Pacific, some from the Indian Ocean, and so on), you would look at them as if they were crazy, right? It's like Allen Barton's (1968) analogy of an anatomist who takes a random sample of cells from an organism and studies them all together. But we often allow statisticians to get us to do just that—to make it hard for us to see structure (if there is any) because we need to make *inferences*. That's their job, and kudos to them. But they're not really there to solve our problems of having a social science.

A statistician, as I've said, can help you figure out whether you have correctly inferred the returns to education on average for American men. You can estimate that perhaps to four significant digits. But it's not much to build a social science on. Why? Because if we divide the country at the Mississippi, the chances that we'll get the same values West and East are pretty low. You might suggest that there are then two "isotopes" of American men; but if we split each of these again, we'll get four different numbers, and so on. These numbers are artifacts, in other words; they are the outcomes of our conventional manipulations of measurements. They aren't actually numbers that reflect something like invariants; rather, they are the result of throwing together all the *real invariants which we do not understand and which might well be mathematizable* (though not necessarily parameterizable). Remember: statistics gets you the best estimates of parameters—whether those are meaningless or not is a totally different story.

Speaking of stories, let me give you one involving a nice beef between two of my heroes, Harrison White and Leo Goodman. It had to do with a seemingly arcane issue, namely how to model the distribution of group sizes that might spontaneously form in a larger setting (such as conversational groups at an embassy party shortly before an Arnold Swarzenegger character crashes through the window in a tank). This nicely encapsulates the difference in habitus between mathematical sociologists and statisticians, and why statisticians can always win. White (1962) began from statistical mechanics and developed a rigorous approach to trying to solve this problem, one that built on and transcended some earlier work done by Coleman, by the way. But Goodman (1964) showed that there were problems in the way in which White had approached this.[16] And the most damning issue was that White had treated the expected number of singletons as a parameter, as opposed to a random variable. That is, he had ignored the fact that it had a sampling variance.

Now I get it, and I basically am convinced, or convinced enough. But a mathematical sociologist wants to be able to mathematize *this* group, with its number of isolated people *right here, right now*. If that doesn't do justice to the population of all possible sets of groups, then so be it. Mathematical sociology *isn't* about inference in this sense of sampling, and we shouldn't let statisticians come in and smash the more delicate constructions that we need to make. Or, at least, that would be true if mathematical sociologists were still making models of structure.

But instead, mathematical sociology began to fall apart with the influx of simulation work, which encouraged a lot of usually young, usually male, usually students to consider themselves mathematical sociologists because they were using a *computer*. 'Tain't so, not nohow. But this stuff flooded the journals, pushing out mathematical work, which is much harder to review. And these people were unable to do that sort of review, so even as the mathematical rigor of much work in sociological fields increased, there was less exciting mathematical sociology. But mathematical sociology should be the grail that we are searching for—it may be a myth, but if we stop believing that there are mathematical properties of social interaction, we should leave off all this number crunching (for an example, see Newman, Strogatz, and Watts 2001).

And so in this book, I'm going to be talking about how to get reasonable parameters, but my position—strongly—is that if there *is* a mathematical *order* to social life, we should uncover it wherever it appears, and if there *isn't*, we shouldn't fetishize inferential models; rather, we should *use* whatever we can to rule out wrong ideas . . . at least, the ones that we *should* rule out. We're going to try to get some principles to do this in the next few chapters.

16. Oh, did I mention that White also developed a theoretically defensible approach to modeling mobility tables, again, based on statistical mechanics, and Goodman blew him out of the water with the loglinear system, which had no real theoretical grounding but behaved much better statistically?

* 2 *
Know Your Data

What Have We Here?

Weakest Links

My point in this chapter is simple—you actually need to know your data: where they came from, who made them, why, what they look like, feel like, and so on, before you can do any decent statistics. Lots of our problems could be solved here, and an ounce of prevention is worth a ton of retraction.

You might think this doesn't belong in a book on statistics, but it does. You'll remember that my main argument is that we need to be able to know what we're doing when we *don't* have exactly the right model, the right specification, and all the other things right that guarantee that our numbers coming out are right. So why *don't* we have the right specification? I can give three reasons. The first is that the vast majority of the theoretical questions we are interested in involve processes (for example, "influence") or states (for example, "political sophistication") that are impossible to measure directly. And even if we *could* conceive of how to do it, the data that you happen to have didn't. The second is we just don't know enough about the way the world works to make an acceptable model of it. Those are pretty severe limitations, and it's sad that we can't do more. The third reason is different: it's *because we are stupid and lazy* and don't even bother knowing what we *do* have.

Okay, sorry to yell, and of course, this isn't true for everyone. But most sociologists, I find, do not actually *know* the data they are using. Unless they've collected it themselves, they don't have a real understanding of where it came from, and by "collecting it themselves" I don't mean simply writing a grant to have someone else collect it. They treat data like a gift that Santa Claus left for them under a tree. And they don't (to mix metaphors) look a gift horse in the mouth. As I wrote in *TTM*, accept that gift horse, but inspect the mouth carefully.

Map: I'm going to start with some cautionary tales, tak
vey data, arguing that we can easily think we have found som
haven't. I'll argue that our working theory of science disserves us
making us believe that we can turn a set of numbers into a theo
term by simply *naming* it such. Sorry, Rumpelstiltskin, no theory o
ence allows you to turn straw into gold. Those measures are what they a
How can we find out what they are? I'm going to discuss three tactics. One is to examine your data descriptively, especially using visualizations. The second is to learn whatever you can about the processes that generated the data. And the third is to consider certain formal aspects of your data—where they sit in a space of possibilities, where the variation and the co-variation are. Finally, I'll suggest what sorts of projects are most likely to lead to conflicts over the quality of your data. And I'll suggest you set up camp right there . . . and do a good job.[1]

Horror Stories

I aim to scare the living f**k out of you. That's the way I was raised. "You want to play with the blender? Great idea! Just stick your hand in it, turn it on, blood will cover the walls and you'll pull out a piece of white bone, sharp as a pencil, and then die of loss of blood." Got it, blender ≠ toy, will not actually try. So the stories I'm about to tell are going to remind you that the only quality control standing between you and humiliation is yourself—not peer review. And this is what I want to be left in your head at the end of this chapter. Cautionary tales.

Lenore Weitzman (1985) used data from a sample taken from a Los Angeles Divorce Court to determine how men's and women's standard of living changed after divorce. Her stunning findings were that while men's standard of living went *up* significantly (42%), women's crashed by 73% (in this metric, 100% would mean having *no* standard of living at all). She thought that this seemed extreme, and asked her minions to check it a few times, but when no one reported an error, she went to the press, and in a very big way. No other study—and there had been others—found anything like this. Finally, Peterson (1996) went back to the original paper forms that had been archived, re-entered the data, and demonstrated that there was no way to get this number from her original data. Weitzman (1996) pointed out that she had made a number of half-hearted attempts to trace the problems and re-create the data set, *but she had never gone all the way upstream to actually look at her original data*. Basically, she said that

1. As this goes to press, Howard S. Becker is coming out with an outstanding book, *Evidence*, that should be required reading for all.

she didn't really keep track of what was going on, and that a lot of the decisions were in the hands of nameless research assistants now long forgotten.

And it isn't just sociologists. Thomas Herndon, a student at Amherst, had an assignment to replicate an economics paper, and so he chose the influential paper "Growth in a Time of Debt" by Carmen Reinhart and Kenneth Rogoff (2010), widely used to justify austerity policies by governments. Unfortunately, Reinhart and Rogoff left out Australia, Austria, Belgium, Canada, and Denmark (A–D, you will notice). Put them back in, and the conclusions change a lot (Herndon, Ash, and Pollin 2013).

You might think that no one makes these mistakes any more, but I just finished reading a paper sent to me for someone's tenure evaluation that claims to examine 2000 cases in the General Social Survey (GSS) and finds relationships far stronger than those *I* get when I look at similar sorts of things in the GSS. Well, the analyses have around 200 more cases than should be there, according to the description of the data. Probably this person recoded a bunch of the missing values as "0" and they drive the results—people with missing values on more than one variable easily hold down a nice correlation by sitting in the bottom left of your scatterplot, when they should be off the page. Now *that* was an awkward tenure letter. . . .

Who Is Your Enemy?

The answer is your theory. Everyone makes mistakes. But we are more likely to go to press with a mistake when it fits our theory. That's not only because it's congenial to us, but because, having our brain in creative mode, as opposed to skeptical mode, we can come up with accessory explanations as to why the pattern of data is a *bit* weird, even though it fits our preconceptions. Unfortunately, most of us are likely to look twice only when we're disappointed. For that reason, competing research programs are more likely to unveil errors than the folks working on their own idiosyncratic questions—another reason to embrace the vision of social science I laid out in the first chapter.

An example comes from the sociology of religion. Here one of the most exciting claims for a while was the idea that affiliation with a particular denomination was best explained using rational choice models, and for this reason, competition between religions was associated with high religious involvement—an exact inversion of the old "sacred canopy" idea that pluralism will undermine the assumptions that support much of religious faith. Finke and Stark (1988) had presented some fascinating evidence that confirmed their view. But Breault (1989a) used a different set of data and came to the opposite conclusion. Finke and Stark (1989) responded that one of their colleagues had done a more complete analysis of the same data that Breault used, and had come to the opposite conclusion. Huh?

Daniel Olson (1998) finally put this to rest. Laurence Iannacone quite decently gave Olson the SAS script that was used to calculate the key piece of evidence, namely a positive correlation between pluralism and adherence rates per county. The formula for pluralism they used should be $1 - \sum_i p_i^2$, where p_i is the proportion of adults in the county belonging to the ith religious group. But the program used $\sum_i p_i^2$, instead of subtracting this from 1. But it fit the theory, and so it seemed right. But Breault never had his own confidence in his results shaken (see Breault 1989b). Why? "Remembering advice I had received early in my methods education, I embarked on the herculean task of checking the validity of my conclusions by examining *every* observation, laboriously computing each of the (3,100+) diversity scores."

Now the debate actually resolved itself harmoniously, in that there was increasing agreement that *no one could use the cross-sectional data to answer the theoretical questions* (see, especially, Voas, Olson, and Crockett 2002). That's because, as we'll see in chapter 6, our linear models don't necessarily work for terms that entangled, as often happens when we have ratios. But it took disconfirmation for people to recognize that.

So when you get findings that agree with your theory . . . look a bit more carefully. And that's also true if you come up with surprisingly cool findings that others have missed. They might have missed them because they aren't there. A good example here is the Bruch and Mare piece we're going to briefly refer to in chapter 9 on simulation. They had some robust results, but the big shocker was that, contrary to what everyone else had thought, this one very well-known model for segregation had results that were the opposite of what everyone thought. But this model is the sort of thing that fledgling computer jocks try their hand on the minute they're done with their "hello world" program. So the chance that *everyone* else had missed this was pretty small. When you have a result like this, it makes sense to pass it around widely before going forward.

Finally, it can also be tempting to go too quickly to the press when you have something that's going to shock. The most extreme version of this is when you try to make what my son calls a "dick move," which in sociology is something like getting a lot of attention by doing something like blaming the rape victims for being raped, saying that racism helps the economy, or that fat people commit more crimes, or whatever. That is guaranteed to get you a lot of attention, but not all of it will be what you want. So you're going to have to check your work *really* carefully. And that can be the case even if you don't think you're making such a move.

For example, John Donohue and Steven Levitt (2001) made the shocking claim that the legalization of abortion across the nation led to a decrease in crime 20 years later, as the sorts of young men who were most likely to cause crimes were less likely to be born in the first place. This was a serious argument, made on the basis of plausible linkages—and in chapter 3 I'll advocate these sorts of compositional arguments. But it struck many people as a dick move largely because it seemed to say that the cause of crime was poor people, not inequality/deprivation and so on. I would say that the most important issue for us, however, has to do with how the sort of variation you have tends to lead you to ask your question in such a way as to predispose you to some sorts of answer and not others. (This is so important, we'll have *two* sections on "Where Is My Variation" coming up!) If we think of crime as linked to perpetrators (and not, say, to communities as a whole, or to victims), chances are, our explanations of the "causes" of crime will focus on the alleged perpetrators, even if this isn't the most important cause.[2] In any case, this sort of claim is one that turns out to be very, very difficult to demonstrate, as it requires coming up with close counterfactuals, which nature rarely provides. Fortunately for critics, however, the errors here were relatively easy to demonstrate, as Donohue and Levitt had omitted some controls and not standardized their key measure by population (Foote and Goetze 2005).[3]

So before going forward with such research, don't listen to your theory, or your friends. Send it to whoever's on the other side. Chances are, they're your real friends—those who will catch your mistakes. One more case: my

2. There are probably plenty of famous, well paid, and generally nonimprisoned men—politicians and industrialists—whom we never think of as being "the cause" of crime. But there's a good chance that had their mothers aborted *them*, crime would have been lower than it was. Perhaps we should start talking about them instead.

3. Donohue and Levitt went back and re-did things so that they could still find the effect, but for reasons we'll appreciate after chapter 7 on Time and Space, there was little way of convincing readers. It requires too many different and difficult assumptions; it is no longer that most critics are sure that Donohue and Levitt's revision has a goof—it's just that most serious researchers don't believe that a relation like this can be shown for a historical change that happened only once, unless it's really, really big and very, very crisp (and this one isn't). Trying to forecast to what extent *Roe v. Wade* affected births is an interesting exercise, though of course here already there is going to be a lot of uncertainty. Trying to peg which of those non-existent children would have committed certain crimes (and which states they would have been in when they did—or, more important, were *arrested for*—the crimes, or arrested for crimes they didn't actually do . . .) . . . hmmm. . . . .

use of real world examples. I sent my first draft to the authors involved, and most of them wrote back, with various degrees of heat, and in all cases helped me give a more accurate treatment. In one case (I still break out into a nauseated sweat when I remember this) I was 100% incompetent, and Michael Rosenfeld gently pointed this out to me, saving me what would have been the biggest embarrassment in my life.

We all make mistakes, but we're better at catching other people's than our own. So get them to help you. Show your surprising work around before blabbing.[4]

Magnified Errors

We've been looking at errors that come from analysts making a mistake. But sometimes the big error analysts make is not anticipating the big effect of the little errors that *respondents* can make. Here, I want to follow Micceri et al. (2009) in speaking of the "Gulliver effect"—when we have a hard time estimating the size of a "Lilliputian" population because of the errors coming from the Gullivers. An example would be "tweens" with exclusive same-sex orientation. No one is quite sure to what extent this notion of sexual orientation makes sense for kids just going through puberty, but studies suggest that maybe around 1% of 9–14-year-olds have this sort of same-sex orientation (Austin et al. 2004); that number will go up over high school, but it's still not really clear how much. Imagine that we were to do a large survey of 9–14-year-olds and ask them whether they have this same-sex orientation. And imagine that everyone has a 3% probability of making an error. We'll find a value not of 1% but of 3.94% (as $1 \times .97 + 99 \times .03 = .97 + 2.97 = 3.94\%$)—not only have we greatly overstated the size of this group but most of the people we include as having same-sex orientation don't. Error leads people in the more popular categories to inflate the numbers in the rarer ones, often so much so that it isn't worth even trying to look at the people in the most interesting classes. Look for the "upper class" in a survey and you're likely to get a bunch of workers who said the wrong thing or who were mis-coded.[5]

Of course, assuming a 3% random error might seem pretty implausible.

4. The stoic philosopher Epictetus reported that his teacher Rufus upbraided him for overlooking an error in a particular syllogism. "Well, it's not like I burned down the capitol or anything," he carelessly replied, rather proud of his own cheek. "Fool," answered Rufus, "this syllogism *is* the capitol." [I have used the version from my memory, which I prefer to the original, *Discourses*, chapter 7.]

5. This is well understood by epidemiologists, because even with rather accurate tests, most of those who are diagnosed as positive for certain rare diseases do not actually have the disease.

But it gets worse, because there's often *non*-random error here as well. That's when our sample includes cut-ups and goof-offs: people who are intentionally messing with the survey.

And unfortunately, such cut-ups probably distort some of our examinations of youth homosexuality: it can appear that there are long-term negative consequences to early expression of same-sex desire, but we can't tell how much of that is due to distress suffered by gay youth, and how much is because some of those who seem to have had same-sex orientation were actually wiseguys: the ones who sit in the back of the class, put down that they had romantic attachment to boys, and a peg leg, and sniffed glue, and—all in all—enjoyed a fun few years before getting pushed into juvie hall.

Cheaper data generally has higher rates of people messing with you. I confess that when I get an automatic telephone poll, if there's no baby around to hand the receiver to, I walk around the kitchen cooking and pressing random buttons when I remember that I'm holding a phone. For every cheap-ass poll that comes up with garbage, that's one more contract for NORC (the National Opinion Research Center) and our graduate students, I figure. But even the highest quality data can have this problem of wiseguyism.

Consider the National Longitudinal Study of Adolescent Health, which is is one of the most important studies of the development of youth; it's been used for many pivotal studies. And the data is of high quality. But they aren't immune to wiseguys. (Here I follow Savin-Williams and Joyner [2014].) Of the 253 adolescent students who reported during the self-administered, in-school questionnaire having an artificial limb, only two admitted to having the prosthetic during a later home interview. Did they, like axolotl, regenerate their limbs? Or, in the us-versus-them context of the school, was it a lot funnier to give wacky responses in school? It seems that a substantial portion of the youth identified in this survey as having same-sex "romantic attraction" were jokesters—and others may have misunderstood the question, as "romantic attraction" seems to be ambiguous to a lot of Americans.

So one has to move incredibly cautiously when making claims about relatively small populations. Let me close with attention to a now pretty famous example. Mark Regnerus (2012) wrote an article using new data collected for the purpose of determining the effects of different types of child-rearing situations on later health. (And, significantly, he talked up the results before checking them carefully—the "peer review" for this piece was famously cursory.) He wanted to see whether kids who grew up raised by gay parents later had worse psychological health. He said they did. But "being raised by gay parents" was, for the cohorts he was studying, rather rare. Extreme care would need to be taken to make sure that the people he was studying were those he was *talking* about, as opposed to bad respondents and tricksters. That care was not taken. In addition to the fact that Regnerus simply confused having a parent have a romantic attachment to a same-sex person (and actually Savin-Williams shows us that this phrasing may be especially likely to lead to a large number of false positives) with growing up with two gay parents, Cheng and Powell (2015: 620) found some of the key cases included the respondent who claimed to be 7′8″, to weigh 88 pounds, to have been married 8 times and have 8 children; another claimed to have been arrested at age 1 and spent less than 10 minutes on the survey. Take out all the questionable cases, and guess what else you take out? The findings.[6]

Reification and Nominalism

Reification

We have seen the errors that come from mistakes of calculation, and from respondent error. But those are just the most flagrant forms of the many reasons we have to think twice before moving ahead too quickly. Even using high quality data, like census data, doesn't automatically exempt us from a skeptical consideration of what we have to work with. Here I want to start with the buzzword "reification"—or, to use Alfred North Whitehead's phrase, the "fallacy of misplaced concreteness." Sometimes those irritating kids in theory class make it seem like reification is intrinsically bad, and something only idiots do. That isn't so—as Hegel would say, if you're not willing to reify anything, you're not willing to develop at all. All of our data is going to involve reifications to some extent. But in some cases, this reification in our measures maps onto the reification in the world. Then the measures seem to work well enough. In others, it doesn't, and so we can't accept our measures at face value.

A great example here has to do with official racial classifications, like those of the US Census. This census data comes from surveys, increasingly self-administered. Many scholars have focused on the changes in how the Census Bureau classifies people. But there are other changes, ones the Bureau didn't necessarily plan, that come from complexities in how

6. As this goes to press, we learn from Sullins (2017) that errors in the National Center for Health Statistics data means that around three-quarters of those classified as same-sex parents from 2004 to 2007 were in fact misclassified opposite-sex parents!

people answer these questions, and that can really wreck certain research programs. And indeed, studies of census behavior in the US, but even more in Latin America, have found that the outcome of race measurements can be a negotiation between the census taker and the resident. Not surprisingly, we also expect that for some questions, the mode of data gathering—mail questionnaire or in person—will make a difference.

Researchers have found that in Latin America, in mixed-race marriages, the wife often "follows" the husband's race—if he's white, she's more likely to "become" white (to identify as such) than he is to become non-white; if he's black, she's more likely to become black than he is to become non-black; and so on. I'm not saying that it's morally or scientifically wrong to reify race, but I am saying that it is a problematic analytic choice, since our subjects sometimes work hard to de-reify it. And just because the government says it, doesn't mean it's so.[7]

Indeed, using the census data on race is something that requires a great deal of care, because both the categories and the procedures used to assess race have been in *constant flux*. Some of the data—especially the mulatto categories from the 1890 census—were basically declared by the Bureau to be junk as soon as they were in. This enumeration was pushed on the Bureau by politicians (Hochschild and Powell 2008), and census workers were asked to determine the proportion of black ancestry to the nearest eighth (Bennett 2000: 166)—but no instructions were given as to how this was to happen. A precipitous drop in the number of mulattoes from 1910 to 1920—decried by the official journal of the National Urban League, *Opportunity* (1925: 291)—seems to be a result of the fact that the Bureau employed a large number of black numerators in the 1910 census, leading to a boost in these numbers (Hochschild and Powell 2008: 70). And a switch from the census worker determining race to having race generally be self-enumerated in the 1960 census means that before and after statistics are going to be very difficult to compare.

7. One time, discussing some issues about data reporting regimes, Ed Laumann mentioned that in 1997 epidemiologists were confused by the data on chlamydia coming from the Centers for Disease Control (which collate the information from localities). Why would Indianapolis have so much when Detroit (!) had almost none? And why did Chicago have a huge burst for a few months? They had to go upstream to figure out where the original data had come from. In Indianapolis, a local doctor and expert in STDs was leaning on all the doctors to test for it so that he could do a study; Detroit, in contrast, ran out of money and couldn't even pay a clerk to process the paperwork. Why the blip in Chicago? That corresponded to the time when CDC officials were in town to conduct training and start up their surveillance program. We'll return to the puzzles of dealing with similarly pooled data in chapter 6. Keep this story in mind.

Reification is a bigger problem than it might seem, because to many of us, it's actually a *solution*. That's because we're taught that the solution to problems of disagreement about terms is that each of us has to "define" our theoretical terms. In too many cases, that becomes equivalent to a commitment never to undo misleading reification. So let me go on and discuss the problems with this approach, before I turn to what we can do about these problems.

The Perils of Nominalism

In *Thinking Through Theory* (2015) and elsewhere, I've emphasized that sociologists tend to *think* ideal-typically, but *work* nominististically. *Nominalism* is the theory of cognition (and often the basis for an epistemology) that says that the only reason that we see generalities in the world (for example, "mammals") is that we have defined them so. If we didn't have an intellectual commitment to the notion of mammals, we wouldn't see them as such. In contrast, *realism* (in this sense) is the theory that generalities are inherent in the nature of the world.

So our *theory* of knowledge (nominalism) is that we "construct," for our own purposes, "concepts" that are like boxes that we can put particularities in. If you "define" as a juvenile delinquent "anyone under 18 who has had an arrest or more than three misdemeanor citations," then, for your manipulations, that's what a "juvenile delinquent" *is*. I'm not here to bust on nomimalism. The problem is that while we *work* in nominalist terms, we revert to realism when it is time to interpret the meaning of our findings. So no matter how you *define* delinquency, and code up your cases accordingly, when you *think* about the causes or effects of delinquency, chances are, you have a particular vision in your head of a *real* delinquent—perhaps Mean Jim who pantsed you in tenth grade—and this may not be typical of the people you include in your category.

For another example, consider immigration. It's common to distinguish between immigrants and their US-born children. Well and good. But I find students often assuming that the "immigrants" are uniformly those who immigrated *as adults*. That's their ideal typical model. But in all but citizenship status, a person who immigrated with her parents when she was 8 months old may be much more like someone born in the US to parents who immigrated four months before the birth, than she is to someone who immigrated as a young adult.

It is vital that you don't just *define* your categories into existence—that is, just throw out your conceptual nets over the world, and then rush to analyze. You first need to inspect very carefully what you have caught. If we don't, we'll be (like many researchers) troubled by the surprising finding that most "children" who die of gunshot wounds are killed by other "children," that most "abusive spouses" are wives, and so on.[8]

And there's one last, I think underappreciated, danger in our tendency to substitute our terms for the data. It's a normal cognitive error for humans to give undue weight to one side of certain bifurcations or continua, because that's what we *label* them (they're cognitively "marked"—see Zerubavel 1997). For example, when we distinguish race as black/white, black is the marked category, and white the unmarked. The random American is assumed to be white, and white doesn't seem to bring information. So when we think about a "race" effect, we tend to imagine that it has to do with blacks, because they're the only ones who "have" race.[9] When we think about "education," we think it comes from the more educated, and is related to the education they got. But whatever's driving your data doesn't know the terms you've applied! Maybe you need to think less, and look more.

Theory Testing

This all might seem obvious, but it's actually a matter of dispute. Many sociologists—especially those who teach methods courses, I think—hold that more important is that you have a clear theory, define your terms, figure out how to translate them into measurements, and test your hypotheses. I think that encourages us to spend too much time with things that might not mean anything in particular. Let me use some of the work that comes from pursuing the (I think misleading) concept of "social capital." One of the key versions of this was developed by James Coleman (1988), who proposed that high school students did better where they were embedded in networks with "social closure." (Coleman lived in Hyde Park, Illinois, and he couldn't help noticing not only that he saw the

8. This is because if we call 16- and 17-year-olds children, we'll see a lot of them killing each other; if we define spousal abuse as any physical striking, we'll find that women do it more often than men (presumably since they are less likely to harm the partner).
9. I confess to an interesting episode of this, provoked by some hierarchical linear models of voting which had random slopes by race failing in some early years. We looked closely at the data to figure out the problem, doing crosstabs of party-voted-for by race in each congressional district in our sample, and learned two lessons we of course already knew. The first was that our models were struggling because before the Voting Rights Act, there were plenty of congressional districts in the South in which we could not find a *single* black respondent in the sample who had cast a vote. The second is that the variation in the race parameter wasn't due to blacks, who, at this time, were voting nearly monolithically Democrat where they were *able* to vote. It was due to *whites*, but because we termed it "race," I first stupidly assumed it had to do with blacks.

same people every day, but that they were as multiply related as second cousins in L'il Abner's world. Your colleague was also your son's friend's father, who was also your neighbor, who sat with you on this town committee and whatever. It's a small, depressing, little town of overinvolved people.) Anyway, Coleman thought that such social closure could explain why the kids of Hyde Parkers, or those in Catholic schools, did so well in school—the social closure of parents knowing their children's friends' parents. You can always track down your kid if he's playing hooky instead of working.

So Coleman collaborated on making a survey (the National Education Longitudinal Study of 1988) where this could be measured: they specifically asked parents to name their kids' friends' parents. Later, analysts attempted to test his hypotheses. Carbonaro (1998; also see 1999) and Morgan and Sorensen (1999) both examined to what extent student achievement could be predicted by the number of a student's five closest friends' parents known by that student's parents (hence "closure"). Carbonaro looked at this as an *individual*-level covariate, and found it mattered, while Morgan and Sorensen treated it as only at the aggregated school level, and found it didn't. This seems like just the sort of theoretically focused debate that we want in sociology.

But (as I also found, when working with a student replicating, and we were confused as to some patterns of results and so we went to the codebook) Hallinan and Kubitschek (1999) showed that it was far from clear that the debate was being waged in the proper terms, because the variation in what both sides were accepting as a measure of "social closure" was being driven by very different processes. Let's dig into the codebook (available at http://nces.ed.gov/surveys/nels88/pdf/10_F2_Parent.pdf; accessed November 27, 2016). Here's how it reads: "Do you know the first name (or nickname) of any of your teenager's close friends?" If No, then we skip to the next page. If Yes, then they are asked to name as many as they can, up to five. Then, for each friend, the parent is asked to mark whether this kid "attends school with your teenager" and whether "you know the parent/s of this teenager."

You need to be able to know your kid's friends' *names* before you can even be asked about knowing their parents. Almost all the important variation comes in the fact that there are plenty of parents who *don't even know their kids' friends' names*. They don't know their kids' friends' *parents*, or their kids' friends' *shoe sizes*, or their kids' friends' *favorite color*. Any of those might have done equally well in the model. Those who *do* know their kids' friends tend to know the parents too (the mean of the number of friends parents named is around 4, and the mean number of parents that parents claimed to know is around 3.3).

The conventional approach to methodology gives the theory of social closure advanced credit for predicting a relation between the supposed measure and some outcome. That's why Morgan and Sorensen felt justified to respond to Hallinan that, because Coleman had *planned* the item to measure closure, that's what they would term the measure. I'm totally unmoved by that. I say, look at the data on its own terms, and try to figure out what, in real world terms, not abstractions, is going on. Is this about social closure? Maybe. Is it about more active parenting? Seems more plausible to me. But if we are talking about social closure as a characteristic of some schools and not others, the question is whether this is just a rough indicator of something about the school environment that we could measure in any of a set of different ways (the number of Rice Krispies Treats sold in bake sales, 1 – the proportion of desks that have "do beers" carved in them, and so on). That calls out for a control strategy that we'll examine in chapter 4. But just in case you're on the edge of your seat wondering whether social closure *does* increase math scores—Morgan and Todd (2009) returned to the issue, trying to use measures that more closely tracked the theoretical concept, as well as the rigorous causal framework that Morgan and Winship (2007) lay out. The more controls one adds, the more the coefficient creeps downwards, until it winks out. Something is going on, but there are far too many correlated variables characterizing contexts to be able to make a strong causal statement.

To sum up, I don't think we should let our training give us confidence in interpreting a pattern in the data the way we want to, just because that's how we defined things. But of course, there are times when it seems we *do* need to do some defining. That's when we try to combine a set of answers into a single number, like when we make a scale. This is where a lot of problems can enter, so let's move a bit slowly.

Measures, Scales, and Indices

Our adherence to nominalism often leads us to confuse all our variables with measures. A *measure* is something that we believe is anchored in a single property of a unit, something that has out-of-the-lab validity. Not all numbers that we attach to our units are measures. For example, just adding up all the things we don't like about someone—and a fair number of our psychological inventories don't seem to me to do much more than that—isn't really a "measure" of anything. Can I philosophically draw the line between real measures and various nominalistic perversities? Not in the abstract. But you should be able to do so in practice. And where you hit something you're not sure about (for example, "*g*," the "general intelligence" factor that emerges from factor analysis), it's best to treat it with a healthy dollop of skepticism.

Here's a recent example. Ferraro et al. (2016) recently reported that, using the MIDUS (Midlife in the United States) study, only around a quarter (27.3%) of American adults were *not* abused by their parents! True, 31.5% had only "rare" abuse, but that leaves over 40% of Americans having experienced frequent abuse.

Ah, but if you look at the "Documentation of Scales" (Institute on Aging, University of Wisconsin, 2004), p. 24, you will see that what they called "abuse" was experiencing "at least once when growing up" that your parent . . .

List A: / Insulted you or swore at you / Sulked or refused to talk to you / Stomped out of the room / Did or said something to spite you / Threatened to hit you / Smashed or kicked something in anger

List B: / Pushed, grabbed, or shoved you / Slapped you / Threw something at you

List C: / Kicked, bit, or hit you with a fist / Hit or tried to hit you with something / Beat you up / Chocked [*sic*] you / Burned or scalded you

Some of these definitely sound like abuse. Plenty of them aren't. And some are downright funny. So if one time you called your dad a stupid fascist pig, and he refused to talk to you, chalk one up for abuse!

This is just an extreme form of a nominalistic problem that we have with scales, which is that we tend to replace in our mind the actual data with what we have decided to call the scale. For example, political psychologists were interested in whether certain political opinions were rooted in the "authoritarianism" of some people, a trait which encompasses (among other things) aggression toward people of other groups, when these other groups are made vulnerable by some authorities (see Martin 2001). To measure this, we make a scale to see how many things-that-an-authoritarian-would-think each respondent agrees with. And then we score people from low to high.

But there can be a big difference between being *higher* on the scale and actually instantiating the concept we had proposed. In some cases, the people who drive our empirical findings actually have only a few of the less central attributes that we've used to define our scale. We need to make sure that the features of our concept to which we appeal to explain some action are indeed present in a large proportion of what we take as the empirical instantiations of this concept. Do those people who both are conservative and oppose affirmative action actually also have enough authoritarian aggression to fulfill our understanding of the concept? Because if they only have "low" aggression (as opposed to none), we might need to rethink our ideas.

The road to scientific hell is paved with unreflective, uncritical methodological choices that seem to have the support of convention.

There's one last way in which our concepts and our scales/indices can mismatch. As I said above, we have a tendency to be distracted by the "marking" of one side of a variable due to our names. If we are measuring "education," we assume that whatever we are talking about has to do with the more educated, when it might be the less educated driving the phenomenon.[10] Making scales can be fine, and it can help deal with simple measurement error, but at the cost of a bigger danger—weakening the connection that the world has over our theoretical concepts, because we feel freer to label the numbers something that they might not be. So we can definitely do it wrong by forgetting the nature of the data and relying on our names for the resulting sums. But can we do it right?

Indices and Scales

The motivation for index construction starts with the recognition that often we measure things inexactly. For example, a classic item meant to measure self-esteem is whether one agrees with the statement "I can do things as well as most people." However, perhaps our respondent is 85 years old, and happens to be thinking of things like sliding down bannisters and running across the street, and, although having quite high self-

10. This problem of markedness returns in the construction of scales and other summaries, where there is often a temptation to weight some sorts of responses more heavily than others. For example, sometimes we ask people to put cases (say, foodstuffs *A, B, C, D, E,* and *F*) into piles based on similarity. We let them decide how many piles to make. One person makes piles [*A, B*], [*C, D*], and [*E, F*]; the other makes piles [*A, B, C*] and [*D, E, F*]. Shouldn't we weight the first's *AB* connection *more* than the second person's? After all, they must be *closer* for him, because he only includes these two in the pile. And he only makes half as many same-pile relations as the second person. For this reason, some methodologists have actually encouraged practitioners to weight their distances in this way—we give the first respondents' *AB* tie twice as much weight as the second's.

But now let's think of the second person's *not* putting *C* and *D* together. Since she has only two piles, this is a greater indication of difference than the first person's not putting *B* and *C* together. Maybe we should double-weight this. But twice zero is still zero . . . so let's count a non-tie as –1. And we'll double *that* for the second person and . . . we're back where we started; the intra-individual differences are now the same. (It's this logic that explains the counter-intuitive implication of the Rasch [1960] model that a simple raw score, weighting hard and easy items the same, can be a sufficient statistic for a latent trait.)

esteem, says "No" to this item. We might not be getting at self-esteem with this one question, at least not for people like this. So if we ask lots of different questions all having *some* relation to self-esteem, and we add them all up, we have a good chance of having the non-self-esteem stuff cancel out, and we'll have a better indicator of self-esteem than if we didn't do this. Many indirect measures, put together, might approximate a direct measure. That's our logic, and it's fine.

But we need to think through the logic of indices a bit further. That's because sets of measures that are *effects* of some underlying attribute behave very differently from sets of measures that are *causes*. The former should all co-vary together. And in fact, many of the statistics that we use to determine whether a set of measurements can be treated as an index *assume* that they are all effects of a common cause. But if we are measuring something by its *causes*, there's no reason to think that they covary. In fact, sometimes they need to covary *inversely*—unemployment and overwork can both lead to stress. But if you have one, you can't have the other.

Wouldn't it be great to have one word for measurement-by-causes (*index*) and another for measurement-by-effects (*scale*)? That distinction is indeed used in some areas of the social and behavioral sciences, but it isn't standard in all of them.[11] But let's adopt that convention, and tie it to a more familiar one, that between *necessary* and *sufficient* relations. We say that *a* is *necessary* for *b*, if without *a*, we don't see *b*. So *air* is necessary for *fire*; there's never any fire without air, but just because you have air doesn't mean you have fire. On the other hand, we say that *a* is *sufficient* for *b* if whenever you have *a*, you also have *b*. So given pressurized air and fuel vapor, a spark is *sufficient* to cause combustion; there's never any spark without the combustion, but sparks aren't the only way that we can cause combustion (a diesel engine doesn't use them).

My gut feeling is that, in general, we're better off measuring phenomena by their causes than their effects, but that's because (like a lot of other folks) I tend to think that real-world causality tends to sufficiency, and not necessity. Thus depression may lead people to miss work, but there are plenty of other reasons they miss work. So measuring depression by days lost is a bit problematic. But if getting fired increases depression, it should always (probably) increase depression.

11. In sociology we tend to use the word "index" when we are thinking a bit more rigorously about the underlying relation of our terms to one another, and "scale" more widely (when, for example, a single variable might be called a "scale" if there are many ordered response options). That's confusing, because in wider usage, index often means any set of numbers thrown together higgledy-piggledy, and in psychology, "scale" refers to something that is seen in more psychometric terms. Some people tend to use "scale" when they are adding items that themselves have multiple response categories.

I can't prove that mathematically, and I admit there's a wrinkle when it comes to making indices via their causes: we can introduce bias because, in most cases, we include only *some* of the causes. For example, many things can cause stress: death of spouse, loss of employment, being convicted of a crime, being subject to witchcraft, business readjustment, and trouble with in-laws. These are a few items from the famous Holmes-Rahe stress scale. Oh, but of course, they don't include bewitching. Why? Because Holmes and Rahe were the sorts of people who aren't stressed out by witchcraft.

Others are—in fact, it seems to be the most stressful experience possible in many societies, such that it can lead to death. And there are some folks in the US who are also stressed out by witchcraft, though we have no idea how many. But because most of us make indices of causes from our perspective, they tend to track PLUs best (that stands for "people like us").

That's the problem with taking a theory-driven approach. It's only as right as your theory. And our theories are just not that good. They don't reach far enough to get to other people, other cases, other times for us to trust them. It's for this very reason that we need to theorize—not to slap fancy words on our results, but to theorize the processes whereby our data were created. I think you'll find that this gets to be way more interesting than the abstractions you toss around in class. Instead of starting from what some guy with a beard said, you start with the data. You work on testing it, like you would test branches if you were climbing a tree, before you put your full weight on it, and then, trying to get an overview of the whole, and what is *in* your data.

How Can I Learn More about My Data?

Genealogical Investigation

Now I hope that you realize that just because it's "data" doesn't mean you can rely on it for whatever you're wondering about. What's the solution? The first thing is to be attentive to where it came from. In most cases, the data came from interviews—just like those in-depth interviews that many survey analysts seem to think are unscientific. Good data is only as strong as the weakest link in the chain, and the weakest link can be the interview itself. Yet many researchers have no idea what went on here, and they don't want to know. (If you're interested, Hyman [1954] is a great beginning on the phenomenology of the interview.)

Our first cautionary tale is the wonderful case of the rapid increase in social isolation in America. McPherson, Smith-Lovin, and Brashears

(2006), in the flagship journal of the American Sociological Association, reported this shocking finding. They had used *very* good data—what I think is the best social survey data in the country, the GSS (General Social Survey). In 1985 this included a module on social ties, asking people about who they talked to regarding important matters. (This was termed the "core discussion network." Whether or not this is a good way of measuring close ties, we'll return to in a second.) This module was repeated in 2004, and McPherson and colleagues analyzed the new data, finding a shocking increase in the number of those who had *no* close ties. It was such a strong finding that they were themselves skeptical, and they investigated a number of possible explanations for this finding (other than that it was true). Even in their (influential) report, they suggested that they weren't sure about the magnitude of the change. But since they couldn't explain it away, they went to the press.

Now in hindsight, that looks kind of silly, and there were sociologists like Claude Fischer who knew enough about the US to be sure that this just *couldn't* be true. It's like when the thermometer says it's 75 degrees but you can see the lake is still frozen.

Anyway, the puzzle was finally solved by Anthony Paik, along with Kenneth Sanchagrin (2013). It turns out that some interviewers were bad interviewers. They communicated (whether explicitly or not) that they didn't want to spend any more time in the interview. Wherever they went, they got people who said they had no friends and skipped entering in all that information. How did Paik realize this, when the analysts and critics hadn't? Paik had been a graduate student at the University of Chicago, and he had worked for the organization (NORC) that collected the data, though not on this particular project. Still, he understood how the data were made, and so he knew what to look for.[12]

But now let's think about the basic prompt itself, and this notion of measuring ties by "core discussion networks." Does it even make any sense? The answer is that it makes about as much sense as any other approach to measuring a totally vague abstraction like "networks." I want to emphasize that a lot of thought went into choosing this notion and the precise wording used, as opposed to just grabbing a conventional term like "good friends." The question writers were doing their best. But it doesn't make sense to treat networks as if they were obdurate, real entities. Most

12. Lee and Bearman (2017) argue against some of Paik and Sanchagrin's evidence, but the pattern of interviewer contact with isolates just seems too improbable to be compatible with a world of identical interviewers. Lee and Bearman present other important information that is relevant to general issues of reported network *size*; Paik and Sanchagrin's work focuses on *isolation*. Both approaches seem to have uncovered something.

people have a number of ever-fluctuating and multivalent relations with others, with no discrete break dividing friends from acquaintances, acquaintances from strangers. When they're asked to report on these sorts of "who" questions, it's actually a pretty tough cognitive task to pull out some names. And it's doubly tough when the item requires a lot of interpretation—what *is* an important matter?

You don't want to stop with *your* idea of what the item means—you want to see how respondents interpreted it. As Bearman and Parigi (2004) discovered when they used results from a poll that asked the same question, but then recorded *what* people had been talking about, some probably didn't think *anything* they said would be important (and who knows, they might be right), while others elevated topics that might strike some as relatively unimportant to the status of important.[13] Further, we can imagine that something that seems important at one time might seem less important at another (for example, a national crisis might make the issue of Lisa's braces seem less important). And indeed, it appears that when there are important political affairs brewing, many respondents implicitly think that important matters are these contentious *political* matters . . . which they might not talk about (Lee and Bearman 2017).

So when it comes to something as inherently abstract as social networks, we're going to always want to put the labels on hold, and try to understand the processes of prompting, recall, and notation that actually produce the data. Even when we can't see these actual social processes, we can learn a lot by looking at the questionnaire. And where we can't convince ourselves that the data production was done right, we need to be very skeptical of our data.

Skepticism

In other cases, the data don't come from well-conducted interviews that allow us to chase a finding all the way upstream to paper forms. And that's increasingly true of attractive by-product data that is made by organizations. Remember: your analysis is only as strong as its weakest link. That means that if you are merging bad data in with good, the result is bad. This is increasingly an issue, as there are all sorts of attractive "free data" that may promise you a quick fix for a real problem. Uncle Bob's advice might have been right—you get what you pay for.

Let me work through a recent example. So if you live in the United States, you've probably noticed that rich people have been leaving work

13. And I have found that my colleagues seem to think that *anything* they say is important, so that makes us *all* part of their core network!

with a hell of a lot more of the money than they used to. People wonder what explains this change. Lui and Grusky (2013) argued that a good chunk of the reason had to do with the *skills* these elites had. For some reason, "analytic" skills were being more highly rewarded than they used to be.

Okay . . . but how do we tell how much skill any job requires? Well, Lui and Grusky used the Occupational Information Network (O*NET) to get their codes for each occupation's skill requirements. First, I should emphasize that a huge amount of work went into the construction of these estimates (see Tippins and Hilton 2010), and there's been a lot of study of their reliability. (But let's remember, reliability is not the same thing as validity, and that's relevant here.) O*NET really tried to get valid responses; they even tried to get respondents to differentiate between *how important* a skill was to the job, and about *average level* required, anchoring this with brief examples.[14] Second, Lui and Grusky are totally aware of the nature of the data, and they discuss some of the issues.

But still, their key argument relies on the validity of the reports on analytic skills. I don't know how good people are at reporting this for their jobs. I feel pretty confident that the professional analysts aren't great at this; in many cases, all they really have is a job title and a paragraph describing the job, and many of the O*NET ratings seem to be based on sheer prejudice. For example, what occupations require the most "deductive" skills? In second place are "judges" and "police detectives" (O*NET distinguishes the *importance* of the skill and the *average quantity* possessed by incumbents; for detectives, deduction is very important but they might not have as much as do urologists). Plumbers need a lot less deduction than either of these. Now somewhat suspiciously, "gaming cage workers" and "gaming and sports book writers and runners" require and typically have the exact same amount of deductive reasoning as each other.

It's not that I think that a gaming cage worker would be better than an analyst at reporting how much deductive reasoning they need. I don't doubt that there are skills required for jobs, nor that people's reports on this are non-randomly related to the actual skills. But when it comes to something very abstract like the components that went into the construct "analytic skills," we have to worry if even the most dedicated respondents and analysts can do much more than recycle prejudices. The fact that they *share* these prejudices isn't going to be reassuring when it comes

14. For example, when it comes to "critical thinking," they suggest that on their seven point scale, (2) would be "Determine whether a subordinate has a good excuse for being late," (4) "Evaluate customer complaints and determine appropriate responses," and (6), "Write a legal brief challenging a federal law." (7) is the highest, by the way: I think it could be, say, "Write a book critically examining the validity of government data."

to accounting for why the higher strata are raking it in.[15] And that can make it tough to know how much weight to put on them when explaining inequalities. The increase in returns to analytic skills is an interesting story, but here's another one: Okay, so in the eighteenth-century Caribbean, Blackbeard and his friends are hard at work composing a database on sailors. "What be th' skills most required for to be a buccaneer cap'n, me hearties?" Sucking thoughtfully on a bone, up speaks Calico Jack Rackham, "'Tis analytic ability that marks the true seadog, I'll wager." "Aye, spoken as Bible truth." Later, the poor landlubbers attempting to explain why Captain Kidd is so flush, dutifully doing their regression, conclude that it is, after all, explained by his high analytic abilities. He wouldn't *be* a captain if he didn't have them. The pirates even *said* they looked for this! I've met some of those higher-up manager types; some of them did have analytic abilities, often on a par with those of a squirrel monkey, and far inferior to those of their subordinates whose pensions they were raiding. Maybe there's something to that old Marxian idea that it's really hard to do good science in any environment of conflict, when "facts" are easier to make by those who have money than those who don't, and where these facts help them hold on to their money. It isn't necessarily true. But it's a hypothesis worthy of examanination!

Finally, when it comes to cross-national research, just reading the translated version of the codebooks isn't always good enough. Many of the commonly relied upon data sets have serious problems due to translation issues. Strangely, the World Values Survey for 2001 reported that 99% of Vietnamese favored military rule! But that's because the translation actually means "rules *for* the military," not rule *of* the military (Kurzman 2014). Even worse, when it comes to things like values, there aren't always clear equivalents across all cultures (though see Wierzbicka 1996)—there isn't even a *right* translation that will fit our conventional methods! Back away.

Triangulate

Okay, let's say that you've investigated the parentage of your data and found no reason to think that they are deeply problematic. What next?

15. Indeed, troubled by the possibility of respondent over-estimation of their abilities (for incumbents *do* tend to rate their jobs' skill requirements higher than do analysts), Liu and Grusky deflate the analytic abilities of jobs where they think that the rise over time is implausible, by regressing them back to a function of analysts' reports. That might be right (and it also might not affect the results much), but we want to make sure we aren't making our test of validity the shared agreement of analysts, upper status job incumbents, and professors that the less well-paid lack certain skills!

Where you can, you conduct descriptive analyses that can help you get a sense of the internal validity of the data.

I remember a job talk at Wisconsin, where someone was using data that was intrinsically two-sided: perhaps something like husbands' and wives' reports as to what proportion of housework each did, and the presenter was studying whether something like whether women's hours of work changed this balance. And it turned out that she had just averaged the reports of husbands and wives, and "hadn't yet" investigated the degree and nature of agreement between husbands and wives. And Jeremy Freese and I looked at each other and imperceptibly shook our heads. Unacceptable. Who would go ahead and start "explaining" stuff before they knew what they *had* in their hands?

So, too, I've been stunned to see people examining reports from schools on two-sided data like fights, bullying, and partner abuse, and being afraid to look at the simple tables of alter's and ego's reports. The first thing that you should do, if you are serious about learning *from* your data, is to learn *about* your data. See what is on the diagonal (where people agree) and figure out where they *don't* agree. Don't sweep disagreement under the carpet—examine it. You can learn a tremendous amount.

For example, the network data that I used (and still use) from Benjamin Zablocki's Urban Communes Data Set is of very high quality. In fact, for network data, I think it's incredibly good. There are some disagreements as to relationships, however. Albert might say that he never sees Betty, while Betty may say she sees him regularly. In one or two cases there might be simple error, whether by respondent or interviewer, but in most of these (still relatively few) cases, these are disagreements of interpretation. What "counts" as seeing? Of course, Zablocki's items were written to be unambiguous (". . . get together *in person*") but still, some folks may think of "in person" as including the telephone.

There are even one or two disagreements about whether people are married. But this doesn't mean that the data is "bad" or that one of the respondents is "wrong." Many respondents treated committed partnerships as common law marriage; others understood the question to mean "piece of paper" marriage. And there might even be disagreements about whether two people ever were, or still are, formally married. If you briefly joined the Jupiterian church of Dolphin Consciousness, and were there married to "Ba," and then woke up and got the hell out of there, you might decide that the marriage conducted by High Priest Io was invalid, that you never were married, and you sure aren't *still* married. But Ba, still in the group, may believe that his soul and yours are forever entwined, until, in the next life, reborn as dolphins, you jump through hoops together at SeaWorld.

In fact, disagreement can be far more enlightening than agreement. Let's take the National Health and Social Life Survey (Laumann et al. 1994), which is really a top quality study that allows for the exploration of the *meaning* of acts, and not simply what takes place. In this survey, as in most others on contemporary Americans, men, on average, report having more sexual partners than women, on average—something which seems to be a priori impossible. This discrepancy has sometimes (e.g., Lewontin 1995) been used to justify a dismissal of the results of such surveys. But rather than being the ground for dismissal, this is a fascinating opportunity to study the ways in which men and women understand sex. First of all, close analysis may find evidence that the reporting bias is less than imagined. If non-response is correlated with sexual activity for women (positively) or men (negatively) than the actual difference will be overstated. The same will occur if sampled men are more likely than sampled women to have sexual partners who are outside the sampling frame. For one example, if men tend to have sex with younger women, more of their sexual partners will fall below the cut-off for inclusion in any study of adults.

Further, the critics seemed to imagine that the reporting difference would exist only if people were *lying*. But these differences can stem from differences in *recall*, differences in *definition*, or forms of *motivated cognition*. This last possibility means that people may classify ambiguous events in ways that help them make particular self-claims. So, for example, when you were 15 years old and went to some keg party and drank five plastic glasses of beer and then accidentally spilled the sixth before finishing it . . . well, how many beers did you drink? When you're talking to your friends ("Wotta great party!"), you say, "whew, man, I drank six beers!" but when you're talking to a cop you say "but I only drank four beers!" since they probably only have 10 ounces in each, or maybe "three" if those cups were, say, only 82% full. The same may happen with deciding how to classify physical interactions.

Fortunately, we can test this and similar explanations by examining whether the discrepancy between men's and women's reports narrows or widens as we specify the type of partner involved. For example, we can see whether the gap between men's and women's reports is smaller when we look at, say, partnerships between men and women of roughly equal age or very different ages. If all the discrepancy was due to differences in the sampling frame, we'd expect the discrepancy to decrease as we moved to these more focused comparisons. Of course, finding this reduction doesn't mean that we can be confident that in the world, there isn't any reporting difference. But if we found that the discrepancy *doesn't* decrease, we can proceed on the assumption that there *is* a reporting difference that has theoretical significance; further, finding that this difference is greatest when it comes to some sorts of relationships and not others begins to suggest reasons why the difference arises.

To briefly illustrate how such analysis might proceed, the NHSLS provided the somewhat surprising result that Catholic men report the highest rate of heterosexual anal sex, nearly twice as high as Protestants and even higher than those with no religion. (Father Andrew Greeley, who was a priest who *also* worked at NORC, when he wasn't writing steamy novels, loved to go around talking about this finding.) While Catholic women report higher rates of anal sex than do Protestants, these rates are lower than those reported by Catholic men, and lower than those reported by women with no religion. This difference persists even when we look at the last sexual event (where we would imagine reports to be most accurate). Now this does not necessarily mean that there is a reporting difference—Catholic men may be having anal sex with non-Catholic women. But they reported on the religion of their sexual partner, and the difference persists even when we look only at relations between Catholics—Catholic men are more likely to report anal intercourse with Catholic women than Catholic women are regarding their relations with Catholic men.

Now the Catholic men and women in the sample are not the partners of one another—so it is possible that both men and women may be reporting accurately. But the likelihood that such a discrepancy could result with a random sample from a population in which there was no discrepancy can be determined from the data. A sequence of such examinations can then be used to get a handle on the nature of reporting differences more generally. In this case, the data starts to run out on us before we can attempt to see whether this might be due to embarrassment (which might be the case if anal sex is being used as a form of contraception), or to different reporting conventions (e.g., some men count an act as satisfying the definition if *some* portion involved anal sex, or/while some women count it thusly only if it did not also involve another form of penetration).

In sum, rather than being afraid of asymmetries between reporters on the same events, this is where you want to start. This will give you the leverage to pull apart things that are normally very difficult to get a handle on using individual-level data.

So here we've been looking at reports of the same events (or the same population of events) from two sides. But in other cases, where people are asked questions more than one way, or surveyed more than once, we can compare their answers over time. Of course, just because the two answers disagree doesn't mean the data is bad. Some people change their minds. But no one re-grows limbs. Remember, that's how the bad respondents were identified in Add Health. If you *can* test for intra-individual consistency and *don't*, you're a shyster. Finally, we can sometimes see in the pattern of individual responses something that alerts us to a problematic interview—for example, a switch to all "no opinions" or consistently "agree" at some point suggests that someone got bored or irritated—or handed the phone to his one-year-old son.

Describing Data

Getting a Feel

Once you have determined that your data isn't trash, you are in a position to switch from disbelieving your data to believing it—and letting it guide you. One of the first things to ask of your data is "what questions can you answer?" You do this through data exploration. And when you have done this enough, you'll start to get a feel for "good data" and "bad data." I can't quite explain it, but sometimes you know even data that isn't faked can be (relatively) bad. Not that it contradicts your ideas, but that there's not all that much you can learn from it. More often, you have *good* data—it firmly contradicts your ideas, and teaches you something different. And the way you learn is almost always to start by making tables. Two-way tables, four-way tables, six-way tables, as much as your overtaxed brain can handle.

From doing this, you will develop a feel for what to chase down. "Hmm . . . 17 out of 92 in *this* table versus 29 out of 108 in *that*. . . . around 19% vs. 27% . . . not worth pursuing. . . . but 17 of 92 versus 37 of 108. . . . worth a look. . . ." And you make initial inferences ("maybe it's really about the *discrepancy* between these two things . . .") which, if not quickly confirmed ("nope!"), leads to a sense of what is *not* a plausible way of pursuing things. And then the data will help generate hypotheses for you to then chase down.

Here's an example I'm working on at the time of writing this. I was interested in whether *jointly* measuring people's support for (or opposition to) both the Occupy Wall Street (OWS) and the Tea Party (TP) movements could tell us something we didn't know about political ideology and the relation to the party system. Let me tell you what I learned from laying out all those cross-tabulations on my floor. First, supporting one movement is associated with *not* supporting the *other*. Surprised? Probably not. Didn't seem worth the effort, does it? Second, being Democratic and/or liberal (versus Republican and/or conservative) definitely predicts *positive* (versus negative) support for OWS and *negative* (versus positive) support for TP. Are you snoozing yet? Are you laughing at my stupid way of trying to look at the data? But . . . make tables of support for OWS and support for TP, and do this separately for conservatives and liberals, and separately for those with and without college education (a 2 × 2 × 2 × 2) table. And *look* at those tables.

Within any ideological category, there's a *positive* correlation between support for OWS and support for TP! Just what you *wouldn't* expect if they were inherently opposed, such that it was inconsistent in some way to support both. This now suggests a pretty interesting theory: both OWS and TP measure a pissed-offedness, and ideology *orders* which one of these is "easiest" to answer first. By bringing in other data, this theory can be tested. (So far, it seems to be true. . . .) Of course, if you just stop with the generated hypothesis, you'd have a tendency to overfit. What you then should go on to do, as I suggested in the previous chapter, is to test your interpretation not by doing any particular *statistical* test, but determining different *substantive* implications of your working theory. Don't worry—there will be plenty of opportunities to reject your insight later. Let's concentrate now on generating it in the first place.

Making It Easy

The secret to this endeavor turns out to lie in cognitive science, not statistics, and it is this: most of us have plausibly good pattern recognition circuitry. We can *hear* patterns, even feel textures, but it seems to be easiest for us to *see* them . . . and certainly, easiest to convince others with visual patterns (Latour 1986). The problem with using this capacity to make social science is that our environment is usually too complicated for us to really be able to recognize the patterns. So in many cases, the breakthrough is reducing the data to some form in which you can literally see patterns.

For this, "visualizations," as they are often called, are vital. I've been emphasizing using simple tables of numbers as visualizations. Tables have the advantage that they are very general, are very quick to produce, and have a stable language once you know how to read them (all you need is a cell value, a row percentage, and a column percentage, and you're on your way).

There are more highly "leveraged" visualizations that can give you much more, but the simplest "canned" version (with preset parameters) can be misleading for your particular data, and so you must be willing to take the time needed to tweak them to your purposes. Some visualizations are famously bad for pretty much everything (e.g., the "pie chart"), and others work for some questions and some data sets, and not for others (the bar graph). Some visualizations are *great* for low dimensionality data structures (most importantly, correspondence analysis), but are misleading if you have a much higher dimensionality. You can learn to use a "rota-

Table 2.1. Age, Period and Cohort Structure of a Table

		Period										
		1900	**1910**	**1920**	**1930**	**1940**	**1950**	**1960**	**1970**	**1980**	**1990**	**2000**
	100	1800	1810	1820	1830	1840	1850	1860	1870	1880	1890	1900
	90	1810	1820	1830	1840	1850	1860	1870	1880	1890	1900	1910
	80	1820	1830	1840	1850	1860	1870	1880	1890	1900	1910	1920
	70	1830	1840	1850	1860	1870	1880	1890	1900	1910	1920	1930
Age	**60**	1840	1850	1860	1870	1880	1890	1900	1910	1920	1930	1940
	50	1850	1860	1870	1880	1890	1900	1910	1920	1930	1940	1950
	40	1860	1870	1880	1890	1900	1910	1920	1930	1940	1950	1960
	30	1870	1880	1890	1900	1910	1920	1930	1940	1950	1960	1970
	20	1880	1890	1900	1910	1920	1930	1940	1950	1960	1970	1980
	10	1890	1900	1910	1920	1930	1940	1950	1960	1970	1980	1990
	0	1900	1910	1920	1930	1940	1950	1960	1970	1980	1990	2000

tion" or successive slices to get a pretty good sense of a three-dimensional projection, but I personally have never adequately been able to see in my mind's eye four-dimensional data, though we could if we were to produce a cube of data (say, via a hologram) that would change as we went back and forth on a fourth dimension. And maybe we could even get a sense of five dimensions if we could move the cube up and down, side to side, and six if we can walk back and forward in our room. But that's still for the future.

The key to a visualization is that it should give you a head start in developing intuitions about your data. Let me give one nice example. Demographers are familiar with what is called the "Age-Period-Cohort" problem. Something (say, divorce) may be changing, and we want to know is whether this is a function of people's *ages* (everyone changes as they get older), of some *period* (the 1970s were just like that for everyone), or a particular *cohort* (the early baby boomers are *always* different). The problem is each of these is a function of the other two (AGE = PERIOD − COHORT), which means that we can't explore them all simultaneously. It's mathematically proven as an impossibility.

Well, that's math. In sociology, we do impossible things all the time, like breaking out of the hermeneutic circle. Let's do it here. Table 2.1 shows how such data might be arranged. The vertical dimension is age, and the horizontal is the "period"—that is, when the observations were performed. The cell values are the cohort—when people were born. We're using 10-year windows and collapsing everyone in this 10 years together. Note that cohorts "travel" up and to the right (an arrow has been drawn

ble 2.2. Age, Period, and Cohort Data Matrix

						Period					
	1900	**1910**	**1920**	**1930**	**1940**	**1950**	**1960**	**1970**	**1980**	**1990**	**2000**
100	11	6	10	10	59	28	10	11	7	13	12
90	10	5	8	7	68	22	12	6	8	10	8
80	6	5	8	6	66	71	7	8	7	6	10
70	7	11	11	11	54	24	12	7	7	11	8
e **60**	7	13	13	5	27	54	9	9	5	5	5
50	8	5	10	13	54	84	8	12	7	10	6
40	9	12	7	9	13	25	7	6	11	10	11
30	9	10	8	11	40	18	12	6	7	7	10
20	12	9	12	12	26	77	9	7	10	6	10
10	5	12	10	8	41	36	7	5	13	13	5
0	10	7	9	5	35	68	10	10	11	6	5

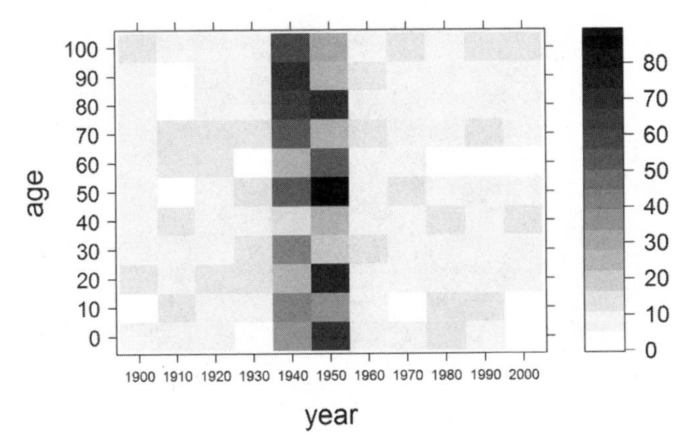

FIGURE 2.1. A Period Effect

to show the 1910 cohort). Then in table 2.2, I present some made-up data (say, it's the percentage of the married women who had their first child within a year of their marriage).

So do a quick exercise: what can you see here? Probably, if you look, you will see something. But you might be fooled. Do you see that the 1940 and 1950 columns are a lot different from the others? If so, you're right. Does it look like Age 70 and 60 is also different? Hmmm. . . . I wonder. Let's turn it into a visualization, by assigning shading proportional to the number in the cell (fig. 2.1).

This is just a quick version made with R and a black-and-white color scheme (R 2.1). But it shows that the pattern is really only a period effect.

Table 2.3. More Pretend Age-Period-Cohort Data

		Period										
		1900	**1910**	**1920**	**1930**	**1940**	**1950**	**1960**	**1970**	**1980**	**1990**	**200**
	100	13	6	7	9	5	13	6	20	8	11	11
	90	5	10	7	9	8	6	20	5	6	9	7
	80	12	7	10	6	9	23	11	6	6	9	9
	70	13	7	8	12	18	13	8	7	8	11	15
Age	**60**	12	11	9	18	7	9	8	12	12	8	11
	50	5	11	22	6	13	12	9	9	23	11	7
	40	10	9	12	10	10	11	6	22	12	9	9
	30	11	13	11	10	10	9	11	6	7	11	12
	20	6	12	9	11	5	17	7	13	11	6	9
	10	7	11	8	10	15	9	9	6	13	10	8
	0	6	10	6	16	8	13	7	5	10	7	11

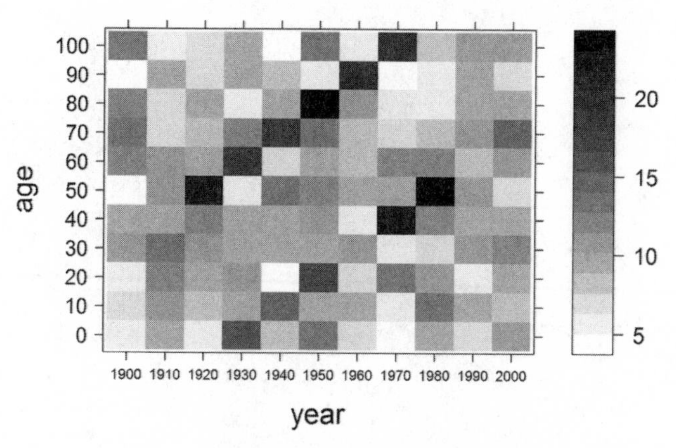

FIGURE 2.2. A Cohort Effect

The seeming pattern you might have looked for in the rows comes because your eye is attentive to the difference between 1 and 2 digit numbers, and the body of the table is random (choosing between 5, 6, 7, 8, 9, 10, 11, 12, 13).

Now that example involved a very strong pattern. All the table values were randomly chosen from a number between 5 and 13, except for the two columns, which were between 5 and 85! You'd have to be blind to miss that. Let's try something subtler—what's going on in *this* pretend data matrix (table 2.3)? Kinda hard to see, right? Let's try the same visualization (fig. 2.2; R 2.2).

A bit clearer, isn't it? If not, squinch up your eyes (an important methodological technique, I have found). There seem to be two unusual cohorts here (born in the 1870s and 1930s). Can I mathematically prove to you that these are cohort effects, as opposed to complex non-linear interactions of period and age? I sure can't. But guess what? I wouldn't be trying to. I'd be off trying to figure out what was going on in these two cohorts in their early years, and see if they had anything in common.

Simple visualizations not only can uncover a thesis, but also can sometimes prove it. A wonderful recent example comes from Voas and Chaves's (2016; also previously found by Hout and Fischer 2014) examination of secularization in the United States. Most sociologists of religion have been impressed with the vitality of religion in the US, and have used this as a key piece of evidence to reject the notion that there is a tendency toward secularization, even in the first world. Voas and Chaves (e.g., p. 1531) present line graphs of various measures of attachment to religion over time, by cohort. *None* of the lines cross. At any age, every cohort in the US is less religious than an earlier one.[16]

Boom. Case proven . . . or as close as we come in social science. Mathematically, it's true that there could be another explanation. But until someone comes up with it, we're going to accept that they're right. The burden of proof just shifted dramatically.

In sum, we want to arrange data so that it gives us a head start in recognizing patterns, and visualization can do this. Some visualizations are totally useless (the "network" hairball that people *still* draw for large networks, just to prove that they can), and nothing works for everything. How do you choose? Put your theory on hold, get a sense of the main contours of the data. Are there clumps? Do some variables have no variation? Are others basically all tapping the same thing? If you have a national-level data set and region really structures the data, always make maps, or separate rural from urban, or whatever. Don't ignore what you *know* about the data and try to "control" it away, which is equivalent to asking what a corn farmer who lived in Brooklyn would think about water policy. Instead, take what you're interested in, or your hunch, and see if you can pursue it in a representation of the data that does justice to what you already know is unavoidably going on.

The Other Side

So I'm very enthusiastic about pattern recognition. But there's a practice that often seems similar, and is deeply problematic. I'm going to call it

16. And yes, new immigrants may bring more religion with them, though, as Voas and Chaves (2016: 1546f) correctly note, this doesn't disprove the thesis at hand; as they put it, the theory that a warm bath cools over time isn't disproved by the ability to top it off with hot water. But in any case, the immigrants aren't religious enough to compensate for the trend among the native born.

"messing around with data" (and I've discussed it a little bit in chapter 8 of *TTM*). This is when we sit down with a new data set, trying one algorithm after another, transforming the data, using cool techniques we've learned about, making pictures, processing the picture of the algorithm of the coding of the picture of the algorithm of the . . . holy smokes! It's four in the morning! This lastest picture *is* really cool, though I'm not sure exactly how I got it. . . .

Induction is about looking for robust patterns that jump out at you . . . not ones that involve a lot of overlaid creative processing of the data. What you are likely to end up with isn't a good grasp of the real data, but something more like pizza topping (often mis-assumed as "meat"). Do you know how they make this? Well, I won't tell you, because you'll barf right on the page. Suffice it to say that they use a lot of processing, and there's not much of what you or I would call meat in it. If you've got to scrape at that thing that hard to get something, walk away. Instead, let me go on to give you some sensible, and reliable, techniques that you can use to get a sense of where in the world of possible data sets the one you have before you lies.

Where Is It?

You Can't Miss What You Can't Measure

When Funkadelic sang those immortal words, it was clear what they were talking about: the fact that social researchers often aren't aware that they are missing the most important explanation for a phenomenon, because it hasn't been properly measured. (Unfortunately, they cut my favorite stanza from the recorded version, where they also point out that you can't miss what doesn't have proper variance in your data set.)

Our methods explain variance *with* variance, like using a rock to break other rocks. That means that no matter what your theory is, your explanation—if you get one—is going to be in something that varies in your data set. If your data set is truncated on some variable (for example, you have a restricted age range), it's going to be harder for that variable to carry a lot of explanatory weight. But even worse, the *level* of variation of your data is going to naturally lead you to emphasize certain types of explanations over others. Unless you know in advance what you are looking for, and how to translate it to your data, if you have individual-level data, chances are, you're going to come up with individual-level explanations. If you have data on criminals, your explanations of crime will have to do with them. If you have data on victims, your explanations will have to do with *them*. If you have national-level data, your explanations will have to do with national-level things.

I'm not saying you can't think of other hypotheses on your own, and even test them, but I am saying, you set yourself up to discover certain sorts of things by the kind of data you have, and where the variation is. More generally, if there's *no* variation in something, it won't appear as a cause, even if it is. So first, what sort of data do you have?

Tops and Bottoms

Once you have thought this through, there are three things to know about your data, especially about your dependent variable (if you have one). These are where the bottom is, where the top is, and where most of it is. Many of our variables are bounded, whether in reality or only because of our measurements. At the top, there is nowhere to go but down, and at the bottom, there is nowhere to go but up. These lead to what we call "ceiling" and "floor" effects, respectively. That means that we can predict a lot of patterns without even running tables.

I predict that, say, the children of the top 1% are likely to go *down* in income compared to their parents, and the children of the bottom 1% are likely to go *up* in income compared to their parents. Does that support a theory of gradual equalization, or deny the Matthew effect (them as gots, gits?). It surely does not, but it does support the vision of statistics that underlies all our procedures (which assumes that there is a tendency for a regression to the mean, and note that this does *not* imply negative correlations between generations!).[17]

More subtly, if there is asymmetry in the actual relations in the world, whether we are up against a floor or a ceiling can determine our findings. In a data set on children's friendships where we only ask about three friends, and almost all kids name three friends, all our variation comes from the very *unpopular*. Even if we *call* this measure "popularity," we won't find it explaining things that are associated with high popularity.

These sorts of floor and ceiling effects produce predictable patterns in our data when we look for heterogeneity in effects. Imagine that we're trying to predict the effects of education on income, and we have three people, one near the income floor, one near the middle, and one near the income ceiling. The person at the floor can go up, the person at the top down, but the person in the middle, both. You might imagine that there's

17. That is, if your mom was 7 feet tall, you're likely to be *less* than 7 feet tall, but still *taller* than the average.

going to be a tendency for the income-education relation to be stronger in the middle. And indeed, that's generally the case.

Define "local" parameters as those that come from repeating the same basic model in different subsets of our data set. We do that when we partition our data, but we implicitly do it whenever we construct interaction coefficients. Floor and ceiling effects tend to lead to U-shaped patterns of different forms in such local paramters. A reasonable rule of thumb is that when we have some sort of bounding of a dependent variable, looking at categories set up according to values of the *independent* variable leads to U-shaped pattern of slopes, while looking at categories set up according to values of the *dependent* variable leads to *inverted*-U-shaped pattern of slopes.

Let's do a simple simulation to illustrate (R 2.3). Sticking with our example, let's say that education (x) does in fact cause income (y). But we don't measure income exactly—instead, only know for each person, whether she is in one of, say, six evenly split categories (or "n-tiles," where n here is 6)—we'll call this *measured* income variable z. That's usually the sort of data we have. Now the true relation between x and y is the same across all values of x and all values of y.[18] But we don't know that, because we don't observe y. There are two questions we can ask: does the education-income effect differ by level of *education*? and does the education-income effect differ by level of *income*?

To answer the first, we split up our cases by their rank on education (x), and regress income (z) on x within each. And lo and behold, the slope will tend to be highest for the middle categories. (See, for example, the solid line in fig. 2.3.) Why? Because at the higher levels of education, even though we're going up in *true* income (y), that doesn't translate to a change in *measured* income (z), because of the ceiling effect. Now you might think, then I just won't ever do that. But suppose instead of splitting up your cases by their rank on education (x), you split them on something else that was *highly correlated* with education (perhaps, in some data sets, certain moral attitudes). There's going to be the same tendency.

But now let's imagine that we do actually measure true income (y), but we bin by the n-tile of y, namely z. We might do this on purpose if we were doing a "quantile regression." But it also could be that you don't really mean to bin by z; instead, you might be binning by something that is very closely related to z. For example, if neighborhoods are highly stratified by resident income, you might get similar results from doing local regressions within neighborhoods (here arranged in terms of their average in-

18. For this simulation, the relation is $y = x + \varepsilon$, where $\varepsilon = N(0,4)$; that is, the true slope is 1.0 and we have identically normally distributed errors of y at any level of x.

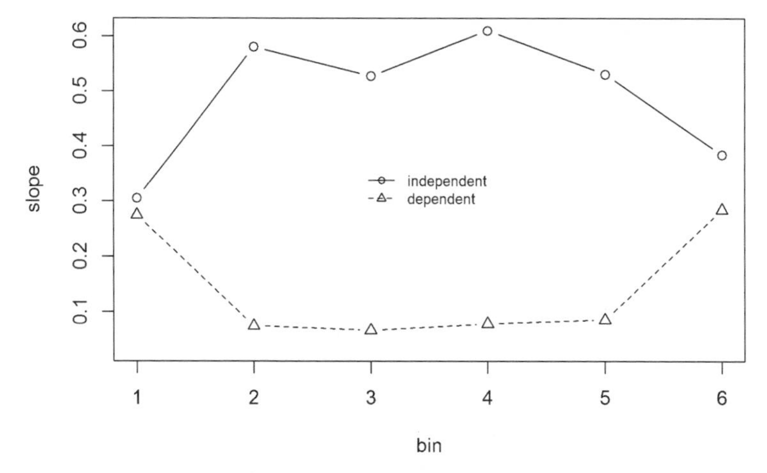

FIGURE 2.3. Floor and Ceiling Effects in Local Parameters

come, say). In this case (see the dashed line in fig. 2.3), we expect the slope to be smallest at the middle bins. Why? Because these include cases where someone with a high x (education) happens to have a negative error ε (that is, this person is in a poorer neighborhood than you'd expect), *and* those cases where someone with a low x happens to have a positive error ε (that is, this person is in a richer one than you'd expect). I'm not saying that a finding of an inverted-U shape is *never* theoretically important, but I *am* saying that the importance of such a finding decreases the more the predictions are actually coming up against hard stops in the data. And I am saying there are a lot of articles that take this fact as a strong piece of evidence for their particular hypothesis, when it isn't.

Where Is Most of It?

Even if we aren't running into floor and ceiling effects, we need to understand where our data are in a space of possibilities. We've already seen that our brains can wander away from where the data actually are. As we saw in chapter 1, if you compose a scale of authoritarianism, and regress it on other variables, because you are used to talking and thinking about "authoritarianism," you're probably imagining the people who are at one extreme end of that scale—those really nasty, creepy, authoritarians. But in actuality, you are regressing deviations from the mean on this scale on deviations from the mean in other variables. You might not have *any* observations at the extreme! Your "oomph"—your explained variance—might be all about people in the middle. Statistics teachers, like good Durkheimians, warn you about those pesky nonconforming outliers—

those with extreme values. But you can also get tripped up by those conformists—the mass of people with non-extreme values whom your model is trying to fit.

For example, Daniel Schneider (2012) had a theory about gender roles in work. Jobs can be ranked in terms of the percentage of those who hold them who are men and those who are women. Thus a job that is 95% men is gendered masculine. Schneider suspected that those who have jobs that are *either* very *consonant* with their own gender, or very *discordant*, would tend to engage in more gender-stereotypical activity *out* of work. So that means that men who are in 95% male occupations, *and* those in 5% male occupations, would do things like play football on the weekend, while those in 50% male jobs would be more likely to crochet.

This implied, or so reasoned Schneider, that one should find a correlation between both the linear and the squared term for %MALE in predicting gender-stereotyped activities. And he found just this. He was excited—though finding that the people in the 95% male jobs did gender-stereotyped activities wasn't the most important part. That would arise if macho guys just did macho work and had macho fun. It was finding that the people in the 5% male jobs *also* did gender-stereotyped activities that fit his conception of identity threat and all that. And enough reviewers for our top journal agreed.

Do you already see what the problem is? It's that very few men are in those 5% male jobs. The quadratic term is struggling to fit the precise shape of the relation at the *other* end of the distribution. It doesn't care much if it misses the few men in gender-discordant jobs. In other words, *the data weren't where the theory was.* Schneider hadn't split the sample to see whether his actual theoretical claim was born out. (He used two data sets; in one there was a negative relation at this low end, and at the other, there wasn't. He kindly did this in response to my request.)

Now to some extent, this supports a point I emphasized in *TTM*, which is that you *don't* want to "test" your theory, in the sense that you learn in statistics class. That's like giving someone a medical degree on the basis of watching them treat a single patient. If you really want to test that theory, shake it, set it against plausible contenders, have it out. Schneider did what we say one should do: test his theory. His theory implied a U shape, and so he tested a U-shape . . . not realizing that he was borrowing from the uninteresting part of his theory to keep the interesting part afloat. How can you avoid this? Easy. Make plots that also show the marginal distributions (called "marginal plots" in many cases)—an example for some simulated data is in fig. 2.4 (R 2.4).

Finally, you also want to make sure that your predictions—especially

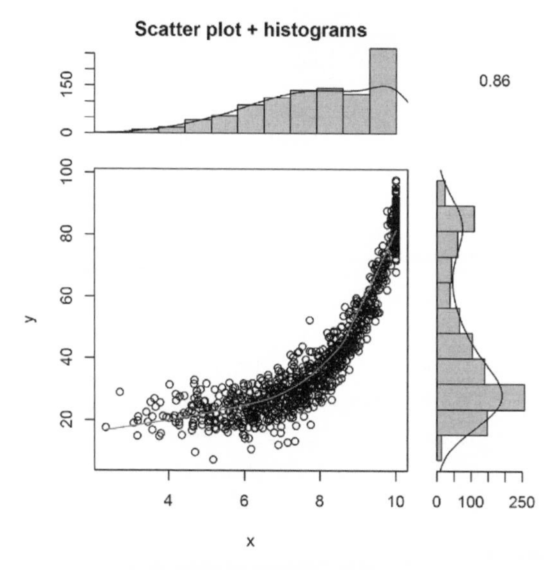

FIGURE 2.4. Marginal Plots

those that you think of as theoretically important—actually are landing somewhat close to your data. Making theoretically interesting predictions that correspond to impossible cases is misleading at best. We're most familiar with this in terms of the funny regression results, such as "this is how much income a college-educated 5-year-old girl would get if she were a ditch digger." But it comes in simpler forms when we just look to illustrate our claims with overly extreme cases. In a very impressive recent paper, Goldberg et al. (2016) have a sample of 601 employees in a firm, and their emails, from which they derive scores of cultural fit and network constraint, which are correlated at $r = .316$. The paper is a great example of how to do text analysis and there's no reason to doubt the findings. But to illustrate the findings (fig. 3; p. 1206), they first come up with four ideal types, two concordant (going "with" the correlation) and two disconcordant, and predict outcomes. The ideal types are formed by looking at those who are two standard deviations above or below the norm in cultural fit, and then those with low and high constraint, being the 10th and the 90th percentile of constraint.

The thing is, with that correlation, they probably have only around two observations in each of the concordant categories, and probably none or maybe one in the disconcordant (R code 2.3). They then show figures that go from +3 to −3 standard deviations in terms of cultural fit. And given that only .27% of a normal distribution is outside of three standard deviations, there's definitely no one there. Could the patterns be extrapolated outwards? Maybe. But it's a risky bet. So I think a good rule of thumb is, never plot "predicted" lines without *also* printing the data.[19]

Where Is My Variation?

Okay, so we've seen the importance of knowing where your data *are*. Now let's go a bit further; you also need to have a sense of where your variation and your covariation really are before you start analyzing. When I am at a workshop and someone presents a paper, I do what everyone hates—I tune out the speaker while he or she puts up PowerPoint slides with a literature review or a theory or something equally irrelevant to the subject at hand and I flip to the back of the paper. And in particular, I look at the correlation matrix. I read it. I get a sense of where the covariation is. And only then am I ready to understand what some multivariate analyses might be. And students are often surprised that sometimes I seem to know something more about the analyses than does the presenter! It isn't always true, and sometimes I am wrong, but lots of times I stump that chump, because he has no idea what his data actually *are*. He knows that some coefficient he wants to prove is significant has a big standard error, but he doesn't know *why*. Yet almost all the information is there in the correlation matrix he printed out.

That simple matrix sometimes tells us that the sort of question the presenter was asking can't really be answered with his data. Maybe *nothing* is correlated with the dependent variable. Maybe *everything* is, and it's all a big mess. But before you can try to *explain* your variation, you need to know where it is. That's especially true for complex data structures, like multiple units observed at multiple times. Is your variation basically within cases, over time? Between cases? Evenly spread around, or weirdly clumped?

Each variable, as Breiger (2000) emphasizes, is itself a way of dividing up cases. So there's a difference between what we *call* a variable, and what it *is* (a pattern of cases). This is, I think, a better way of thinking about these issues than "multicollinearity" for most of us, and definitely better when we have qualitative variables or few cases. Let me give an example (also illustrated by Breiger 2009 in a piece that should be required reading for *everybody*). There are a lot of quantitative analyses of the 18 OECD (Organisation for Economic Co-operation and Development) countries in western Europe, looking at the relation between this aspect of their political structure and that. When it comes to polities, there are basically three

19. People who print a line by extrapolating from a regression coefficient, and then conclude from this, that the relation is linear, should be shot.

types of countries in Western Europe, and they turn into three "clumps" in the variable space. You can *call* them anything you please. But that's what you have to work with. So we have around 80 different theories, all using the same evidence (that there are these three clumps), but *labeling* them differently. We don't want to confuse those labels with the underlying structure of the data. Those who have a sense of the variation are far less likely to make major mistakes.

One last thing here: I want to reiterate a point made wonderfully by Stanley Lieberson (1985). He spoke in terms of proximate versus basic causes, while I've talked similarly (in *Thinking Through Theory*) of conditional versus unconditional explanations, and here I'm going to pull out a different way of thinking about the same point. Lieberson argues that we often confused an explanation of *variation in* some phenomenon with an explanation of that phenomenon *itself*. His simple example was different objects falling; his more complex one has to do with occupational stratification. Explaining why *this* person gets the $12,000/year job—or no job at all—while *that* person gets the $12,000,000/year job isn't the same thing as explaining why there are $12,000/year jobs and $12,000,000/year jobs. Sometimes, the story you give "explaining" the stratification, if you confuse these, even if it passes your standards as a "causal" story, is still a whitewash.

Scale

As you look for variation, keep in mind that this can obscure a different aspect of your data, namely, the *scale* of different variables. The same correlation can have very different real-world implications depending on the scale that your variables are measured at. Knowing your scale will keep you from doing that irritating thing of reporting a coefficient as ".000***."[20] But it can also help you get the right coefficients, too. While OLS analyses are impervious to issues of scale, because they start with the correlation matrix, that isn't true of non-linear models. Many of the more advanced statistical routines for these models aren't quite as idiot proof as your more common ones in commercial packages. They can start off in their search for the best estimates caring quite a great deal about the scale of your variables. It never hurts, and sometimes helps, to use for your variables a scale that is similar to those used in the statistical routines.

Of course scaling might seem like a "small" issue—but without a sense of scale, I suppose we don't know what "small" means. And indeed, I have

20. Our silly author is also likely to report "$p = .000$," or even better, "$p < .000$." Why these are both crazy is left to the reader as an exercise.

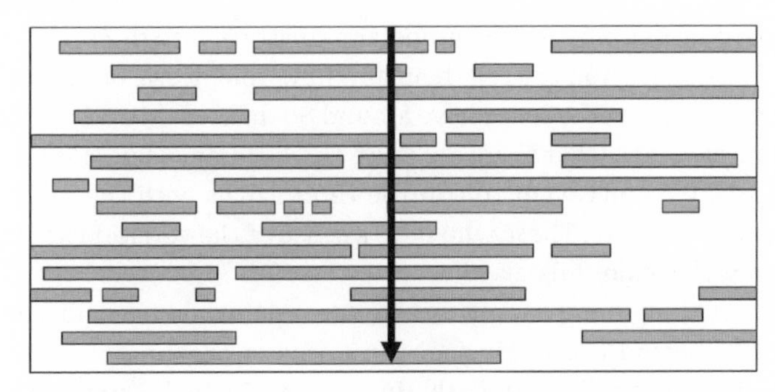

FIGURE 2.5. Sampling Relationships

seen more than one job candidate reduced to blathering incoherence when it became clear that she had never thought through the scale of her parameters. One in particular had made strong claims about social processes, when people in the audience could figure out that she was discussing results that would be reached in around 1200 years.

Lifespan

Finally, in many cases, you also need to know about the "lifespan" of the elements you are sampling. Imagine that you are conducting a sample of persons—you're only interviewing the people who are alive, presumably. That means you don't have a random sample of the *cohorts* that the people came from—all the dead folks are gone. That isn't a huge bias in a world with low death rates, so usually only gerontologists worry about this. But the same problems arise in other cases.

For an example, imagine that you are interested in marriages, and trying to make a table on the rates of marriage or cohabitation between different ethnic groups. Romantic relationships have extremely variable *lifespans.* And while every person who is alive now had his or her one life, there are some persons alive who have had marriages, though those marriages are no longer "alive." Now the census data is collected every 10 years on a specific day (nearly). Fig. 2.5 schematizes this. Each row represents a person's life, and each rectangle represents a romantic partner relationship, with time proceeding left to right. The arrow is when we happen to sample. We get a lot of the long-lived relations, and very few of the short ones. That means that our number of, say, interracial partnerships could indicate that there are a few very long-lived ones, or lots of short-lived ones (we only get a few of these).

In other words, an instantaneous type sample of variable-duration events is sampling not on the events, but on event-days. That isn't bad if that's what we want, and it isn't terrible even if it isn't, so long as we know when the events started. But even when we know when they started, we don't (at first) actually know when they end. It's a fun game to play, estimating the lifespan of each, and sometimes, to really get the estimate you want, you're going need to play that game—and win.

Trust No One

One last thing. I've started with the importance of knowing your data—how it was made. Well, you need to also know the nuts and bolts of what you are doing with it . . . at least as far as you are able. You aren't responsible for errors built into Stata, but you *are* responsible for programming errors *you* make. Programming has gotten a lot easier, which is great, but it means that it can be very easy to make big mistakes. That's especially true if you are a basically self-trained programmer . . . as almost all practicing sociologists are. You can write a clever program to do something like trace through a network, simulate the spread of a meme, say, regress this on something else, and boil it all down to a cool program. Remember: big data mean that minor errors lead to major f*** ups.

But would you know if you had made a mistake? You might, if you *really* knew your data. But the bigger the data set, the harder it is to get a feel for it, and the more complex the procedures, the harder it is to know you've made a mistake.

So what do you do? It's easy. Take your data and make a subset—depending on the sort of data, it might be as small as 10 cases and two variables. Something small enough so that you can print out and see at a glance all the data, as well as everything that is being done with it. Visualize it. Make the program spit out intermediary information. And *check it by hand*. Of course, there are some things you can't necessarily check by hand, like a model that requires iterative fitting (though you can check it with a separate routine). But take a turn at being John Henry and see if you can beat that machine. Sometimes, you'll find interesting things as a result. Once when I was checking the QAP (Quadratic Assignment Procedure) routine I had written for multiple groups and non-linear models (Martin 1999a), my results didn't agree where they should with the most widely used program for doing such tests at the time, UCINET IV. SPSS agreed with me, and so I did all the calculations by hand. There was an error in how UCINET dealt with missing values. It happens to the best of us. . . .

Once you're sure your program works well for the teeny data, scale up a bit. Does it work for 100 cases? What about 1000? Still seem right? Doing this can give you a feel for what to expect and what not. For another example, again from that QAP program, at a certain point, my *p*-values didn't feel right. The program hadn't changed, but the statistics seemed like they had. They were too extreme. I couldn't explain it, but given what I knew about the data, I didn't believe it. I looked into the code, and I admit it took me a long time to find the problem—I had a line that "reset" the seed on the random number generator by using the current time's hundredths-of-a-second counter.

That should have been called once at the beginning of the execution, but as I had moved code around, it had gotten embedded in the loop, which didn't make any difference—until I moved to a faster machine in which the entire analysis of one permuted data set was occurring in *less than a hundredth of a second* (which I had never imagined as a possibility), and so I was actually having the same permutations happening more than they should. Instead of 1000 different permutations, I was getting more like 280.

How did I tell? I had the program spit out exactly what it was doing and why, so that I could look at each permutation—where each node had been moved in the network—and notice that they were the same. You can do that. If you care about being right. And you don't mind wasting hundreds and hundreds of sheets of paper, because the odds are, if you try looking for it on the screen, you won't catch it. You often need the paper all over the floor.

Again: are you merging files? Take a few cases—*not* the first!—and trace them through. Is everything happening the way you expect? (Don't take the first case because many errors, especially that have to do with arrays being understood as a dimensionality different from what you're thinking, won't appear there.) Is this too much work? I guarantee you will sleep better after a long evening of this checking than if you don't.

Conclusion

Social statistics is a craft, like others. If you want to be a carpenter, you need to learn about wood. What can it do? How do the characteristics of the grain affect its capacity? What's the difference between different trees? And so on. You get the idea. Your data are your wood. You might have an awesome set of plans for a cute desk, but before you start in, you need to choose your wood carefully, look it over, and keep an eye on it. We're going to go on now and think about how, with decently good data that we understand, we can ask and answer some questions.

* 3 *
Selectivity

Selectivity Introduced

Variables, Causation, and the World

In this chapter, we're briefly going to deal with issues of selectivity. "Selectivity" is a key concept for those who are bound and determined to try to squeeze estimates of causal effects from their data—something I'm suggesting is rarely a good use of your time. But selectivity turns out to be a great way of beginning to think about the big problem for most statistical social science, which is the relevance of unobserved variables to our interpretation of our estimated parameters.

Why? Because from a statistical point of view, there are an infinite number of things you aren't taking into account in your model. How can you be sure that you haven't omitted a relevant predictor that is correlated with your included predictors? You can't. Is all hope lost? Not according to the pragmatist conception I laid out in chapter 1. We're trying to make our knowledge better, and good data should be able to help us. Rather than think about the infinite number of possible confounders, we want to think about the *likely* ones—whether or not they are in our data set.

How can we orient ourselves to this question? Selection—how people screw up statisticians' attempts to get causal estimates by getting up and *doing* things—turns out to be a great way. That's because *we're* people, and we can start with the question, "what would I do?" and go on to "what have I heard about or seen others do?" When, instead, we just try to think of "other things that might matter," we tend to spin out long lists of generally irrelevant factors. (Omitted predictors that aren't correlated with what you have in the model don't mess up our conclusions.) Selection orients us to the ones that are likely to be most fatal for our interpretation.

Map: So what I want to do is briefly lay out the classical understanding of a causal estimate, how selectivity enters, and how this wrecks causal interpretations from non-experimental data. Then I want to back up, and, in place of the causal modeling strategy that many textbooks now favor, suggest a way of thinking in terms of decomposition taken from demography, one which we'll use throughout the book. We'll then be ready to turn to the "control" strategy in the next chapter.

The Experimental Model

The classic experiment involves four crucial elements:

1) a *set* of at least two, and usually considerably more than two, instantiations of the unit of analysis, which we'll call a "sample";
2) the *random assignment* of one portion of our sample to a "treatment" group, and the rest to a "control" group;
3) the *application* of the suspected causal factor to the treatment group, but not the control group; and
4) the *observation* of the outcome in the two groups.

For example, we might take a set of adorable, furry kittens, and randomly give half of them an injection of makeup remover to see if they sprout tumors. This produces data like those shown in table 3.1 (for now, look only at the italicized row and column labels). The number of observations in any cell are given as f, with 0 meaning the control/no effect cases, and 1 meaning the treatment/effect cases.

Now let's try to apply this logic to a case in the social realm, this question being whether dropping out causes crime. Imagine that these are dichotomies: either you drop out or you don't; either you commit crimes or you don't. The format of our data look just like what we had before for our bunny tumor case . . . but now, we aren't applying the treatment—we're just observing a 2 × 2 distribution (and now look at table 3.1, but reading the non-italicized row and column names). We tend to focus on the symmetric relation between the rows and the columns, which I've (2015) called our "casual causal" idea: the putative cause "definitely maybe" causes the effect, in the sense that the "odds ratio" $(f_{11}f_{00})/(f_{10}f_{01})$ is greater than 1.[1]

This clearly doesn't fit the experimental model, because people, unlike the kittens, weren't randomly allocated to the treatment and control groups. Some folks say that we should use the notion of causality only where we can at least conceive of that sort of allocation. I get that point,

1. I thought that I was the first to make this obvious dig, but it seems I was inadvertently preceded in this by Woodwell (2014), where, due to a copyediting mistake, all figures accidentally (if wonderfully) replace "causal theory" with "casual theory."

Table 3.1. Classic Experiment

	No tumor No crimes	*Tumor* Crimes
Control In school	f_{00}	f_{01}
Treatment Drop Out	f_{10}	f_{11}

and I supposed we get to choose what our words are going to mean, but it isn't obvious to me that this is really going to do very much for us. That's because the whole idea of "cause" really only comes down to "what I consider a decent answer to a 'Why' question that I have." People were talking about causality long before some sicko had the idea of injecting the kittens.

But more important than arguing about words is to see *why* the experimental model fails. Rather than this being a source of disappointment, it can be the source of us starting to get concrete about what's going on in social life. First, we want to think through what sort of relations we might want to interpret as causal, and then we're going to let that point us to the problems of selectivity.

Getting Clear on Causality

Asymmetries

We know that we can't interpret the number from that table as just "the" measure of the causal effect of dropping out on criminal activity. For one thing, are we sure that dropping out causes people to do crimes? Or could it be that staying in school causes them *not* to do crimes? It might seem the same thing here in our 2 × 2 table, but it's not necessarily the same thing in real life. For example, maybe dropping out *causes* crime because people look for a job and find they can't get one because they don't have a diploma and so they start a life of crime. Or maybe being in school *suppresses* crime just because it gives you something to do all day—maybe people who drop out are no more motivated to commit a crime than those in school, they just have more opportunities.

Second, it is also true that our cross-sectional methods lead to a way of thinking that tends to assume that causes are *reversible*, and that they are the same as their inverses, only upside down—that is, that if *adding* a cause *increases* our effect, then *subtracting* the cause will *decrease* it. But some causes are *irreversible*. Humpty Dumpty had a great fall; he went down from the wall to the ground and broke. But moving him from the ground won't put him together again—not even all the king's horses and all the king's men can do that.

For a more sociological example, while there are characteristics that predict political party identification, it seems that it is harder for people to *renounce* a party allegiance and become independent than it is for unaffiliated people to identify with a party. This means that it is harder for the weak Republican to become an independent than for the independent to become a weak Republican. If we were to imagine that increasing income is positively associated with one's distribution on a line going from strong Democrat to weak Democrat to independent to weak Republican to strong Republican, this means that an increase in income of $20,000 has less of an effect in moving someone from weak Democrat to independent, than a *decrease* of $20,000 has in moving you from independent to weak Democrat.

That sort of asymmetry throws many of our conventional ways of analyzing data into total chaos. For example, Jacob Habinek, Benjamin Zablocki, and I (2015) were looking at how local relational structures—who your friends are, and who their friends are—affected how people formed new friendships, or let old ones lapse. We found that if person *A* moves closer to person *B* in geographical space, then the more friends they have in common, the more likely they are to form a new friendship, which makes perfect sense. (If you want to look at the results, a few are in chapter 9.) But if they already have a friendship, and move away, then the more friends they have in common doesn't make them any less likely to *drop* their friendship! But most of our methods assume that this is the case, and so we just tend to assume symmetry out of laziness (for a different approach, see York and Light 2017).

When we do pursue the issue of asymmetry, sociologists often focus on the issue of "reverse causality." And we try to solve such problems by ordering our relations in time, and forcing an asymmetry. The problem is that sometimes—at least, so far as our data is concerned—causality can flow backward in time. So let's return to our running example. Maybe it's not that dropping out causes crime, but that getting into a life of crime makes you more likely to drop out. That certainly makes sense—why waste your life in school if you've decided to become a thief? You might say, no problem: we'll just let temporal ordering answer this one for us. Which comes first, dropping out or entering a life of crime? But all our measures on crime might come after the measures of dropping out, either because initial delinquencies were too minor to leave a record, or that the dropping out was necessary for our would-be thieves to free up sufficient time for crime.

For a happier version of this dynamic, we might imagine that graduating high school is a cause of attending college. But the reverse is also true. That is, if a student can't attend college, why graduate high school? So when a Maryland millionaire, Stewart Bainum, pledged to pay for college for any kid in a certain class of a poor high school who got into one (this was in 1988), high school graduation rates there went up. Going to college is later than graduating high school. If we looked at the data, we'd be misled into thinking that this association was wholly explicable by high school graduation being an independent variable and college attendance a dependent variable.[2]

One last, and particularly devilish, form of this backflow comes when there is what we can consider *systemic* behavior. By *system* I mean a set of relations or variables that, when perturbed, has a tendency to return to its fundamental arrangement. In the last chapter, I discussed the importance of knowing where your variation is: are you explaining (say) between-person outcomes of stratification *within* some system, or are you explaining the system itself? Lieberson (1985) points out that sometimes you can't actually focus on the first and ignore the second, because you get tricked into putting explanatory weight on apparent causal factors that are in fact effects of the outcome—if *these* particular "causes" weren't present, something else would be. In Lieberson's wonderful words: "Those who write the rules, write rules that enable them to continue to write the rules" (167).

A great example is the introduction of the personal essay in college applications. At first, nice elite colleges could figure out who was Jewish because they had names like "Moishe Lieberson," which meant that you could make sure there weren't too many of them screwing up your nice elite world. But when Jews started changing their names (adopting Anglo ones like Stanley and Martin), the top colleges were getting flooded (or so they felt) with Jews who lacked that "character" that is so important in an elite environment. So they invented the essay question. If you write a moving essay about your struggles to break your thoroughbred horse so that you could lead the polo team, well, clearly, you're our kind of guy. If you write about playing stickball in the shtetl, well, maybe we don't really

2. Sure, of course you can give me an ontology in which it was really the *present* expectation of the future benefits that changed the kids' behaviors. That's fine. But we've been talking about whether, when we have a matrix of data **X**, with columns as variables that are associated with different times of observation, the ordering of those columns in time determines which ones can be causes of which others. The point is, it doesn't.

want you after all. So if you asked the question, "Why are there so few Jews at Harvard in 1920?" you could do a study and conclude, "Because they do poorly on the essay question." But we should instead ask, "Why is there an essay question?" And the answer would be, "Because Jews do poorly on it."

This sort of backflow, whereby causes are there to have the effects that they produce, is inherent to functionalism. You might think that functionalism is a bad theoretical approach, and you may well be right. But that doesn't mean that there aren't pockets of functionality in social life. If you ignore them, and tell a straightforward causal story, at best, you're telling only half the story. And at worst, you're being a patsy.

Causes and Causal Powers

So we've seen some of the complexities that come from a casual causal analysis, and we can appreciate the reason to try to push our analysis closer to an experimental model. And we're going to explore the issue of selectivity from this perspective. But before we plunge in, one more brain break—let's for a moment think about this issue of selectivity from a *different* approach to causality. Imagine our experimenter finds that the make-up remover *does* cause tumors, and is sitting in the lab excitedly dictating a research report. But one rabbit, gifted with language, vehemently disagrees, calling out from his cage: "What caused my tumor, you ask? *You* injected me with that crap, and so *you* caused my tumor!"

What our crabby rabbit is doing is invoking a different way of thinking about causality, one often termed a "causal powers" approach. He's looking around to see what has this causal power, and what can get things done. I think that makes a lot of sense for sociology. In most cases, we're going to be better off bearing in mind that our causes need to have some causal powers than if we just focus on the statistical issues.

This helps prevent us from implicitly attributing causal power to *attributes* (such as implying that your education could somehow cause you to get money without you doing something about it) or, even worse, to *direct objects* (like when an experiment proves that the cause of hiring discrimination is the race of the discriminated). But there's something else—something truly wonderful—about switching to thinking in terms of causal powers. When we started from the experimental model, selectivity was what *wrecked* the attempt to identify causality. People were going around doing stuff. That's no good! But from the causal powers way of thinking, that's precisely where the causality is and should be. What we are going to want to do in our research is play these two off one another. If your statistics uses variables, think about what might happen if some other jerk went around trying to manipulate those. How would people respond? This points you toward the omitted predictors. And that's what we do when we theorize processes of selectivity.

Selectivity

The Problem of Selection

From the causal powers perspective, it's probably a good thing that sociologists don't take a sample of 500 families with grade school children, and randomly make 200 of them undergo a divorce to see whether this will lower the kids' self-esteem. But from the conventional perspective, it's a problem, because it means that families that *do* get divorced probably differ from families that don't get divorced in lots of ways. Most obviously, families that get divorced may be families where there is a lot of tension. And maybe it's the tension that hurts kids' self-esteem (if this is found to be the case), and not the divorce.

Note that while we often focus on *self*-selection, we can also be selected by others. Some divorced people don't want to be divorced. It wasn't randomly allocated, but it wasn't a choice either. Either way, this selection can wreck any application of the experimental model. In general, far from waiting in a cage for us to apply some treatment as do the bunnies, people are busily self-selecting themselves into one sort of fix or another, like the unfortunate human beings they are.

There have been four ways of responding to this problem. One—still dominant—has been to attempt to "control" for selectivity. A second has been to try to model the selectivity itself, and to adjust the results for this. A third has been to try to figure out how *fragile* our results are, given non-random selectivity. This can involve seeing how strong an unmeasured confounder would need to be to send our results to zero (e.g., Harding 2003), or how many potential subjects would need to be missed due to sample selectivity to invalidate an inference (see Frank and Min 2007). And a fourth has been to try to focus only on situations where we think we won't get screwed up by selectivity.

The first approach is so obviously statistically inadequate that fewer and fewer serious books will even tell you how to do it. I'm actually going to be arguing against ignoring this approach; since it's what we mostly do, we probably should do it as well as we can. We'll be looking at this in the next chapter. Regarding the second, as you might expect, where the model of the selection process is good, the results are usually good, but where that model isn't very accurate, this approach can make things worse. (But to be fair, *all* the methods I'm discussing can make things worse.) There's no guarantee that an imperfect model with *more* relevant predictors will come up with better estimates than will an imperfect model with fewer. It's a nice thought exercise, but I stand with Lieberson (1985) here—if we learn that there is a selectivity problem, it makes little sense to try to go ahead and still be determined to *get* that gosh darn causal effect, no matter how hard it squiggles away from you. Instead, we want to study the selectivity process; that's probably the more important thing.

I think that the idea of looking at the fragility of our results (the third option) is great, and something that can and should be done more often than it currently is. But the fourth approach tends to lead to a "look under the lamppost" strategy of analyzing only the naturally occurring experiments we happen to find. These *are* interesting, and it's great to compare the results to those that arise in a multivariate analysis, but they're too few and far between for us. We can't just be lazy and wait for trivial findings to fall into our hands. (And if you've read *TTM* you'll know why I don't think experiments are a general solution for this problem.) Instead, let's focus on how we can explore selectivity, and, to start with, determine how fatal it might be to our capacity to use data to answer certain questions, starting with the propensity approach.

Thinking in Terms of Propensities

One of the reasons why some statisticians have dropped thinking in terms of "control" is that they realized that if we're really interested in only *one* causal effect, especially if it's dichotomous, then it doesn't matter whether we really try to specify a full model. At least in the simplest cases, all we need to do is estimate the probability that any unit would get the "causal treatment," known as the propensity (Rosenbaum and Rubin 1983). Then we can compare those who *did* get the "treatment" (e.g., dropped out) and those who *didn't* in terms of their propensity. Our data really can support a causal claim only when there is some overlap of these distributions—that is, we need a fair number of people of basically the same propensity who did and didn't get the treatment.

As many skeptics will correctly point out, the assumptions that go into the propensity score method are no different from those that are required for the classic control strategy; while the propensity score has some technical advantages where the number of control variables gets large, this is rarely our problem. So why do I think we should take this approach very seriously? Because it tends to push our attention to where it should be: selection processes, and how much "support" they leave for making the key comparison of interest to us (which is basically a weighted version of that 2 × 2 treatment-by-outcome table). It's possible to do this using conventional controls, but it's astoundingly easy when we've reduce all those variables to a single propensity score: we simply make sure that we have enough of an overlap in propensity between our pretend control and pretend treatment groups. And the results of this examination are often enough to make a serious analyst immediately give up the search for a causal model altogether.

Further, boiling things down to the propensity can be of great use in exploring effect heterogeneity. For a wonderful example, Diaz and Fiel (2016) were interested in the effects of pregnancy on teens' completion of schooling. As you can imagine, those who get pregnant aren't as likely to finish school as are those who don't. Can we interpret this as a causal effect? Not really: pregnancy isn't quite randomly allocated, though it isn't a simple choice either. You could use propensity scores for becoming pregnant to try to get back an estimate of the causal effect, but what Diaz and Fiel do is more interesting: they take slices of the data at different propensities to re-estimate the model. And what they find is that the (possibly) causal effect is *larger* for those with *lower* propensities.[3]

For purposes of simplification, we'll imagine that there are conventionally "good girls," and conventionally "bad girls." The finding is that there is a stronger specifically *causal* effect of pregnancy on good girls. There are three interpretations of this sort of finding. The first is that we accept that this causal heterogeneity exists and that we are finding something that is really about differential susceptibility to a classic treatment. Thus we might say that the causal effect is larger on the low-propensity girls, because it is more of a surprise, and these are the girls who had a lot more to lose.

The second interpretation is that the reason there isn't a causal effect on bad girls is because they aren't being "treated" in the same manner; they are the more willful ones, who deliberately threw caution to the wind. The estimate for the good girls is close to what a sadistic scientist would get who randomly replaced half of a set of sexually active good girl's birth control pills with placebos. But we haven't actually gotten all the selectivity out of our comparison.

And a third, more extreme interpretation, is just that we didn't get the propensity model right. We're underpredicting.[4] We've said that that good-but-somehow-pregnant girls were going to have a lot less education than we'd expect, because we're comparing them to the good-but-not-pregnant girls. But maybe we just couldn't quite identify the bad girls

3. Note that Turney and Wildeman (2015) find something similar for a different case of family disruption; also see Brand, Pfeffer, and Goldbrick-Rab (2014) for another interesting analysis.

4. I am pleased to find that this point has just been made rigorously and formally by Breen, Choi, and Holm (2015).

well enough. We thought that this B+ student was going on to a four-year school, because we never got to see her terrorizing the neighborhood on her Harley Sportster and getting into fights with bouncers at clubs.[5]

Finally, we can also see this sort of apparent differential treatment effect due to the Gulliver effect. Imagine that we have a good predictor of being a good girl versus a bad girl—whether, in the third grade version of Dante's *Divine Comedy*, the kids chose to play Angels or Devils. Because this taps the propensity, we could use it to see whether there really is a causal effect. In a nice little simulation (R 3.1), where, by construction, the effect of pregnancy on dropping out is the same for both the Angels and the Devils, we falsely conclude that the effect is much greater for Devils than for Angels.[6] Why? The tendency of the Angels to get pregnant and to drop out is so low that lots of our reports are due to "error" and so the signal is comparatively weak compared to the noise.

In any case, my discussion here is in no way a critique of this work on looking at effect variation by propensity. In fact, I think this is the most important do-able avenue for improving attempts at causal analysis that we have right now. But I do think that if we're excited about using propensity scores to get causal effects (for example, by weighting our cases inverse to their probability of treatment), we're going to need to explore this heterogeneity carefully.[7]

Now I noted that we need to restrict comparisons to those of roughly equal propensities, but for many problems, we find that the propensity distributions of those who did and those who didn't get the "treatment" barely overlap! Even more problematic, it can be that if we restrict our attention to those in this region, we may not be able to generalize much. Let's return to the case of whether dropping out is a cause of crime. We can imagine some folks who are right on the fence regarding whether or

5. What if we were to see that it was those of *higher* propensity who had a larger effect? Does it mean we're "overpredicting"? Maybe, but in practice, it again points to the possibility (not the necessity) of omitted variables, especially intensifiers. For example, some people have the resources to cope with many different kinds of adversities. Life is such, however, that these folks are the least likely to *get* those adversities in the first place. So those with a low propensity are more resilient, and have a lower treatment effect.

6. The equation producing pregnancy is $d_i = 1$ iff $.1 + .7Devil_i + \varepsilon_1 > 0$; $\varepsilon_1 = N(0,.25)$. The equation producing dropping out is $c_i = 1$ iff $d_i + \varepsilon_2 > 0$; $\varepsilon_2 = N(0,9)$. The odds ratio (the ratio of the diagonal cells in table 3.1 to the off-diagonal ones) is 3.40 for the Devils, as opposed to 1.42 for the Angels.

7. Some of the exciting new methods, like inverse treatment propensity weighting (e.g., Sharkey and Elwert 2011), are much more robust than other techniques that we commonly use . . . but if there is effect heterogeneity, they *can* introduce or exacerbate bias (see Imai et al. 2011). And it stands to reason that heterogeneity is the rule, not the exception.

not they are going to drop out. They're all basically the same on the variables that affect their propensity to drop out, but for accidental reasons, some do drop out and some don't. These are going to be the people who help us identify a causal effect, true enough. But there's no reason to think that *their* dropping out is the same as the dropping out of a solid 0.0 GPA kid who has been waiting to drop out since the first day of kindergarten. Chances are, we can't really support a good counterfactual for him.

Now most statisticians will shrug their shoulders and say that's just as it *should* be. It's those fence-sitter types who are the only ones for whom the idea of dropping out as a possible cause of crime even makes sense. But still . . . if the kids are wishy-washy about dropping out, maybe they're wishy-washy about other things. If so, we're going to start finding that for them, chances are good that *anything* is a cause of *y*—rain showers, the Red Sox losing the playoffs, whatever. Sometimes I worry that we can go pretty far in having a social science that only studies the wishy-washy.

In and Out

Now I'm going to be arguing that selectivity is a good way of thinking about a wide variety of problems, not only causal ones. Causal questions do have the advantage, however, of making it easy to guess as to what we need to worry about, since its selection into a particular treatment category. But not just selection *into*. *Out* is just as important. A nice recent example was the very surprising claim by Nicholas Christakis and James Fowler (2007) that if a person's friends got fatter, she or he was likely to get fatter. Well-schooled in the finer points of causal modeling, Christakis and Fowler attempted to take possible selectivity into account. The most obvious form is that if fatter kids are less popular as friends, because of anti-fat attitudes, then they might be more likely to find that it was only other fat kids who were interested in being their friends. Christakis and Fowler did a good job, or as good as seemed humanly possible, at taking this into account.

But they forgot that we don't just select *into* friendships . . . we select *out* of them. Further simulations by Hans Noel and Brendhan Nyhan (2011) showed that some of their results could be explained by a process whereby kids who got fatter lost their non-fat friends. In this case, the puzzle was solved rather quickly, because the focus on a single causal parameter (weight of friends → own weight) immediately oriented researchers to possible sources of selectivity. Once we get used to thinking this way, we'll remember that, say, it isn't just that depressed people may be less likely to be *offered* jobs, but they may be more likely to be *fired*, and so on, complicating any examination of a causal relation between the two.

(We're going to return to some of this work in chapter 8, when we deal with the complexity of dealing with social network data structures.)

Here's a second interesting recent example from the *American Sociological Review*. Christin Munsch (2015) used data to analyze how the proportion of income that husbands and wives bring in affects their infidelity. She has an interesting finding—it appears that the greater a man's earnings, the *more* likely he is to have an extramarital affair, while the greater a woman's earnings, the *less* likely she is. Munsch (p. 481) considered the possibility that there was some mediating factor having to do with marital quality and took this into account. Her results were robust (p. 483). So she concluded that the reason that the reason women's infidelity decreased with their earnings was that they are trying to stabilize the relation, given the inherently destabilizing effect of female-dominated earnings stream (p. 489).

See the problem? It has to do with the marriages we *don't* see, because they dissolved. Let's imagine that there are two women, one a high earner, the other a low earner. Each runs into Mister Charming—cute, but little earning potential. The high earner ditches the husband and runs off, because she doesn't need hubby's paycheck. The low earner, however, stays with Mister Doofus, but enters into a extramarital affair. When we count up the number of affairs, we have more among low-earning women . . . but it might be different if we took selectivity out of marriage into account. How can we adjudicate between the different possibilities? We can't, at least, not without a lot more information on marriages and marriage quality. Something interesting is going on here, but without the right data, and attention to selectivity, we aren't sure what. As in many cases, we also need to make sure this isn't about a difference in class cultures or something analogous.

Increasing Selectivity with Time

Causal reasoning helps us identify this issue, but so far, selection has appeared only as a discrete presence or absence in our data. More generally, we can have forms of selection bias that increase over time. (This is often called "survivorship bias.") Imagine data on workers' salaries, and you're interested in how these change over time. Every second that someone can drop out of a data set is an opportunity for bias to increase, and the people who stayed in the longest are the most different from those who entered the data structure early.

What does that mean? Often it means that what conclusions you reach may depend whether you sample at the beginning or the ending of the stretch of time you're interested in! You might claim that you're not doing either of these, but rather doing both at once, by only looking at those who have "complete" data. In most cases, that isn't really correct, because in practice, it means sampling on completion. Many people started some process, some don't finish (and leave the data set), but some do. If everyone who finished *had* to start, then, if you only include complete records, you're sampling on the final period. And if you're interested in timing until this final state (for example, how soon officers become promoted to two-star general in a database on generals), you're sampling on the dependent variable. If you *can* change your question to one that is about some process *conditional* on being at the final stage of a process, then that's what you want to do. But otherwise, you need to try to figure out how your conclusions may be affected by selectivity.

Further, in many cases, what may appear to be temporal change is due to one particular type of selectivity, namely, a shift in proportion between two latent classes. For example, let's return to the example of data on workers' salaries. Suppose you find that initial salaries of workers are low at many firms, but they go up over time. You might think, "well, perhaps employers aren't as bad as everyone says. True, the starting salaries are unlivable, but if you stick with it, they really stick with you!"

But that assumes no unobserved heterogeneity in your class of workers. Suppose there are actually two different types of workers (as in theories of a "dual labor market"). One set, we'll call them the "in-crowd," have privileged positions; although they may be paid hourly, they have good salaries, good schedules, and benefits. The other, the "out-crowd," have low salaries, intermittent and fluctuating schedules, and no benefits. Further, the company actually *wants* them to have high turn-over: they don't want people around long enough to organize, and they'd rather burn them out through overwork, have them quit, and replace them with new hires. Members of this class will have low mean tenure, while the other class will have on average longer tenure. That means that if you look at income by years-of-experience, it will go up steadily, only because the proportion of cannon fodder workers is steadily decreasing.

This puzzle is formally identical to one famously pointed out by the great O. D. Duncan (1966), namely that intergenerational stratification processes look really different if you sample on parents than if you sample on children. And the problem is intensified if we go to multiple generations. Not all people have the same number of children. Indeed, in some cases, the most important aspect of inequality is *who gets to leave living descendants* in the first place! (Here see the great work of Song, Campbell, and Lee 2015.) In this case, sampling on descendants can greatly understate the degree of intergenerational transmission of inequality.

Here's a great example where what seems like paradoxical opinion change may be due to survivorship bias. Sullins (1999) was interested in how people of different religious traditions, and with varying degrees of religious commitment, changed over time regarding their opinions on abortion. Using the General Social Survey from 1976 to 1996, he found (as others did) a convergence between Protestants and Catholics; it used to be Catholics who were least supportive of legalizing abortion, but their attitudes became more permissive while those of Protestants became more restrictive.

Yet most of the rise in Catholic prochoice sentiment appeared right where you'd be least likely to expect it—among Catholics who regularly attended mass, who are the *least* likely to be prochoice! Those who only attend every now and then—who seem less committed to doing what the church says, and who are less exposed to anti-abortion messages, didn't change as much. Why? Here's Sullins's explanation. The thing about being a committed believer is that you stay with the church even when you disagree. Catholics all over might be becoming more prochoice over time. We have two Catholics, say, one who is a regular attender and committed (call her Mary), and the other who attends infrequently (Marsha). Both become prochoice. Marsha goes, listens to the prolife sermons, and quits with disgust. Because she (and others like her) exit the class of occasional attenders (and become non-attenders), when we examine the class of occasional attenders, we don't see them becoming more prochoice. When, in contrast, Mary listens to the prolife sermons, she disagrees, but bears in mind that the true Church is the body of believers, not the authority structure, and keeps going to mass. Therefore, we see the proportion prochoice of the more regular attenders increasing. No causal argument is being made here. But we still need to be very attentive to selection in and out of our categories.

To remind ourselves, while people *are* caged overall—that's why life sucks, remember?—they're *not* in cages when it comes to treatment and control. They at least get up and move from one depressing cage to another. So we need less work on how to adjust standard errors for this or that and more studies of what the mobility patterns are between the cages. Thinking of selectivity is great because it reminds us that people are not passive, and it forces us to be more *concrete* when we think about the challenges to our causal interpretation, while "control" strategies tend to bracket our understandings of *actors* and instead focus on the relation between the *variables* in our data.

And so now, after piggy-backing on the causal modelers, I think it's time for us to clamber down. Because they're going to take a different turn.

Using Imaginations

The Asymmetries of Regressions

We've been using the basic causal framework to highlight the problems caused by unobserved predictors. But I've dissented from the current orthodoxy regarding how we respond. Most of those who are attempting to solve these difficulties numerically begin with a counterfactual approach to defining causality. The causal effect of dropping out for any person (A) is the difference we'd see in her criminal record if we somehow twitched a dial from low to high and made her drop out. Because we observe A only once, and either she did or didn't drop out, this requires thinking "counterfactually"—contrary to fact. Would this person have committed a crime if she *hadn't* dropped out? Even if we have someone in the data set who is identical on all observed covariates as A, except she hasn't dropped out, we can't be sure that her value on crime is what A's would have been had *she* not dropped out. Much of the current work in sociological methodology involves attempts to correct for this, so as to get the proper estimate of the true causal effect, most commonly, the "average treatment effect" (ATE). That's because this is what we'd be interested if sociology were a biomedical science in which we were figuring out how many Pop Rocks™ make your throat explode if you mix them with Coke.

But that isn't what sociology is all about. And refining an ATE almost never makes very much sense. I'd like to walk through a different way of thinking about what our general models are, one that gives us different criteria of what "good" means in our practice—and one that doesn't see regression as a "second best" for an experimental determination of a treatment effect. It's going to involve a few steps. First, we're going to need to rethink what a "model" is, and what we're doing with our statistics. To do that, we're going to need to think about *conditional* distributions and conditional probabilities. With that, we're (second) going to review some simple demographic approaches to explanation, and marvel at how deficient the current sociological imagination is in this regard. And then, third, we'll be able to contrast this to the counterfactual approach in terms of how we combine reality and imagination.

So when you take sociological statistics, you'll learn the regression model, which basically is an equation that we write in the following form $y = bx + c + \varepsilon$. For any value of the independent variable x, we produce a predicted y; c is a constant, and b our slope, with ε some error. But statisticians often think of it a bit differently. They define $\mathrm{E}[y \mid x]$ to mean our *expectation* of what y we should see *given* any x. In sociology we get all hung

FIGURE 3.1. Non-linear Relation

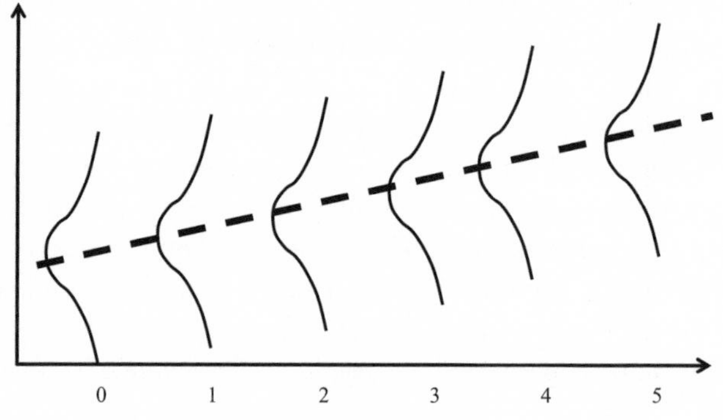

FIGURE 3.2. Linear Relation

up on the *b* coefficient, and do our best to forget about the underlying distribution of *x*. But if we didn't, we would realize that we're just predicting a distribution of *y* based on a distribution of *x*'s.

Perhaps the single best way to get a handle on what we are really doing is to start thinking in terms of $E[y \mid x]$ and not *b*'s. Thus consider two different schema representing the relation between some *x* (horizontal) and *y* (vertical) (fig. 3.1 and fig. 3.2). In each, we have six values of *x*, while *y* is continuous. The distribution $E[y \mid x]$ is drawn as a 90-degree rotated distribution for each value of *x*. In both cases, given any value of *x*, our degree of accuracy in predicting *y* is the same. The only difference is that in the fig. 3.2, the conditional means of *y* are a linear function of the values of *x*. That makes things easier, to be sure. But it's not always a very big deal.

As an aside, learning how to see a scatterplot in this way is vital for bearing in mind the *asymmetry* in a regression.[8] Take a look at the final

8. And I don't know where else to put this, but if you are interested in this issue, you need to read Berkson (1950) and have your socks blown off.

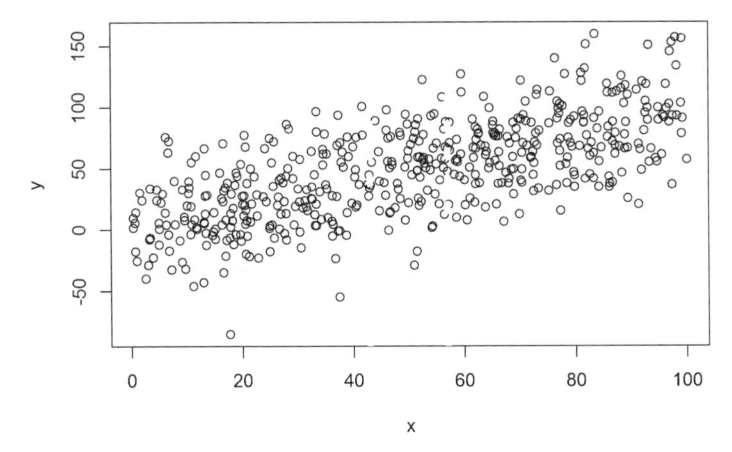

FIGURE 3.3. Truncated Range of Independent Variable

example, fig. 3.3, above. The overall slope here is about 1.0 (it's generated from a simple relation $y = x + \varepsilon$). But what would the slope be if we looked only between the two dashed vertical lines?

If you're assuming it's going to be *less* than 1, you're reading the chart wrong; you're looking at it as a correlation, symmetrically, when you need to read it as these successive slices along the x axis. The answer is that (except for sampling error, which will make the coefficient smaller *or greater* than 1), the slope is the same no matter where we put the dashed lines. Now that we're thinking asymmetrically, let's go on.

Demographic Designs

Let's now use the same basic way of notating and thinking to shift from continuous to discrete outcomes. In particular, let's think about conditional probabilities. Read "$\Pr[y = y^* \mid x = x^*] = a$" as "*given* that the x variable takes on the specific value x^*, then the probability that the y variable takes on value y^* is the number a." We call this a *conditional* probability, as it is the probability of one thing happening *given that* we know something else. So for example, the probability of a person committing any crime in any year may be .10. But the probability may vary by age, as follows

$$\Pr[C = \text{yes} \mid \text{age} = 0\text{–}10] = .0$$
$$\Pr[C = \text{yes} \mid \text{age} = 11\text{–}20] = .2$$
$$\Pr[C = \text{yes} \mid \text{age} = 21\text{–}30] = .3$$
$$\Pr[C = \text{yes} \mid \text{age} = 31\text{–}40] = .1$$
$$\Pr[C = \text{yes} \mid \text{age} = 41\text{–}50] = .0$$
$$\Pr[C = \text{yes} \mid \text{age} = 51\text{–}99] = .0$$

This sort of presentation of data not only tells us who the bad guys are likely to be, but it allows us to predict how a change in the age distribution, *all other things remaining constant*, might likely change the crime rate. What is our best guess as to the total number of crimes that will be committed? It's going to be a sum of the number of crimes committed by each age group, right? Well, what is our best guess as to the number of crimes committed by any age group? It's going to be the product of the probability that our person is in age range k, times the probability of committing a crime, *given* that one is in that age range. (Or, $\Pr[C = \text{yes} \mid \text{age} = k] \times \Pr[\text{age} = k]$.) In other words, the total number of crimes will be bigger the more 11-30-year-olds there are, and lower the fewer these are. That means that the crime rate will go up and down as the *composition* of the population changes . . . even if the same sorts of people do the same sorts of things!

This is then a form of *forecasting*. It's often used in demography to predict the likely effect of changing age structures. Imagine that instead of looking at crime, we are interested in who has babies. So we calculate conditional probabilities of child-bearing by age, and we see that the probability of having a child is concentrated for age of woman. That means that a bulge in the age distribution where women are most likely to have babies means . . . expect more babies. That relation in the United States these days is actually pretty flat from around 20 to 35, with around 1% of women having a baby in any year. But let's simplify, or take the white pattern, which is to have a greater likelihood of birth for 20–25-year-olds.

So let's say that for some reason, we have a lot of 20–25-year-old women in 2000. We can predict that in around 2006, school enrollments will start rising (so start new building construction now), that in around 2015 street crime will start to increase, and that around 2020 birth rates will *again* go up, as these kids have kids of their own, crime will go *down*, and we're likely to have either a relatively good time (with more gainful workers and fewer babies and old people to support), or a relatively bad time (with too many unskilled workers if there is a job-skills mismatch). But most pleasingly, this pulse is like a wave that has its own momentum, though it will eventually die out because of the degree of variance in mother's age at birth (see Guillot 2005, e.g.).

Let's take a brief pause, and use this basic demographic principle to correct a major weakness in common sociological practice—a way in which we tend not to think sociologically. In a nutshell, sociology is about "muchness," and that means *variance*. But we tend to have a Durkheimian way of thinking, in which all variance ("error") is bracketed as long as possible. And we often have such emotional distance from most of those we're talking about that we tend to do two things. First, we tend to simplify them all into a single abstract average person, which is what our statistics are based on anyway. And then we try to explain changes based on a narrative about this Mr. and Ms. Average. As a result, we frequently confuse compositional changes—changes in the makeup of what kinds of people we've got before us right now, or *changes of* people—with some sort of subjective changes, *changes in* people.

For example, consider the remarkable changes in American culture in the early twentieth century. What caused all this? Many of these changes were explained by sociologists and historians by making recourse to the "experience" of this average imaginary American, who had gone to the war, had come back, and had a new outlook on life. Those kinds of stories are always pretty appealing; they make sense (too much sense, I'd say) to us, and tie up lots of loose ends. Of course, they're hard to "test" in any systematic sense. But, you might reasonably ask, what else can we do?

Well, one thing that we can do is *forecast* what we would expect on the basis of the conditional probabilities in some y given a change in x. A forecast isn't a prediction—it's an attempt to see what would happen were only one or two factors changed. And it isn't a theory. It's really a null model. Rejecting the null model of a forecast means something interesting is happening, something worth explaining. In sociology we often use *randomness* or *independence* as our null model: *nothing* predicts your vote for president, or your vote is *independent* of your income, say. Well, these forecasts are a better null model, better than complete randomness. If we reject them, then it looks like we should dust off our explaining tools. If not, then, "nothing to see here."

For example, farmers (not peasants) tend to save a lot, if they possibly can. Until the rise of agribusiness, credit was ruinously expensive for farmers, often intentionally and sadistically so. So if a farmer *could* save, he'd save. And they tend to marry late. And they are often pretty wary of premarital sex, because sudden marriage can interfere with their inheritance plans. This all has to do with their relation to the land (their means of production). Clerks (i.e., desk workers) and many other non-farm workers often don't save at all. They have a steady paycheck and can borrow on it, and, starting in the early twentieth century, they lived in areas with stores with easy credit terms (Levin et al. 1934: 81, 257, 108; Leach 1993: 124, 299). Unless they're the type slated for upward mobility, they tend to marry early or have more disorganized courtship patterns.

Well, there was a steady decrease in the proportion of American workers who were farmers since basically whenever. There isn't necessarily a tipping point, but in the early decades of the twentieth century, there were, for the first time, not only more non-farmers than farmers, but more people in urban than rural areas, and hence not tied to farm economies.

Without saying that it's all simple, I do think that in this case, when we think about it, perhaps we'd find that most of the social changes in the US were pretty much what you'd expect given the changing composition of the population. In sum, our *stories* generally have to do with changes in conditional probabilities (whether or not we think about it this way), and we ignore the changing distributions.

For another example, we often hear about how ignorant the average American high schooler is, which is pretty scary, because, by definition, half of them are even more ignorant than that! High school seniors can't find Australia on a map—even if it's just a map of Australia. They can't count past eleventy or spell *ain't*. Even more, the average college student is far less educated than was the case 80 years ago. Shockingly so. But before coming up with an argument as to why "we" have gotten stupider, it's worth noticing that there's been an explosive increase in the proportion of all people who get higher education, with high school nearly universal and college becoming much more common. It stands to reason that this wouldn't happen without the achievement changing.

So it might well be that the knowledge of the top 20th percentile in academic readiness hasn't changed at all, but the average college student has, because we're adding in new people. Actually, the top 20% *are* more ignorant (even if they're just as smart), but that's a different story, and it's relatively happy one. Instead of spending all of college reading stupid books and going to stupid classes, they get to have a pisswhacker of a good time! Who's against that? Not me. Go out and play. But first, rethink your idea of using regressions to get at ATEs.

ATEs and Where the Imaginary Is

Let's now combine the ways of thinking from the last two sections: the first, in which we set the value of one variable (x—say, divorce) to some value, and see what that does to the value of another variable (y—say, income), and the second, in which we are attentive to the changing proportions of the population in different categories (which we can express as a categorical variable $Z = z$—that is, we use the first to index all possible combinations of, say, age/race/religion/region/sex/cohort, and the latter to be one specific combination). Assume for the moment, though this isn't necessary for our main point, that Z includes all possible information on our persons. And let's say we're interested in answering some question about the relation between education and income, and we are bound and determined to see it as a causal one. And we've decided that the most scientific way of doing this is to compute the best estimate of the "average treatment effect" (ATE).

Here's the rub. Imagine, for a moment, that x really is a cause of y, in whatever way you want to define cause, but that it isn't constant across all persons. In that case, the ATE is just an average, the same way that "average height" is just an average (and not, as Quetelet had first imagined, a *law* for the production of people). That means that we expect that causal effect of x on y to vary within each category of Z, right? So let's quantify this effect as a slope within any category z of Z as b_z. Now even though, by assumption, the causal effect isn't constant across all persons, it could still turn out to be the same, on average, in each of the categories of Z. In that case, we'd expect all these slopes to be the same (and $b_z = b \; \forall \; z \in Z$), or at least, compatible with a model in which they're drawn from a real world in which the population values are the same. In that case, you're in luck! You don't even need a completely random sample to get an unbiased estimate (your sample can be stratified by the variables making up Z).

But what if there *is* heterogeneity? That's simple, you think. If we have a perfect random sample, we don't have to worry about that at all, because we'll automatically get people in the Z categories proportional to their representation in the population! More or less true (we'll only *tend* to get this), but the whole joy of taking the counterfactual argument seriously is that it helps us determine what the results of a treatment on our population *would* be—how much income would the average American lose if she or he got divorced?[9] But the problem is that if our average comes from combining heterogeneous shares of the population, the average changes as the population changes. Even if that each of our b_z's is perfectly estimated and stable over time, our ATE will drift away from the true result of an intervention. As, say, the population ages and there are more 72-year-olds and fewer 23-year-olds, we need to adjust our weights and our resulting number accordingly. As the fraction of adults with college education goes up, we have to change that weighting also. And so on.

Okay, now imagine that it takes five years from fielding the survey to our article ending up in print, which is pretty optimistic. And imagine that the true Average Treatment Effect will shift over five years by around 3%. Would you think that it makes sense to spend a lot of time increasing the precision of our estimate by, say, 3%? I sure don't.

If the ATE is inherently labile, why do we spend so much effort trying to nail it down? In most cases, it has basically zero theoretical significance. It's simply a mishmash, and the best that can be said about it is that, if we lived in a very different world (which we don't), in which we could (say) experimentally allocate divorce to people (which we can't), this is our best guess as to how the income *would* have changed, *given* the

9. Note, not how much would Americans lose if they *all* got divorced!

kinds of people that we had in the population when we made our sample, which has now changed. I don't blame you if you want to spend your life doing something else than estimate these. And you can.

A Different Approach to Heterogeneity

Causal modelers know that contrary to many of our implicit assumptions, regression analysis doesn't actually produce values corresponding to the ATE if there is heterogeneity (e.g., Aronow and Samil 2015). That's important stuff to know. But the way that causal modelers approach this is to try to push the data as close as possible to a classic experiment—to find, in the data, a good example of someone just like person #4119, only who didn't drop out. That's why they call this approach "counterfactual"—you can only come up with a causal estimate if, for at least some of the people in your data set, you can find someone who is a plausible example of the person #4119 who *doesn't actually exist*, the one without the college degree. In other words, the so-called counterfactual approach isn't really very counterfactual. It *starts* with the logic of counterfactuality, but then runs right back to the data, and just *hopes* that it can approximate the counterfactuality with factuality.

In contrast, the demographic decomposition approach can be used to give a much more coherently counterfactual analysis. Let's say that we are wondering, as did Deirdre Bloome (2014), whether the pattern of racial inequality in income in recent cohorts can be explained by that of the *previous* generation . . . plus the normal correlations between parent's and children's socioeconomic status. If you took a conventional causal approach, what you need to do is reframe the question: Can we treat this as if it were an experiment? Can we somehow adjust things so that we can, in effect, find, for every kid born to poor parents a nearly-identical replica, but who was born to rich parents? Instead, Bloome used population projections to see whether the current state of income distributions among whites and blacks was basically what you would expect given transition rates between quintiles in the income distribution for different types of people.

Why use this sort of indirect method? I think the great virtue is that both the causal analysis and the population projection require that we use our *imaginations* and create pretend cases that aren't in the real world. But the causal approach requires that we have the imaginary stuff *in our data*—we need someone who is just like the imaginary version of some person *a* (who, say, had poor white parents) but who was born to rich black parents. We mix up the real and the pretend together so much that all the king's horses and all the king's men can't get them separated.

In the demographic approach involving forecasting, all the imagination stays on the imaginary side, and all the reality on the reality side. We compare them *as a whole*, imagination versus reality, and see if they match up. Most simply, we fix the conditional probabilities, and try to shift the distribution; there are of course more complex ways of making forecasts. The key thing is that if the forecast tracks the real developments well, we think, maybe this imaginary situation gets at what was really going on. But the causal account requires that our imagination versus reality comparisons happen piecemeal, scattered throughout the data. You can do this right. But if you don't, it's a lot harder for you, or others, to tell. And this gets to the 9th rule of sociology: everything is somewhere. Including imaginary things. You have an *N* of 4000—do you know where your imaginary friends are? Have they scurried among these?

Conclusion

We're fortunate that a number of methodologists have been pushing us to worry about selectivity. Although I don't think we should only be trying to neutralize it, we want to think about how any direct interpretation of a coefficient might be wrong, and a result of selective processes we don't have measurements on. That orients us to our true problem, which we call "unobserved heterogeneity"—the almost certain fact that our cases are different in ways that we haven't measured, and which are correlated with both our dependent variable and our included predictors.

That means that we need to remember that people who *seem* to us interchangeable (because they are the same on all our measured variables) still differ in ways that can be important for our investigation. Take the example of attempting to determine the effect of divorce on children's self-esteem. According to the experimental logic, the only way to really tell would be to get a bunch of families and randomly force some to divorce. That doesn't seem to have much to do with what we mean by "divorce," but we'll put that to the side for the moment. The problem is that to make the causal argument, we need to assume that the families that are divorced *don't really* differ all that much from the non-divorced family in any way . . . at least, in any way we don't know about.

If that's not true, then we're not really comparing divorced and non-divorced families: we may be comparing divorced *and* unchurched *and* poor *and* conflict-ridden families to intact *and* religiously observant *and* well-off *and* happy families. Most obviously, one factor encouraging divorce might be certain people realizing that they don't like having a family very much. Divorce is only the final insult in a parenting style of avoidance and rejection. In that case, we certainly might imagine that their children, feeling constant rejection or disinterest from a parent, have low self-esteem, though this precedes the divorce.

Our response to such a scenario is usually to quickly promise to measure whatever it is raised as a possible confounding factor. But there's *always* some unobserved heterogeneity that will slip through any measure we construct. Should we take into consideration time spent with kids? But we can't distinguish between the parent who has to work and lovingly spends all available time (little as it is) and the parent who, gritting his teeth, pushes a carriage to the store to get the latest issue of *Road and Track* and reads that while the baby screams all the way home. We need to remember that two people who *look* the same to us aren't the same, and there's a really good chance that some of the ways in which they're different are consequential for our results.

So you should be thinking about selectivity even if you aren't trying to get an estimate of the causal effect of this treatment. And indeed, trying to nail down that effect is probably not going to be a good use of your time. You are, I think, a student, and my guess is, you're not pursuing an MA in statistics. *Those* kids are going to stick pretty close to some well-understood problems and well-understood data sets. That's great, and that's the way most science is done. But you are likely to be a bit further away from city-center, as it were, and you may have a lot of loose ends in your project. You probably are exploring four or five possible "stories," and unsure of which one to try to chase down. You probably present your work at four conferences and get six "take-accounts" (as in, "you need to take *z* into account") to take back home with you.

You are not in a position to optimize, and to worry about bestimates. You need to figure out which paths are dead ends, relatively early, if you are going to get anything done. That's what multivariate regression methods are surprisingly good at—even though there is never any proof that they'll work in any case, and even though we can't actually justify them as defensible models of anything in particular.

Maybe that means our methods are pretty bad and we should do something else. But our "abstinence only" education has not been working. I'm taking for granted that you're going to do regression, and I don't want you getting your knowledge from the older delinquents on the street corner, peddling their filthy *American Sociological Review* mags to you. Consider the next chapter a manual on safe regression, trying to get your mortality risk as low as is compatible with having a good time.

Coda: Causality and the Linear Model

After all this discussion of causality, you might start thinking that we use *only* regression type models to explore causality. But that ain't so, not nohow. So to inoculate you, let me just walk through an nice example of a linear model that I'll discuss again later. It's called the Bradley-Terry model (1952), and was developed for data arising from paired comparisons (like, "which do you prefer, Coke or Pepsi?").

The basic data for such a model are counts at the dyadic level: thus dyadic observation x_{ij} contains the number of occasions on which drink i is preferred to drink j (or, as I used it in a paper, child i dominated child j). We imagine that every object or individual has some status along a vertical hierarchy, even though we do not directly observe this status, and propose that the odds of object i being preferred to object j (or person i ending up dominating person j) in any given encounter is a log-linear function of the difference between these two individuals in terms of this status. That is, any two bouts are independent conditional on the difference between the statuses of the persons involved. If we denote the latent status of object/person i as a_i and that of object/person j as a_j, this can be modeled as follows:

$$\ln(x_{ij}/x_{ji}) = a_i - a_j \tag{3.1}$$

Now this turns out to be something that you can fit with a logistic regression, if you come up with the right design matrix. But I hope you'll note that there really isn't any causal interpretation here. What we're trying to do is to use the model to test a reduction of the data—basically saying that a set of data on the order of N^2 really arises from a process that only involves information on the order of N. And now think about using it to express the dominance status of children—using this model is completely compatible with a vision of human action that denies that it is ever caused at all (if you care about that for some sort of reason). We don't need people's parameters to be *allocated* to them in order for the results to be interpretable. They're free to *select* their position. All that matters is that we can *estimate* their position.

That's a legitimate interpretation of a linear model. And there are plenty like this. So bear in mind that we can use regression type models for non-causal generators of patterns in data—and in many cases, we'll *still* need to think about selectivity!

* 4 *
Misspecification and Control

"It was all just another control project" (or words to that effect)

NEO, one of the terrible Matrix sequels

Chaos and Control

In the previous chapter, we focused on the advantages of rigorously thinking through selectivity problems. Most of those I drew on want to go on to *adjust* their causal estimates to deal with these problems. But that makes sense only for the few cases where we not only have a clear causal question, but are confident that we've got the right model. When should we have this confidence? Almost never. It's for this reason that in practice, most sociologists rely on the less elegant "control" strategies.

Map: In this chapter, I'm going to discuss different types of control strategies, and then focus on the one that leads to the most problems for us, namely "partialing out." I'll show some problems with our general usage of this to adjudicate between models. The key problem is that our current strategy usually works only if all our measures are perfect and our model is correctly specified. And that's never. I then go on to demonstrate that the problems with the control strategy are even greater when we include interaction terms in our model.

The Idea of Control

The idea of "control" starts from the assumption that there can be more than one "cause" of something like "doing crimes," and we're interested in a few of them, although we're not sure what all of them are. When it comes to any causes of interest (say, dropping out and presence of both parents in the home), we partition *other* causes into two sets—those we have to worry about and those we don't have to worry about. What gets a cause into that second category is that it is *uncorrelated* with either our causes of interest or with our dependent variable. So maybe boys are more likely to do crimes than girls, but boys are no more likely to drop out than are girls or to come from two-parent families. In this case, the fact that boys are more likely to commit crimes doesn't really affect our test of the dropout-crime or the family structure–crime connections.

When we don't include these uncorrelated predictors in our model, we don't worry much. We call them "error," because they weaken our ability to predict but they don't throw off our conclusions. So we might guess that person #7449 won't commit crimes because person #7449 is in school, but person #7449 is a boy and so commits a boy-related crime. We missed that one, and we miss lots of others, but we are still right to think that dropping out causes crime. Our model is misspecified, but in a way that I'll call "properly specified misspecification"—it's just the kind of misspecification our methods are pretty good with. Most important, we aren't deeply misled by our results. Our theory is right, but partial.

I'm going to be *endorsing* this somewhat nutsy way of thinking, arguing that it gets to the problems that should be keeping us up at night, namely the unincluded predictors of our dependent variable that are correlated with our included predictors. That's the improperly specified misspecification. All statisticians will admit that if you have that, they can't be held accountable for your results being wrong. And you probably always do have that. The project of control isn't about absolutely fixing your errors, but about trying to get a sense of how robust your conclusions are.

I don't deny that there is a way in which the control project is *inherently reductionist* and, indeed, anti-sociological in its extreme. That is, the best way of disproving that x leads to y is to stratify your sample by z and show that, within levels of z, x no longer predicts y. And the best way of *countering* this disproof is to find some w so that, when stratifying your sample by the joint distribution of $w \times z$, x *once again* predicts y. This is *valid*. It is also *paradoxical*. Why? Because we use one tool (a new independent variable) to fix another (a previous estimate), when we know perfectly well that this tool can break what we're trying to fix in a new way, requiring still *another* tool, and that we have no mathematical way of demonstrating that this process ever converges (Lieberson 1985). That's a logical problem—though we don't yet really know how big a practical problem it is—with contemporary methods. If you don't like them, don't use them, and join me in trying to come up with other ones. But if you do use them, don't be a baby about it. If someone splits the sample and shows you up, don't sulk and name-call ("reductionism"). Get back to work.

From Correlation to Causality and Back Again

To work through the logic of control, it helps to imagine that we are starting with an established correlation, and wondering whether we can use it to a support a claim of casual causality. Do, for example, theories that give a lot of explanatory oomph to dropping out when analyzing crime have merit? We might not yet be clear as to whether we think that this means that changes in the former will lead to changes in the latter, or that an intervention would lower crime, but we might be able to find good reasons to rule out *all* of these ideas even without completely nailing down what we mean.

First, note that we *start* with a correlation. By this, I mean that if the relation *doesn't* appear in the bivariate statistics, in 98% of cases you should stop there. The chance that, sans theory, you'll uncover for the first time a true relationship that is suppressed by a third variable is lower than the chance that you'll produce a false positive. There are only two types of cases in which I think it makes sense to keep going to a multivariate set up. One is when there are *universally recognized* confounders (call one of these z) that sociologists have found can be partialed out. For example, we are used to the idea that, at least in some populations, we need to make sure we aren't confusing *age* with *education*. So even if we don't see a relation between education and, say, a certain attitude in the bivariate statistics, we want to go ahead and see what happens after we control for age. (We also do this because we've learned that the correlation between age and education isn't *so* high that we can't pull them apart.) The second time that we proceed without bivariate statistics is when you actually aren't interested in x itself, but in x-conditional-on-z ($x \mid z$), and you can pick that z in advance. For example, we sometimes aren't really interested in test scores, but in test-scores-conditional-on-age. That's because we think that what measures sharpness is how much they're ahead of their cohort.

It's worth deciding early whether you are really interested in this $x \mid z$, because that's often what you end up with. And the phenomena-after-controls aren't always the same things as those before (Lieberson 1985). For example, let's say we are trying to gauge the effect of wealth on students' high school achievement, and we want to separate three things that we expect may be combined. One is the effect of each students' parents' wealth. A second is the effect of all the parents' wealth of all the *other* students in the school. And a third is school quality. We might therefore try to control for the proportion of rich families in the school when we examine the first coefficient (own family's wealth). But if rich kids are almost invariably in rich kid schools with other rich kids, the rich kid not in a rich school isn't a rich kid . . . he's a weirdo.

Now my point doesn't have anything to do with what for a while was a distracting false criticism of regression—how we interpret a constant (the person who was just born with 0 income and no height, etc.). It's not just about the one rich kid in a poor school—it's pointing out that our entire capacity to identify this individual-family coefficient requires looking *against* the grain, as it were, of the data. The main data might be a narrow blimp, indicating a high correlation between family wealth and average wealth. Pulling these apart is looking orthogonal to the main nose-to-tail axis of this blimp. That's necessary, but it can mean that we're actually looking at a different phenomenon. We'll be able to see this more concretely in some future examples. For now, we need to bear in mind that the control strategy can have this unanticipated effect of transmogrifying something that we might understand (the wealth of a kid's parents) into a different bird that is less familiar to us (the kid's parents' wealth conditional on the average wealth in their school).

In any case, the control idea works reasonably well if we start with an association, and are proceeding by attempting to *falsify* a particular interpretation of it. We can throw out—and so we *should* throw out—the claim that dropping out leads to crime if including some other variable z sends this relation to zero. (Of course, if this is because dropping out leads to this z, and then z leads to dropping out, in which case z is an intervening or mediating variable, we haven't *falsified* our interpretation; we've elaborated our understanding.) If adding one control doesn't kill off our finding, we have survived to the next round. The number of rounds is determined by the number of plausible alternatives (or alternatives that reviewers can convince an editor are plausible).

So how are we actually going to "control" these things that we can't actually control? Generally, not all that well—but we have four basic approaches, namely *stripping*, *shredding*, *fixing*, and *adjusting*. Regarding the first: Sometimes people say that they "control" for various alternative causes by setting up a scenario in which these causes cannot operate, stripping everything away but the factor of interest. For example, interested in the relation between task competence and stratification, they might make it impossible for group members to see one another, thereby being able to rule out visual appearance as a basis for judgments. But if we wanted to see whether orange juice increased health, we wouldn't want to test this by feeding people *only* orange juice, would we? But that is exactly what we often do.

This is related to the statistical procedure I'll call *shredding*. Here the notion is that we divide our data into smaller fragments in which the effect of the variable of interest is going to be more visible. A good example of this is found in Durkheim's (1951 [1897]) justly classic *Suicide*. Durkheim was wondering whether those with children are less likely to commit suicide than those without. It seemed like they are. The problem was that, in the nineteenth-century European data he had, most people with children were married, and vice versa. Maybe it's really about being married. So Durkheim cross-classified the two cases, looking at widowers with children, as well as married people who hadn't had children. This isn't foolproof—those married people who end up with living children are different from those who don't in many different ways—but it's a start.

Regarding the third tactic (*fixing*), sometimes people say that they "control" for a potentially troublesome variable by only looking at one level of that variable. In some cases this makes a great deal of sense. In most it doesn't. If you were worried that an examination of the effects of poverty-induced stress on health might be contaminated by malnutrition, it wouldn't be very smart to look only at malnourished people. But that is exactly what we often do. We think we control for race by ignoring non-whites, for example.

But it's the fourth (*adjusting*) that will be my focus in this chapter. We often say that we control for one variable by entering it into a linear model, and around seven days after the end of our Stats 1 class, we forget what this means. Partialing out variance with linearity assumptions isn't the same thing as "controlling for"—because to say you "controlled for" *z* implies that it *worked*. That's as if doctors called giving medicine "saving the patient." Still, here I'll use "control for" to mean *attempt to control for*. But I'll go over some of the ways in which it fails, and where we can patch it up.

How (and How Not) to Use Control Strategies

Guns Don't Kill People

Unfortunately, in addition to the problem that there is no guarantee that adding controls makes our estimates better as opposed to worse, there are other paradoxes that often arise when we start trying to walk down the control road. We start saying things that don't make any sense. For example: suppose that we find that girls do worse than boys on mathematics tests in twelfth grade. Someone argues that this shows that girls have lower mathematical ability than boys. But someone else adds "semesters of higher math" as a control, and finds that the difference disappears. This person argues that the real cause of the difference isn't sex, but training.

Now that's a reasonable claim. This implies that were we to increase the average girl's mathematical training by enrolling her in higher math courses, the sex difference would disappear. But it may be that rather than taking math courses leading to ability, ability leads to choosing to take math courses. Or perhaps one's degree of interest leads both to taking courses and to ability. In this case, forcing people to take math won't necessarily help them. We can't *assume* a direction to the chains of dependency that fits our attempted control strategy. If we can test this directionality, we should, but if we can't, and both ways are plausible, there's a good chance we have to leave things undecided. There's no reason to add controls if the results won't help us throw out at least one interpretation of the data.

For another example, control projects along chains of determination can lead us to control for intervening variables without understanding that this is what we are doing. As a result, we can lose more distant—and often, to us, more important—predictors and replace them with more proximate, often trivial, ones. I call this the "guns don't kill people, bullets kill people" argument. (Here you can see the fuller discussion of "dependent arising" in *TTM*, pp. 232f). And don't think people don't make this mistake—while writing this, I found an article that "disproved" that "religion" "leads to" "intolerance." But that's because religion "leads to" a decrease in social liberalism. And social liberalism "leads to" tolerance.

So sometimes overlaps in variance can lead us to replace interesting ideas with trivial ones that are really no more true. But they can also lead us to ignore important findings. For example, sometimes you can have a perfect predictor of at least one state of a dichotomy. (That is, $z = 1$ might predict $y = 1$ perfectly, but $z = 0$ is compatible with y being 0 or 1.) You can't include it as a "control" where you look at the relation of x on y. So what do you do? You might feel that you simply have to throw z out, like Fearon and Laitin (2003; for an appreciation of this article see *TTM*).[1] But if you do that, you can't ignore it when it comes to conclusions. And you'll have to take a stand: do you think that this is really a strong predictor (deterministic or near deterministic) in the population? If it's very skewed, then you can perhaps convince the readers that the perfect prediction is a fluke. But if not, you are really coming up with two sets of conclusion. The first one is that you have a strong, perhaps deterministic, claim, since you have a perfect predictor. There is some x that is necessary (say) for y.

1. They found that membership in the Soviet bloc perfectly predicted a nation being at peace and so they dropped it, noting that this finding was compatible with their key theoretical arguments, since this could indicate support for an incumbent regime. But they never really discussed it in their conclusions. Who wants to say that the solution to the problem of war is a one world communist dictatorship, anyway? Not me.

Within the cases that are positive on *x*, however, you have a second claim, one that is perhaps stochastic.

There is, however, a second route, which is to use a Bayesian approach, and assume that the deterministic finding in your *sample* is overestimating the true degree of relationship in the *population* (see Gelman et al. 2008). While I am a Bayeskeptic, I think this makes a great deal of sense, and is better than the first approach when your deterministic predictor only deterministically predicts a relatively small subset of your cases.

In sum, yes, you probably need to use controls. But you need to envision this as part of a coherent strategy. How you start off down the control road can make a huge difference—but so can how you stop.

Stopping Points

The simplest path down the control road is trying to "take everything into account." Sometimes we make the mistake of just jumping into a single regression equation throwing all our controls in at once; more often, we work toward such a model seriatim, as readers (or reviewers) come up with more and more possible "coulditbe"s. For example, imagine that we find that in fact, there is statistical support for our idea that dropping out causes crime. But someone else suggests that our result could be due to *boys* being more likely to drop out *and* more likely to commit crimes. And so we redo the work adding the control for sex. Our first coefficient was 1.93 (let's say, just making something up), and it was considered to be statistically significant with a *p*-value of .015; with the control, we might find our coefficient is basically unchanged at 1.91 ($p = .016$).

But now someone *else* however suggests that it could be due to parent's income—maybe kids from poorer families are more likely to drop out *and* more likely to commit crimes. Cursing under our breath, we type in the other model, close our eyes, hit the enter key, and . . . we peek. Thank the Lord! Our result survived. Now the coefficient is 1.72, and $p = .042$. Still under the .05 conventional cutoff!

Unfortunately, given the finitude of our sample sizes, you *can't* control for everything, and usually not even everything that people think of. You get long strings of wonky coefficients, large standard errors, and a model that is telling you "I don't know!" Even if in the population, some of these are valid predictors, you can sap the strength of your model too much by adding too many independent variables. Thus the "right model" isn't always the same thing as the "true model."

But there's a simpler, and worse, problem: this is a way of working that inherently gives you an incentive to do a bad job. Here I'm going to divide

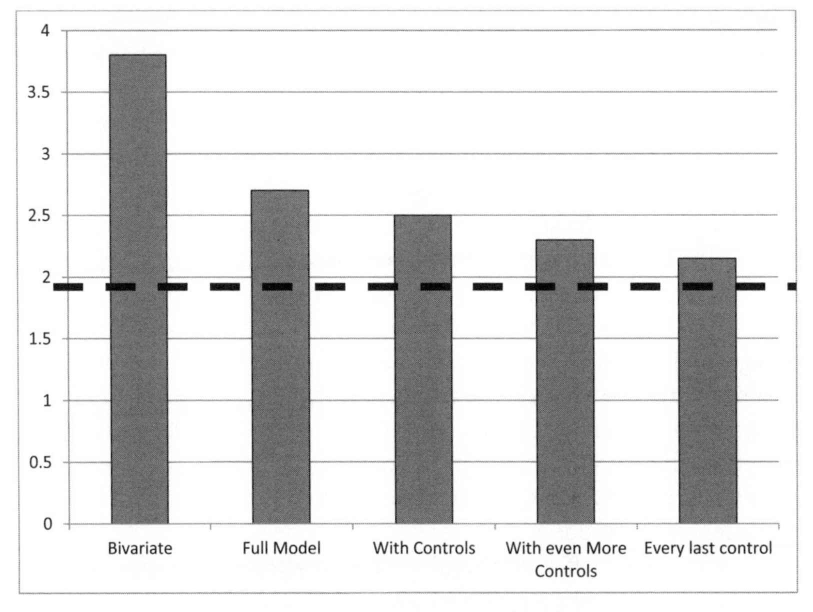

FIGURE 4.1. A Downwrong Set of Analyses

the implicit incentive structures inherent in different methods into two categories. The *upright* are those that reward you for more effort and more rigor. The *corrupt* (or the "downwrong," I suppose), are those that reward you for avoiding data, increasing error, and in general, doing a lame-ass job.

One of the nice things about the current approach to statistics, with its "rejection-of-the-null-hypothesis," is that usually the null hypothesis *isn't* yours, and you *want* to reject it, which leads to an upright incentive structure. We have an incentive to collect a lot of data, most obviously. But when it comes to control strategies, the incentive structure is not to look hard, because looking hard is looking for trouble.

It's not uncommon to see something like the chart in fig. 4.1: the bars show the strength of the coefficient we're interested in. Each bar represents the same coefficient (say, b_1) from a particular model. Here the effect is scaled to a *t*-score, and the dashed line indicates 1.96 (significant at the .05 level). Our researcher has dutifully added in controls for things that critics have suggested might be an alternative explanation, and the coefficient remains significant. Sigh of relief. Off to the presses.

This notion—that as long as we stop while our *p*-value is <.05, we can still make our claim—is completely wrong. I defended the use of *p*-values above (pp. 23–25), but this is one time when there's no excuse for the practice. That would mean that depending on where we stop our research, we would or would not get to be "right." In particular, the less we know—the less work we do to take other things into account—the more likely we are to be able to get to *say* we know (that is, "I was right!").

But the corruption is more fundamental: any way you mess up the process of control *helps* you. Misspecifying the functional form of one of those controls, for example, probably makes it more likely that your treasured coefficient stays significant. Using noisier measures will help too. So if throwing in a rough measure of family background like income, treated as a linear interval variable, leads your main coefficient to become 1.72 (p = .042), there's little chance that a *proper* specification of parental family background wouldn't decrease that coefficient a hell of a lot more. The idea of taking the final significance as evidence that you are right only makes sense if you assume that you have the totally correct model and everything measured without error. But almost certainly you don't.

So if successively adding controls successively decreases your treasured coefficient, then it doesn't matter where it ends up. Assume that if you didn't stop, your coefficient would melt away altogether. The data is trying to tell you that whatever you're interested in is too bound up with the competing explanations for you to be able to pull it out.

In contrast, when adding controls leads our coefficient to bounce around, as opposed to trending down, we aren't getting a strong signal. So if I actually cared about being right, I'd go to the presses with something like the results in fig. 4.2, though the final model here isn't as significant as that in the previous figure.

So the key take-away is that if you go down the control road, if you don't want to end up being a fraud, the game *isn't* "I get to be right if I stop controls before my effect shrinks below t = 1.96." Instead, the game is "I get to be right if I have seriously examined plausible counter explanations and they don't rock my finding."

Absence of Evidence does not constitute Evidence of Absence. You didn't see it. Maybe it's not there. Or maybe you did a crappy job of looking.

Controls and Error

Controls with *Error*

So we can't think through the implications of our control strategy without also thinking about the nature of error. One of the problems that we have in thinking through our statistical practice is that we use a single word, "error," to indicate very different phenomena. Further, the kind of error that statisticians deal with most often is an extremely tame and tractable kind. That sort of error, also termed "residual" error, comes from

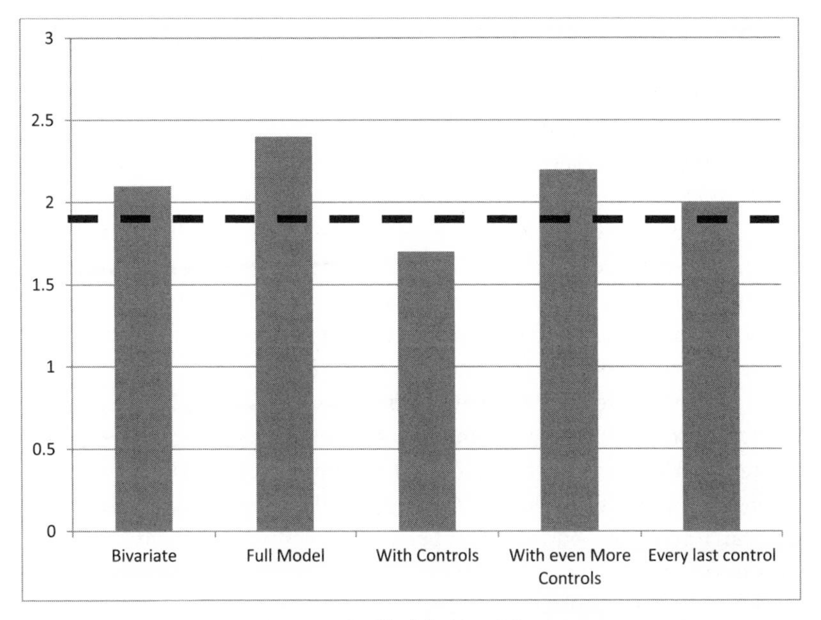

FIGURE 4.2. An Upright Set of Analyses

the difference between any case's value on our predicted variable and our predictions:

$$\varepsilon = y - \hat{y} \tag{4.1}$$

It's only the madness of nineteenth-century French statistics that makes us think there's something *wrong* if people deviate from the average! That's not an error—that's *distribution*, that's *variance*, and that's life.

An error is when something goes wrong, and in our statistical practice, there are really three types of error, and two qualities of each. The qualities are "biased" and "unbiased." The types of errors are *sampling* error, *specification* error, and *measurement* error.[2]

As we recall from chapter 1, statistics as a science was historically oriented around questions of sampling error. The nicest version of this comes when we have simple random samples, and while we have error, it's unbiased (we're as likely to be too high as we are too low in some estimate).[3]

2. Note that these correspond to errors of inference, interrogation, and measurement respectively.

3. It is important to emphasize that the kind of sampling error *process* that I am calling "unbiased" can still lead to the bias in certain estimates. In particular, imagine that we sample people by choosing towns randomly, though in proportion to their population, and from each town, we choose people randomly. It's still the case that, considered

Specification error is when we have errors, usually errors of omission or substitution, in our predictors. As I emphasized in chapter 1, that's our biggest error, and no formal techniques can help you much here.[4] You've got to actually switch on your substantive brain.

And I think social scientists are, quite reasonably, focused on such omitted predictors. But I think they often are less interested in measurement error . . . perhaps because talk of measurement seems all fuddy-duddy, associated with the squares who wanted to be all sciencey. Still, your control effort requires that your measures be accurate. It doesn't make sense to rely on controls and not care about the quality of the measures used. We've tended to assume that random errors just lead us to be "conservative," as they bias relations toward zero. So we don't worry about them. But random errors in a *control* mean that we aren't succeeding in the control strategy—it's *not* conservative for our theoretical claim. (Random errors in the dependent variable aren't as big a deal—they just decrease the fit, and can't be separated from "properly misspecified specification error.") This isn't something you can ignore.

Let's take the simplest kind of measurement error—respondents honestly forgetting the right answer. You ask them how many cousins they have, and someone who has six says five. There are also errors that come after this stage that (unless we have multiple interviewers, say) can be indistinguishable from reporting error. The respondent does correctly say "six" but the interviewer writes down "5." Or the interviewer does correctly write "6" but the data entry worker types in "5." Or the analyst going over a spread sheet means to hit the "left" arrow but hits 4.

We've become very complacent about this sort of thing because we're convinced that these sorts of errors cancel out on average. What we mean is, we *wish* and *hope* and *pray* that they do, or, if they don't, that it isn't a big deal. But here's a point that C. S. Peirce (1986 [1870]) made, in his close analysis of errors of observation in his work for the Coast Guard, which he found to deviate considerably from normality. His example is a data entry technician (in today's terms) writing 53 instead of 35. In a classical theory of error, this is extremely unlikely, less likely than typing 40. But if data is

singly, any person is equally likely to be picked as any other. But two people from the same town are likely to share things, and they aren't *independent* in their likelihood of being picked—they tend to go in or out together. That means that the degree to which we're likely to be wrong is greater than if individuals were sampled—we'll jerk a bit more violently around the true value. Therefore a naïve estimate of the *standard error* of our main estimate is biased, biased downwards (towards zero).

4. In a perfect world, the over-dispersion of the data compared to our model indicates an omitted predictor, and statisticians can help you with that . . . but you need to be right about a lot of other things in your model, such as your functional form.

being entered on a keyboard, as opposed to being read off an analog output (like a thermometer), we expect that 450 is more likely to become 45 than it is 387, and so on.[5]

Further, some kinds of error like this aren't unbiased. For one simple example, in many cases, people are more likely to *drop* digits than to *add* them. That's because if you press a key too lightly, nothing registers, while pressing it too hard usually doesn't lead it to print twice. But even where the errors are unbiased, they can lead to bias in estimates of parameters—in fact, they generally lead parameters of association to be biased downwards, toward zero.

I noted that these sorts of random errors aren't conservative if they are present in our controls. Let me give a simple simulated illustration (R 4.1). Imagine that an unobserved variable, say, "rambunctiousness" (z), affects days-skipped-in-school (x), and then both of these affect number-of-delinquencies (y) (to the same amount, which I'm setting at .5). So the proper b_x should be .5. If we don't take rambunctiousness into account, we get the over-optimistic b_x of .86. If we take rambunctiousness into account, then we get the correct estimate. But what if we use a scale that we *call* rambunctiousness, but it has error in it? Fig. 4.3 shows what happen as we slide a parameter for the proportion of our measured rambunctiousness that is random noise.

It isn't surprising, but it's worth noting that the bias gets noticeable rather quickly. The lesson is, when you have error in your controls, they under-control. If they make your b go down a *bit*, understand that in the real world, it probably goes down a lot *more*.

Finally, there are also some sorts of measurement error that are *amplative* instead of *attenuative* (I guess I made up those words, but we'll keep 'em). This is most often the case when other *people* are doing our measurements for us. If we measure "delinquency" by days absent, it might predict dropping out, but not as well as measuring "delinquency" by asking principals to list the kids with a bad attitude. That doesn't mean that the second is a better "measure." It could be that instead of measuring a personal trait, we're tapping into a social *process* that we don't com-

5. A related problem is often known as heaping, especially by those (e.g., Roberts and Brewer 2001) who have models to help us "un-heap" the data. How many birthday parties did you attend by the time you graduated high school? 10? 20? 30? 40? 50? Chances are, you're going to take a guess, and chances are, that your guess will end with a 0. I notice that students often criticize quantitative research for turning everything into a five-point scale, but that's what most of us as respondents will do anyway in our heads, and we'll label those points with convenient even numbers. That means that our errors are clumped in a particular way. People whose true value is near 100 are likely to have much less error than those whose true value is 86.

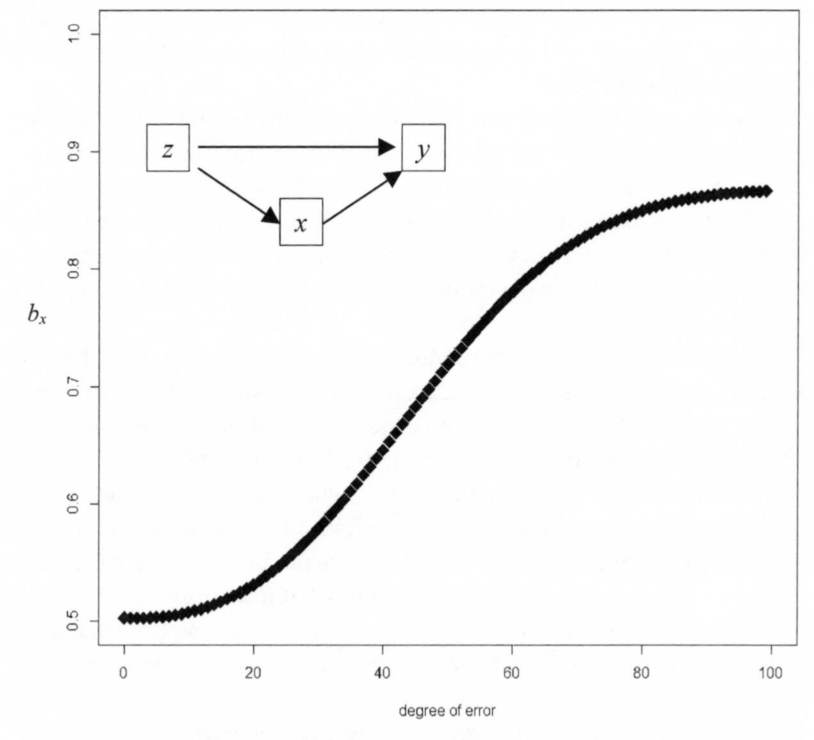

FIGURE 4.3. Imperfect Observations of Controls

pletely understand. And doing this tends to *increase* the association between individual-level covariates, because those seemingly intra-personal associations are caused by people (like police officers, principals, potential spouses) who are attentive to and interested in the outcomes we're studying. Thus we need to make sure we're not inserting the equivalent of transistors in a circuit and boosting the gain of an often weak and unreliable signal. This is such a big deal that I'm going to devote a big chunk of chapter 8 to it.

Correcting for Errors

You may not disbelieve what I'm saying, but you might not appreciate it, either. One of the problems with the wide-scale abandonment of interest in structural equation modeling (SEM) is that students are far less likely to start with the presumption that their measures have error in them. The problem with SEM when it first became hegemonic in the social sciences was in part that it was interpreted as promising quick fixes to the probable presence of error in our measurement. We should deny SEM's solutions, at least in many cases, but still adopt its focus on the problem.

Why jump off the SEM train? It suggests that we can pool data to correct for errors, and come up with better estimates of the true, underlying, variables, than if we don't do this! And that's right . . . if all of your assumptions are correct. Let me give a classic example. We ask people about who they are planning on voting for in the next presidential election. And then we ask the same people the same question two months later. Some give a different answer. And we say that they have changed their minds. Indeed, Philip Converse (1964) found that so many people changed their opinions between two reasonably close survey waves that he decided that many of them really didn't have opinions at all—they were just answering randomly.

Maybe. But maybe the reports have errors. Some misheard the question and gave the wrong answer. Others gave the right answer, but it was changed by a partisan data-entry technician. We might be over-stating the degree of change if we don't consider the possibility of measurement error. So Christopher Achen (1975) pointed out that this was assuming perfect measurement, and if we fit a measurement model that allows for error, we find that the data are actually compatible with a high degree of attitudinal constraint among the populace. Converse's reply was wonderful—he didn't doubt the math, fancy as it was for the time. He just said, "I can't forget what the data looked like *before* they went into this."[6]

Now I have every reason to side with Achen here—not only is he one of my favorite social scientists, but the methods used to make these corrections were the ones developed by my mentor and friend Jim Wiley (Wiley and Wiley 1970). But I think Converse had a point too. The way I'd phrase it is this—for many processes, all we can say is that *someone* is screwing up. We can't tell whether it's the respondent, or us. If we really want to make a statement about one of them, we're usually going to have to work out a special research design just to do this. So more important than undoing error is knowing about it . . . and not actually deliberately *increasing* it. Which people often do.

Collapsing

For some bizarre reasons, social researchers sometimes deliberately collapse their data into fewer categories than were used in the measure-

6. Although he can't forget that, I can forget *where* he said it, and indeed, despite serious searching I have yet been unable to find this reference.

ment. There can be good reasons to do this, if this allows you to be more flexible in relating a predictor to the predicted. For example, if you turn years of education into NO HIGH SCHOOL, HIGH SCHOOL GRAD, SOME COLLEGE, COLLEGE GRAD, AND MORE THAN COLLEGE, and enter four of these five dummy variables as predictors, you may do a better job than if you simply used "years of education." But if you aren't going to do this, you would need some extremely strong reasons to throw information away.

Because throwing information away is increasing error. And that means that you can end up under-controlling, which (assuming the case of positive correlations), biases your preferred coefficient *upwards*. Let's take an imaginary example (R 4.2) in which education (x) affects job status (y). But both of these are affected by parents' income (z). (It doesn't have to be that z causes x; all that matters is that there is an association between the two.)[7]

$$x = z + \varepsilon_1 \tag{4.2}$$

$$y = .5x + z + \varepsilon_2 \tag{4.3}$$

So we're interested in the slope of job status on education (b_x). If we do a conventional regression of job status on both education (x) and parents' income (z), we'll reproduce the true coefficient (.500). If we leave out parents' income (z), our estimate of the effect of job status is too big ($b_x = 1.166$). But what if we include a "grainy" version of parents' income? That's what will happen if we collapse the values into bins. Fig. 4.4 shows how increasing coarseness starts to make our estimate of b_x increasingly optimistic.

You might not deliberately coarsen your data.[8] But guess what? It's also a problem if *you* didn't do the collapsing, but it happened anyway, because the data-gathering instrument used rough categories (for example, five categories for income). In either case, the further we collapse parents' income, the further its capacity to knock b_x down decreases.

And that's assuming that there is no error in the binning. But as we've seen before, error really leads our controls to underperform a bit more quickly. What if we are collapsing error-prone data? Then we're going to see something more like what is portrayed in fig. 4.5 (R 4.3), which also

7. Here I simulated 1000 random points uniform on the unit interval, and made $\varepsilon_1 = \varepsilon_2 = N(0,.2^2)$.

8. Fascinatingly, in chapter 5, I'll show how something that often *seems* identical—a kind of binning—actually has the *opposite* effect, and *inflates* our R^2 dramatically!

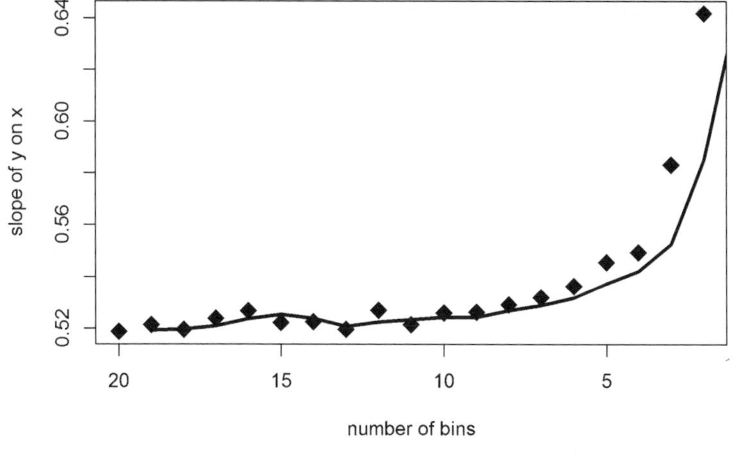

FIGURE 4.4. Adding Grain

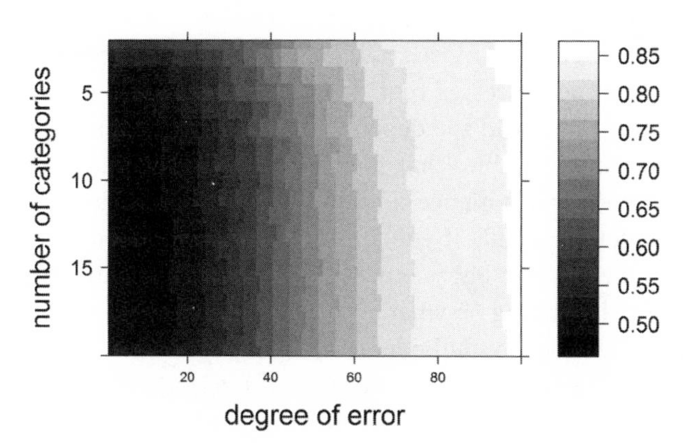

FIGURE 4.5. Collapsing and Error

graphs the value of b_x where we decrease the number of categories (from bottom to top of graph) and increase the amount of error (from left to right)—only the bottom left corner has the right value of b_x; as the figure gets lighter, the bias increases.

If you compare to the results for degree of error above, you'll see that the collapsing intensifies the bias in the parameter, especially for the measures with relatively low degrees of error. Of course, we can't ever be sure, but most of the time, if you're putting in your six dummies for "income," you aren't in the bottom left here, you're more in the center—biasing other parameters upwards. Still, many errors in the *z* variable don't effect the eventual bin into which a case is placed. For this reason, the results are, not surprisingly, likely to be much worse if the error comes not in the underlying *z* variable, which is then binned properly, but in the binning process itself. In the first case, someone with an income of $48,000 reports it as $50,000, and goes into the 50–59K category instead of the 40–49K category. In the second, instead of checking the mark "4" (for the fourth category), the interviewer marks "6." With this sort of mistake, the *scale* of the error is likely to be larger.

So errors in our variables are going to make the control process go awry, because we can undercontrol, and therefore walk away with confidence where we shouldn't. But the same can happen if we misspecify the functional form relating our controls to our dependent variable.

Model Specification and Linearity

I'm actually not going to present simulations here showing how getting the functional form wrong undercontrols, because we'll actually do that—for more interesting cases—in both chapter 7 and chapter 8. Here I want to instead focus on what we can do about this.

There are three all-purpose strategies. The first is to spend some time puzzling over the processes that you think will lead to a link between the control variable and the dependent variable. We sometimes do this when it comes to variables that sociologists treat as interval only by the grace of laziness, convenience, and tradition. "Years of education" is only an interval-level variable if you are interested in the number of school lunches someone has been exposed to. As Winship and Mare (1983) pointed out, depending on what aspect of schooling may predict our dependent variable, we might be interested only in the presence of degrees (and so a "half year" difference of 12.1 and 11.6 may be more important than the four-year difference 11.6–7.6), or we might be interested in total skills learned (output, not time input). Sure, we often have nothing but years of education. But no one said you have to just dump it in the model as such. If you can't rely on the findings of others, you might try breaking up your variable into three to five categories and entering these as dummy variables to get a sense of a proper functional form.

And you might need to do this even when it comes to variables that really do have a strong numerical quality to them. Money is a good example. There aren't really natural break points when it comes to income (tax brackets don't seem to matter for most people). But that doesn't mean that there will be a linear relation between income and something that is equally numerical. Again, it's worth thinking it through, and trying different specifications if there's a good reason to think that it will matter.

The second all-purpose strategy is to *visualize your data*. Look at the scatterplots of the relation between the control variables and your dependent variable. (Looking at residuals is generally what you are taught to do, and you should, but it often makes sense to start with the bivariate relations, especially when you have a lot of variables in your model.) Draw a curve that expresses the overall relation between each control and the dependent variable using, say, loess regression (this is easy and automated in all stats packages). How linear does this look?

And the third strategy is to use non-parametric techniques. Here we make statistical adjustments for some or all of our predictors that treat them as having only ordinal (and not interval) characteristics. I think we'll be seeing more and more sociological work using this as the "go-to." But not enough of us know about these—and this includes me—for me to steer you to how they can be used robustly. (Some are developed as machine learning devices that are really oriented to within-sample prediction, and not the test of whether population parameters are non-zero, which is what we usually want.) So until such techniques are the new workhorse of sociology, it's worth taking the time to explore different functional forms where it matters.

Model Specification: Interactions

What Is an Interaction?

So we've seen some of the ways that the control strategy can go awry. The problems here get magnified—literally, multiplicatively—when we include interactions. One of the confusing things about interactions is that, in a real sense, they're in the eye of the beholder. One researcher's direct effect is another researcher's interaction. As we'll see in chapter 9, "high debt" is an interaction between "high expenses" and "low income." I can't tackle that theoretical question; here I only want to think about interactions between measured variables.

I remember once showing my mentor Jim Wiley an analysis I had done, and he just smiled at a model I was showing him, refusing a direct comment, instead saying, "Well, there's two kinds of people in the world; those who love interactions and those who don't."[9] I loved them then. I also liked when we wrote down equations as models. That seemed theoretical. I didn't like analysis of variance and all that sort of stuff, because

9. Terry N. Clark has told that once at Chicago when Goodman was working on the log-linear system, Daniel Bell asked Goodman for an explanation of it. The early Goodman system involves interactions at all possible levels: the researcher then tries to move to a more parsimonious model by setting some interactions to zero and then interpreting the remaining dependencies. When Goodman was done with his explanation, Bell decided, "I like it. It seems like a very Jewish approach to statistics."

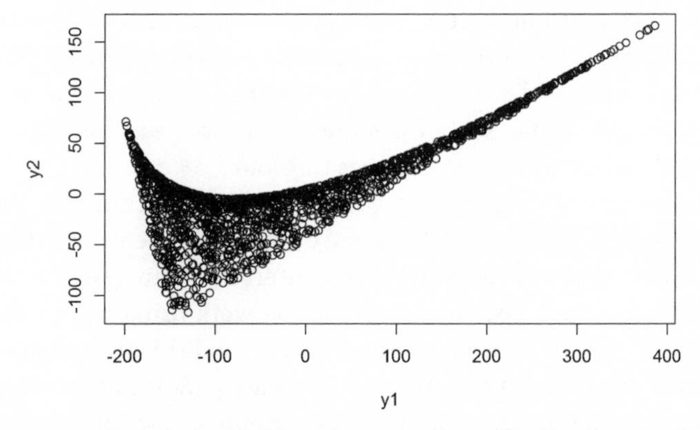

FIGURE 4.6. Predictions from Differently Centered Variables but Same Coefficients

it seemed petty, complicated, and boring. But because I wasn't thinking in those terms, I didn't necessarily know what I was talking about.

Let's say that I think that the effects of church attendance on sex role attitudes increase with one's level of doctrinal orthodoxy, so I write out

$$\hat{y} = b_1x_1 + b_2x_2 + b_3x_1x_2 + c \tag{4.4}$$

and test it. And then I interpret all the coefficients for a set of listeners. Someone asks me if I am using mean-centered variables, and I reply haughtily that it doesn't matter for my *theory*, that's only something for the grubby little fitting machine to worry about. That someone laughs and laughs at me.

If my dismissal was justified, the predictions from two models with the same variables, and with the same coefficient values, only one raw and the other mean-centered, would be the same. But fig. 4.6 below shows (R 4.4), they aren't (the interaction computed using raw values is the horizontal axis, that using the mean-centered on the vertical).

What's going on? If we were to expand out the models, and do a little algebra, we'd realize that the slopes for the *non*-interacted ("constituent") parameters are necessarily different, depending how our variables are centered! (If you are interested, see appendix 4-A.) If all you care about is the coefficient of the interaction, then no worries; but if you want to interpret the slopes of the constituents, then how you zero your variables is weighty. And this also means it's almost never reasonable to claim that your "theory" implies that these should be zero (here see Brambor, Clark, and Golder 2006). Either the predictions or the estimated slopes have to change. Lesson #1: Plot out your predictions.

Interactions and Misspecification

We've seen the importance of not misspecifying an interaction; unfortunately, it's also the case the misspecification or bounding of other variables can make it *appear* as if there is an interaction. Most importantly, we see such ghost interactions due to floor and ceiling effects.[10] So, for example, if you are investigating something with an intrinsic bound, like a depression scale, and you have two very strong predictors—say, suicide of a family member, and diagnosis of a serious disease—there's a good chance that in a linear model, you'll find a *negative* interaction between them. You might be excited that people can be protected against depression after the suicide of a family member if you also tell them they're dying of pancreatic cancer. But it isn't so. It's just that the model is asking a question, "how much more depressed can you get?" and the answer, in classic Spinal Tap fashion, is "none more. None more depressed."

You might think that these sorts of effects become stronger the more bounded the nature of your variables—if you have a few discrete categories, then sure, this is a problem, but not if you have some unbounded scale, like that composed via summing up other variables. But actually, in practice, these problems are *less* likely to arise with discrete outcomes that have only a few categories, . . . because in this case, you're more likely to be using some form of a log-linear model, and these do better (though not, in practice, perfectly) at dealing with floor/ceiling effects. On the other hand, most seemingly unbounded variables are, whether because of the data-gathering regimen, the construction of the variable, or the nature of the world, de facto bounded.

Finally, we often see this problem arising in multilevel models, which we'll talk about in the next chapter. The key is that here we are combining individual-level factors (e.g., being from a rich family) and contextual level factors (e.g., living in an area where most families are rich). Again, we often here see negative "cross-level" interactions, and they seem quite

10. A very impressive example of this sort of ghost interaction was uncovered by the great Art Stinchcombe (1983–1984). Otis Dudley Duncan had proposed that in some cases, we may be able to discover classes of unusually ideologically consistent responders from data on political opinions. Duncan suggested that the excess of persons in the "all answers conservative" or "all answers liberal" cells compared to predictions from a model allowing for bivariate dependence could be used to estimate the number of consistent ideologues in the population. But Stinchcombe showed that this identification required that we correctly understand the metric of the underlying trait and its relation to the items. If we misspecify *this*, we're going to have wrong predictions, and they'll probably get bigger at the tails. That means we'll think there are these M-way interactions, when there aren't.

theoretically exciting. But, given that there's no reason to believe that our linearity assumptions are justified, it isn't worth going to the press with an interaction unless you've really investigated whether the interaction is found in cases that aren't bumping up against your floor or ceiling.

How do you do this investigating? Certainly, you might want to move to a non-parametric approach to do your modeling. But there's something simpler: visualize the data. You need to make a three-dimensional plot with your two potential interactants on the x and y axes, and represent the outcome as height or color (and read Berry, Golder, and Milton 2012). As we'll see in chapter 6, where we walk through a very interesting example, the results can be very different from what you might have imagined. Lesson #2: Beware of ghost interactions.

Being Too Lucky

Interactions and Overfitting

So Jim Wiley had gently hinted to me that while I was predisposed to like interactions, he thought differently. He was right. Usually young scholars like interactions because they save theories that shouldn't be saved. Perhaps you were convinced that some variable x should matter for some y. Your first prediction was wrong, but you only had one chance on that (you predicted $b_x > 0$). But with K other variables in the model, you now have K more chances to find *something*—just interact each one of them with x. Chances are, you can find *some area* in your data where the explained variance of one variable by another is bigger than other areas, and, by using an interaction, you can parameterize your model so that you get to put an asterisk by *something*. And if you don't even have a focal variable, then there are "K choose 2" possibilities (which is $K[K-1]/2$).

There are of course well-known (especially since Goodman 1969) corrections for statistical significance when carrying out multiple tests. That requires, however, that people are honest about the number of tests that they do. And they aren't going to be, for two reasons. The dishonest people aren't, because if they admit that they actually did $K[K-1]/2$ tests, they'll have to admit that all their $p = .034$'s are more like $p = .500$'s. The honest people aren't because they did more like 500 tests, really working to throw their initial finding out, not by doing *random* tests but by examining *plausible alternatives*. Their p of .034 is indeed roughly in the right neighborhood, because if it weren't, chances are good (though not perfect) that they would have figured it out, because they've learned how to read tables.

But most of us don't have a good sense of how to read interactions, which is—literally—polynomially more difficult than interpreting bivariate relations. So now I would say that any interaction that isn't strongly predicted in advance isn't worth examination, unless it can be found in totally different data sets (not pooled panels of the same one). Do you think that this isn't a major problem for sociological practice? Think again. I'm going to give sustained attention to an apparently successful research program that has been built on interactions.

> "WASHINGTON (Reuters) - Three genes may play a strong role in determining why some young men raised in rough neighborhoods or deprived families become violent criminals, while others do not, U.S. researchers reported on Monday.
>
> . . .
>
> People with a particular variation of the MAOA gene called 2R were very prone to criminal and delinquent behavior, said sociology professor Guang Guo, who led the study.
>
> "I don't want to say it is a crime gene, but 1 percent of people have it and scored very high in violence and delinquency," Guo said in a telephone interview.
>
> His team, which studied only boys, used data from the National Longitudinal Study of Adolescent Health, a U.S. nationally representative sample of about 20,000 adolescents in grades 7 to 12. The young men in the study are interviewed in person regularly, and some give blood samples.
>
> . . . .
>
> And a certain mutation in DRD2 seemed to set off a young man if he did not have regular meals with his family.
>
> "But if people with the same gene have a parent who has regular meals with them, then the risk is gone," Guo said.
>
> "Having a family meal is probably a proxy for parental involvement," he added. "It suggests that parenting is very important."
>
> . . . .
>
> Guo said it was far too early to explore whether drugs might be developed to protect a young man. He also was unsure if criminals might use a "genetic defense" in court.
>
> "In some courts (the judge might) think they maybe will commit the same crime again and again, and this would make the court less willing to let them out," he said."

Yow! Genetic science that courts could use to keep people locked up! Did you know sociology was doing important work like that? Let's take a look at this paper (Guo, Roettger, and Cai 2008). Now, first of all, you need to understand that Guo et al. (henceforward GRC) were using the Add-

Health data set, a nationally representative data set for the United States. Ooh, that's good. No, that's bad. Why? Because genes are closely related with what we might call biodescent—things like race and ethnicity—which are tangled with everything else. You can find plenty of genes that predict all sorts of things, such as having had great-great-grandparents who were slaves. Slavery *was* genetic, in the literal sense that it was there from birth by virtue of law and guns. But the clue to it wasn't in the genes, and same goes for lots of other things. We're actually better off examining genetic predictors using data from more homogenous populations. If some gene doesn't lead Danes who have it to be different from Danes who don't, the "genetic" explanation is wrong.

In any case, as the Reuters story says, attention is focused on two genes, MAOA and DRD2. Let's start with the first, which is one of two genes that makes a certain enzyme that turns off some neurotransmitters. A genome contains a particular version (a variable number tandem repeat, or VNTR) of this gene, with a certain number of repeated fragments, 2, 3, 3.5, 4, or 5 copies (referred to as 2R, 3R, etc.). Other researchers had previously found that people with 3R and 5R copies (call these the Bad R's, or BRs) don't do as well making that enzyme as those with 3.5 and 4R (call these the Good R's, GR). So the GRs are better at getting rid of these neurotransmitters including epinephrine, dopamine, and serotonin, than are BRs.[11]

Past research has pointed to 3R and 5Rs as being related to behaviors and outcome, but it's those with the 2R variant—so rare that it's not included in either the BR or GR groups—which has been the focus of a number of Guo's papers.[12] Actually, not even those with 2Rs, since they only look at males. Why? Because for females, there's absolutely no effect (note that in the news story, the research is inaccurately described as "only looking" at males, when it is "only finding" in males). So given 5 VNTR's times two sexes, that's clearly already ten tests; a bit worrisome. But it

11. Some research found that BRs have higher rates of violence than GRs, at least in some situations, but this was only for a study of maltreated children in New Zealand. On the other hand, a replication with the data set being used by GRC, published three years before their own, reports *not* finding a significant different here (Haberstick et al. 2005), though this study was more interested in an issue of whether this mediates maltreatment as a youth.

12. For example, Guo, Ou, Roettger, and Shih (2008) find that, contrary to the implications of the previous research, the difference between GR and BR is in the wrong direction. Those who have the variant that is supposedly bad have (insignificantly) lesser delinquency than those with the good version. I frankly read this as "we tested and rejected the genetic hypothesis." But the authors found a way to save it.

gets worse. There are actually only 11 individuals in the Add Health data set with this variant, and all of Guo's papers on the variant use the same data set—the same 11 individuals. These are the suspects—we could try to get to know them—or we could try to control it all away. And that is going to be key, since MAOA variants aren't randomly distributed across different groups by biodescent—and hence by socially ascribed race (SAR). Which means that . . . it's time for a control project.

And GRC follow the path where, depending on when you stop, you can or can't make your claim. They find (for males) an increased delinquency in the raw data of 3.25 (2R vs. everyone else). After putting in controls for age (?) and race, this has gone down to 1.72—just about $p = .05$, it appears. So we know that with an error-laden measure of SAR, the effect is halved. What would it be with a *perfect* one? Probably not much left. In any case, GRC go on to show that for the handful of students with 2R, repeating a grade is more associated with higher delinquency than it is for other students.

But the interaction that got the press I opened with is that associated with another gene variant, DRD2 (variant 178/304 versus others). Are you wondering why they thought that this would be associated with delinquency? They certainly don't cite anyone other than themselves as having found such an association (and indeed, later work looking for the effect of DRD2 variants and externalization found an impressive effect of as close to zero as one can hope for in this world [Nederhof et al. 2012]). Certainly, we would have to hope that they didn't just start trawling though all 20,000 human genes! Perhaps it is a lucky guess—that this variant will be associated with delinquency. Still, how on earth would they come up with the idea that this genetic propensity would be turned off by *having family meals*? I can't imagine anything other than that they tried a number of different interactions with variables in their data set, and published what worked. The chance that this is a real effect has got to be extremely low. Certainly, it's a bit premature to start testifying and suggesting that we lock people up.

You might think that I'm flaming out here, but for too long, the sociology of the United States has had the same philosophy as its business elite—that we should have a right to make risky gambles, and walk away with a big treasure if we win. But if we lose, we should never be held to account. This leads to intellectual bankruptcy as troubling as our financial bankruptcy. Those who ignore this are complicit, not "nice."

In sum, if you have a single variable, your statistical significance coefficient may not be exactly what it's supposed to be, but it sits in the same world as the statistical theory. If your variable has a set of J categories, and you allow yourself to arbitrarily choose one to contrast to the others (as in Guo and Tong's [2006] examination of the DRD4 gene[13]) that gives you $J - 1$ chances instead of 1. And if you interact this with your K other variables, you have $(J - 1)K$ tries. That means that a p-value has no meaning.

Now remember, I've *not* been saying that you shouldn't comb through your data. We have a lot of learning to do. But finding the seemingly significant interaction isn't the end of the story, it's the beginning. We need to push it, put it in face-offs against other variables, and, most important, derive *other* testable implications.

Disguised Interactions

Finally, we can be conducting interactions, and setting ourselves up for biased conclusions, when we split our sample. It doesn't necessarily look to us like we are misspecifying a model with interactions, but that's in fact what we're doing. Let me illustrate this with another paper by Guo, who seems to have had success in saving hypotheses by finding subsets of the data in which they hold. Guo, Li, Wang, Cai, and Duncan (2015) claimed to have found another example of gene-environment interaction, this time, between the overall-genetic propensity to drink alcohol, and whether college students are placed with a roommate who has drinking experience.

Now this study combines a number of different errors that it is able to illustrate for us, and I'll return to this in the next chapter as a great example of a different problem. But, to prevent you from scratching your head and imagining I must be misrepresenting the logic, I'll deal with any other weirdnesses in footnotes. So they had a purported measure of genetic propensity for alcohol use, and they argued (incorrectly) that it is such a strong predictor, that other factors, such as the peer influence that they were interested in, can influence the outcome (binge drinking) only when this variable is at the middle of its range.[14] So their claim (again) rests on an interaction, though in this case, it's that people in the middle of the range of propensity for one variable have a susceptibility to the effect of a second one.

There is, of course, the problem we already encountered, and this is that genetic variation is highly associated with social traits that might also be

13. There can be from 2 to 11 repeats, and of these 2 and 7 had previously been shown to be different from 4. But guess what? *This* article focuses just on those who have 3.
14. We'll see why this was wrong in the next chapter; given that the article itself presented the data that demonstrated this, I don't know why no one else seemed to notice. For a discussion, see Martin (2016b).

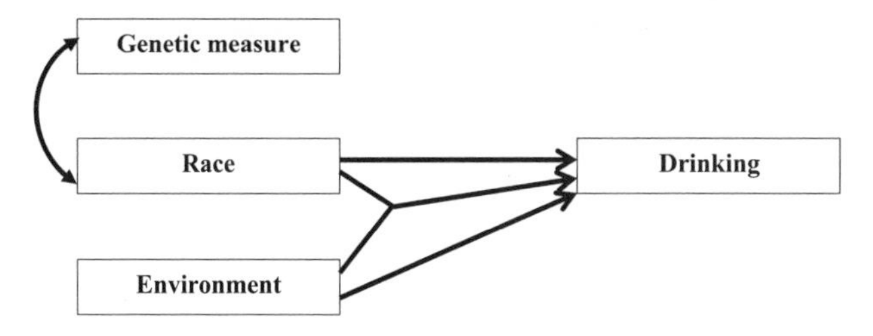

FIGURE 4.7. Possible True Model for Simulations

related to behavior, such as race/ethnicity.[15] So then one has to control for ethnicity and biodescent/SAR. How did they do this? They split the sample into three (unequal) tiers of *high*, *medium*, and *low* propensity, and then regressed alcohol use on their variable of interest and controls for SAR.[16] You might think that this tactic—assuming that SAR is perfectly measured and specified—will take care of the problem. But it doesn't. If you split a sample by some x_1 to establish an interaction, it's actually just like you had the interaction coefficient for every coefficient in the model with x_1. You *aren't* adjudicating between your variable of interest (x_1) and a different one (x_2) that is correlated with yours, because you aren't letting the other variables interact freely.

To make this clear, let me draw a plausible scenario between our three variables, the genetic index ("Genetic measure," x_1), biodescent ("Race," x_2), the roommate's characteristics ("Environment," x_3), and then the outcome, binge-drinking (y). That's given in fig. 4.7. Then I'm going to simulate some data that fit this world (R 4.5).

The genetic score (x_1) and socially ascribed race (x_2) have a correlation

15. This is especially true in the current case because, unlike most genetic studies, which focus on specific genes, at least some of whose function is understood, Guo et al. put all they had on genetic variance into the computer and asked it to predict alcohol use, and then called this "genetic propensity." The funny thing is that *first* they used a technique to adjust statistical significance given multiple tests—just the sort of thing that would have eliminated the findings about the "sit down dinner." But rather than use the results to examine the issue of genetic propensity, yes or no, they just used it to choose the strongest predictors to throw into a stepwise regression which then busily went to work undoing all the conservative aspects of the first procedure!

16. Why this particular binning? Well, there isn't any inherent mathematical reason to draw the lines anywhere in particular. So they put twice as many people in the central category, which allowed them to claim statistical significance for effects that are actually not larger in the middle, and therefore—falsely—claim to have demonstrated their claim of an inverted-U shaped effect. Even so, another of the experimental treatments—for example, whether the roommate is wealthy, is significant for the *high* category and not the *medium*, which is a falsification of the claims.

Table 4.1. The Effect of Misspecifying an Interaction

	Model 1	*Model 2*
Genes (x_1)	-.015	.161
	(.119)	(.197)
Race (x_2)	.885***	.742***
	(.055)	(.092)
Environment (x_3)	.917***	.856***
	(.094)	(.157)
x_1x_3 interaction	-.027	2.222***
	(.203)	(.296)
x_2x_3 interaction	2.011***	
	(.088)	
R^2	.784	.392

(Here and below, $^{***}p < 0.001$; $^{**}p < 0.01$; $^{*}p < 0.05$; $\dagger p < 0.10$.)

of .5. The first doesn't directly predict the outcome, while the second does, but it also interacts with a third variable, environment (x_3). Thus the true equation is $y = b_2x_2 + b_3x_3 + b_{23}x_2x_3 + \varepsilon$, and here I'll assume $b_2 = b_3 = 1$; $b_{23} = 2$, and $\varepsilon = N(0,1)$. When we have the interaction between the proper variables, we find that there is no main or interaction effect for x_1 (see table 4.1). But when we omit the x_2x_3 interaction (Model 2), we wrongly conclude that there is an x_1x_3 interaction.

We might imagine that this problem goes away if we split the data by genes (x_1), but still control for race (x_2) in every category, as done by Guo et al. But this is not so (see table 4.2). We see that the value of the regression coefficient for x_3 varies dramatically across the categories of x_1, even though there is no interaction between x_1 and x_3, and we are controlling for x_2.

Interactions are not something to use lightly. They can ripple through your models in ways you didn't anticipate. In particular, they change what it means to "control" for one thing, and they privilege one variable over others. They dramatically increase the possibility of false positive results.

If you are going to be a serious social scientist, you need to develop a feel for when you are "too lucky"—when you are getting polynomially "better" results than you should. This is a major problem because, as we'll see in chapter 9, there are some methodological procedures that basically *count* on you being this lucky! But when it comes to control projects, you can also be too lucky—if you assume that your particularly convoluted specification, which leads to significant interactions, is the correct one.

Table 4.2. The Effect of Misspecifying an Interaction via Splitting a Sample

	Groups formed by splitting sample		
Coefficients	Low genes (x_1)	Medium genes (x_1)	High genes (x_1)
Race (x_2)	.785***	.700***	.675***
Environment (x_3)	-.535	.682**	2.466***

Conclusions: Get Smart!

We need to be prepared to go down the road of control strategies. But we can't be overly complacent, nor can we trust that if our treasured coefficient stays significant, we can go ahead and interpret it. What we want is not to get "the" right model, but to use a succession of models to chase down potential findings, testing their implications, and eliminating false conclusions. At each stage, we look at our model, examine surprising or important findings, and then treat each of these with the same skeptical consideration that a causal modeler does the key coefficient for an attempted reproduction of a causal estimate from observational data. That helps us flesh out a more concrete understanding of what might be going on, which we then examine with a succession of models.

Let me give an example. When I was analyzing the Young Men's Health Study (which focused on gay men in the San Francisco Bay Area), I found that the use of the recreational drug "poppers" (amyl nitrate) was *negatively*—robustly negatively—associated with seroconversion to HIV+ status. That's sort of interesting because a wacko at Berkeley, Peter Duesberg, had been arguing that these were the *cause* of AIDS, and why gay men were coming down with it (as opposed to the cause being the HIV retrovirus). Even with "controls" in, I had a cool finding—that the more illegal drugs you do of a certain kind, given any level of unprotected sex, the better your health. I could have been "controversial" and published it. But it is almost certainly false that poppers protect you against HIV.

What was going on? Well, I spent some time looking at the different folks in the data—looking at the suspects before trying to hypothesize the motives—and became convinced that in the (Bay Area) sample we had, the use of poppers was strongly associated with gay identification. The more gay-identified men were, the more sex they had (increases HIV risk), but also the more exposed they were to safe sex messages, the more they saw these as applying to them, and the more likely they were to have sexual encounters with people they knew and were comfortable talking with about safe sex. That greatly decreased their risk.

This suggests that if one is proposing to use a regression to estimate the effect of some behavior on seroconversion, one needs to take seriously the heterogeneity between more and less gay-identified actors. But of course there is no perfect measure of this "identification"—which is probably a multidimensional and occasional phenomenon, anyway. So one can't just toss in one's best control. Instead, we need to use a succession of different models to generate a plausible interpretation of the relation between this heterogeneity in identification, largely unmeasured (on the one hand), and the dependent, independent, and other control variables, as well as their relations (on the other). We pay attention to whether the relations between other variables change as you would expect given the inclusion of items that should tap this heterogeneity.

Overall, one might say that we need to bear in mind that we don't *control*, we *try to* control. And if you want to fail at that, you can. But even when you're trying, you can be defeated by puzzles that have to do with different levels of analysis, and different patterns of variation. And that's what we go on to investigate in the next chapter.

Appendix 4-A

Let's expand the model for mean-centered variables, which I'll denote with a star (*) by them.

$$\hat{y}^* = b_1^* x_1^* + b_2^* x_2^* + b_3^* x_1^* x_2^*$$

$$= b_1^* x_1 - b_1^* \bar{x}_1 + b_2^* x_2 - b_2^* \bar{x}_2 + b_3^* x_1 x_2 - b_3^* \bar{x}_1 x_2 - b_3^* x_1 \bar{x}_2 + b_3^* \bar{x}_1 \bar{x}_2 \tag{4.5}$$

We can group all the constants and like terms together and write this now as

$$\hat{y}^* = (b_1^* x_1 - b_3^* x_1 \bar{x}_2) + (b_2^* x_2 - b_3^* \bar{x}_1 x_2) + b_3^* x_1 x_2 + (b_1^* \bar{x}_1 - b_2^* \bar{x}_2 + b_3^* \bar{x}_1 \bar{x}_2)$$

$$= (b_1^* - b_3^* \bar{x}_2) x_1 + (b_2^* - b_3^* \bar{x}_1) x_2 + b_3^* x_1 x_2. \tag{4.6}$$

From this, we can determine the relation between the coefficients in the two versions of the model. And we find that

$$b_1 = b_1^* - b_3^* \bar{x}_2 \tag{4.7}$$

and

$$b_2 = b_2^* - b_3^* \bar{x}_1. \tag{4.8}$$

* 5 *
Where Is the Variance?

Where Is My Variation?

We've seen that our quest to learn from data boils down to an attempt to see whether the patterns of observed covariation and partial-covariation are compatible with our interpretation. We began with relatively easy cases in which we could treat all the cases as independent. From here on, we'll be looking at cases that get progressively more complex in terms of our data being embedded in contexts. I'm going to be pointing to ways in which we can come to wrong conclusions because of the complexities in where the variation is located in such a data structure.

Map: In this chapter, we're going to focus on the simplest case, when we sample units containing more than one individual case each, like classrooms with sets of students. I'm going to start by discussing how our variance can be distributed between and within our units, and then what happens when we collapse certain variables across these units. I'll then turn to a series of cases in which we—deliberately or not—partition our data in such a way that we leave more variance in some parts than others, and come to incorrect conclusions as a result. I'll then generalize to thinking about hierarchical data structures, and discuss what hierarchical models can do for us in such situations. I'll close by talking about some easy-to-mix up issues of cross-level inference.

Back to School

Let's return to the issue of *where* our variation is. If you are like me, you didn't pay enough attention to ANOVA (analysis of variance) in stats class (see chapter 4). Pull out your old text and refresh yourself on it. When you have data like "students nested in schools," you can decompose the variation into that *between* schools, and that *within* schools. That's straightforward. What we don't necessarily appreciate is that measurement error can lead us to be unsure as to whether or not some variable is really at the level of our students, or at the level of the schools.

Consider this nice example from *Academically Adrift*, by Richard Arum and Josipa Roksa (2011: 68). They have a data structure of students nested within colleges. They want to know how well colleges are doing at teaching students. One way to get at this is just to *ask* students; in this case, subjects were asked whether they agreed or disagreed with three questions all beginning "Students at my institution . . ." and ending ". . . have high academic aspirations," or ". . . help each other succeed," or ". . . work hard to succeed academically."

Arum and Roksa (2011: 68) note that "after controlling for individual-level differences in academic preparation and social background, we found that 25 percent of variation in students' reports of peer aspirations and 22 percent of variation in reports of peer support occurred across colleges." They emphasize that institutional differences were around twice as powerful predictors than all other things combined. They were, remember, asking students to report on their institutions. If the students' reports were perfect reflections of a single underlying reality, and there was no error in the data, 100% of the variation would be across colleges. Twenty-five percent of the variation explained by the school level after purging individual differences might at first strike us as being on the low side.

Yet there are way more students than there are schools. There's going to be a lot of individual-level variance, even if the processes are *mostly* at the school level, if there is individual-level error (for example, the respondents have different thresholds at which they think an aspiration is "high"). Let me walk through a simulation (R 5.1) to demonstrate this. I'm creating 50 schools, each with a mean climate of academic rigor, which varies randomly from 0 to 10. Then, from each school, we pick 10 students to survey on the academic rigor. They all add some individual-level variation around their school's mean. For example, fig. 5.1 gives the results from one simulation (with an error variance of just 2). Each school appears as a boxplot, a line indicating the median report of the students, with the central 50% of cases in the box.

In this case, it turns out that 83% of the variance is between the colleges, and 17% within. Naturally, as the error variance of the individuals goes up, the proportion of the variance that is between groups goes down (see fig. 5.2). If we consider a model predicting the outcome for each student using her school's mean, we'll partition the variance into the explained (between school) and the unexplained (within school), which will be *residual* variance—the variance remaining after the model.

It is worth emphasizing that, by construction, the only *interpretable* variation here is between-schools. All the within-school variance is ran-

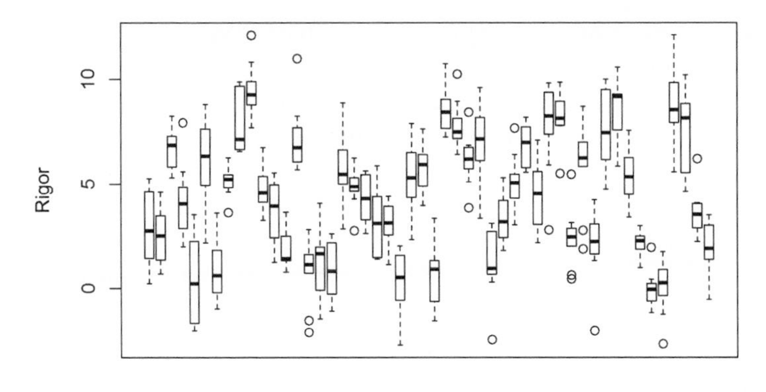

FIGURE 5.1. Distribution with Variance = 2

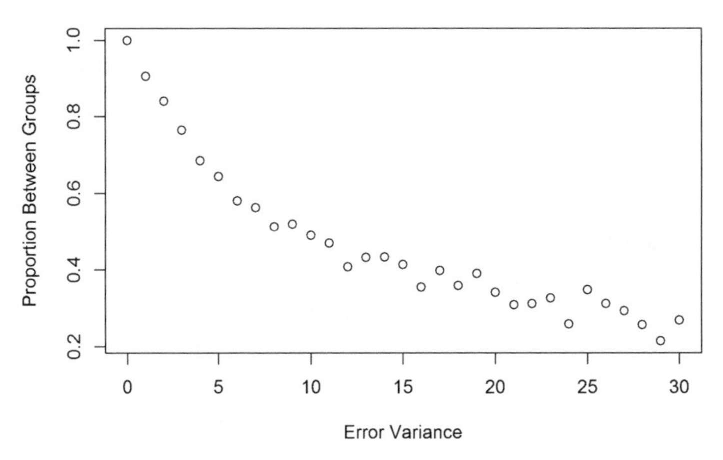

FIGURE 5.2. Between and Within

dom error. Thus it is possible for a very low proportion of explained variance to be between-schools variance to be compatible with a world in which there is a single "true" measure of school quality.

But it also may be that there isn't "a" context at most colleges. There are different pockets and students have radically different experiences. The statistics alone can't decide it for us. But they can make it harder or easier. Once Arum and Roksa account for selectivity (presumably an underestimate), the proportion of variance across schools is 6% for the question "faculty are approachable," 9% for "faculty have high standards," 6% for "faculty have high expectations" (p. 176) and 12% for "students have high expectations," 5% for "students help others succeed," and 11% for "students work hard to succeed" (p. 180). My feeling from this is that we probably have to reject either the idea that schools have unitary contexts, or that students can measure the degree of this treatment, or both.

Looking at *proportions* of variance explained is a good place to start, but it can mislead us in comparisons. It's a problem we'll see again in the next chapter—when it comes to ratios, we often assume that change is happening in the numerator, and forget about the denominator. For example, imagine that we are interested in the whether the educational system is largely egalitarian. We're interested in modeling the achievement of individual children in school as a function of their family backgrounds and the quality of their schools. Students also vary in the degree of effort they put in, but we don't have a direct measure of that.

Now imagine that the variance across schools decreases—for example, someone increases federal funding for schools so there is less variation between different districts. The proportion of the variance in outcome that is caused by family background will *increase* . . . and possibly mislead us into thinking that there is *increased* inequality, when in fact it's decreasing. (Using a regression slope as opposed to percentage of variance explained *might* help us here . . . but only if we have a very stable longitudinal data frame.)

Thinking this way, one of Arum and Roksa's findings might not be as depressing as it seems. They want to know how much going to college actually teaches people, and I think in general, the evidence falls on their side, namely that colleges are teaching less than they used to. But as for how much students learn . . . Arum and Roksa show that on one particular test, students go up only .18 standard deviations in a year (2011: 35). As they put it, this means that a kid who is at the 50th percentile of the sophomore class is only at the 57th percentile of the freshmen. But remember, this could equivalently mean that there is a *lot* of variation across freshmen to begin with—and that schools are doing really well at being open to many different types of students. If we want the change from one year to another to be bigger in terms of *standard deviations*, the best way to do that is to reduce incoming variation, by, for example, denying admission to weaker students. We definitely want to be aware of where most of the variation is, but that's our starting place, not our ending place.

Residual Variation, and How to Disguise It

So we've seen that residual variation, when it takes the form of within-group variance, might well represent noise, as opposed to meaningful variation. But even so, it's important to remember that the between-group variance was still variance among individuals. And so it's easy to get mixed up when thinking about between and within unit questions.

Sticking with the example of schools, you sometimes hear the triumphant news that it turns out that SATs don't predict college success! This might well be true, but hardly relevant, if the issue is whether the SAT scores should be used by colleges for *admission*. Why? Because the comparison of variation in individual scores *within* a school that has *already* selected its students on the basis (at least in part) of SAT scores is irrelevant for the issue of how well students would do in a particular college that has *not* yet made that selection. That is, if a college only takes students who score 1400–1550, precisely where they are in this range might not be important for their success—especially if the college has been attentive to *other* factors that they weigh along with SAT scores. In fact, the successful admissions process should lead to a zero correlation *within* the school of practically *any* criterion![1]

And so even if we don't think that there is much information in the residual variation, better to present it (as did Arum and Roksa) as opposed to making it disappear. How do we make it disappear? We compute school-level means, and conduct our analyses only on that level. Why would we do that? Because it actually can make us feel more secure in our findings. Let me give a simple example (R 5.2). Imagine that we are comparing people's income (y) to their effort in school, measured as percentage of homework they completed on time (x). The pretend data from 1500 people is shown in fig. 5.3.

First, let's take a moment to remind ourselves about how our statistics works (look back to fig. 3.1). We have that cloud of points, and we want to predict the y value with x. In this case, the predictions form a line, but they don't have to. All that matters is that for any value of x, there's a value for our best guess of y. But the observations aren't at that one value; they're scattered above and below. That's the residual variance from this model. It isn't necessarily *error*. The error comes only when we use the same value to make predictions for all these different observations.

Let's say that for some reason, we want to actually collapse our data into only 100 cases, instead of the 1000 that we have. We bin cases according to their position on x, assigning the first 10 lowest cases on x to one value, the next 10 to a second, and so on, and then assign each case its mean value x and y. That's exactly the sort of thing that happens when, say, colleges accept people on their value of x and people only accept the

1. Why all the grousing about the SATs? Well, you might wonder why a school would make SATs "optional," and encourage those with terrible SATs to apply anyway? And you might wonder why schools would "super-score" the SATs, allowing you to submit your *highest* score? And you might then ask yourselves, are colleges ranked on the basis of how many people apply, and how high their average SAT score is? And then you might find the answer wiggling its way into your head. . . .

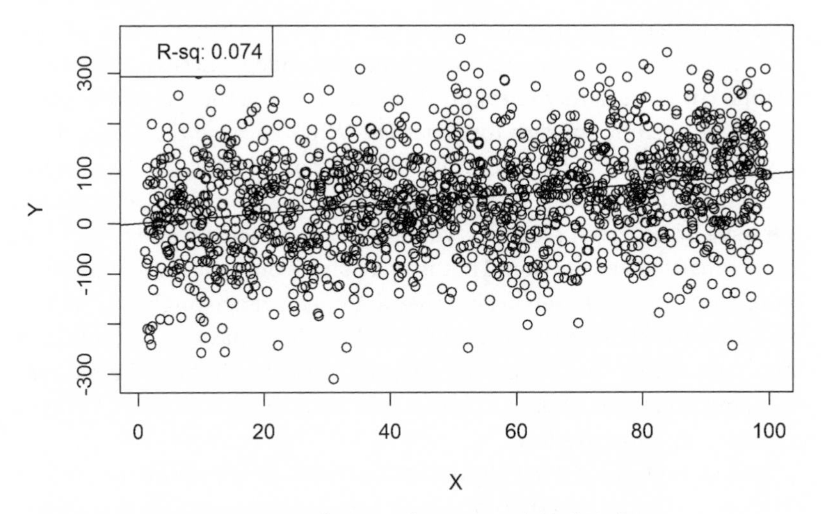

FIGURE 5.3. Typical Scatterplot, Low Explained Variance

highest ranked college that admits them. Then we could imagine examining the distribution of these new collective averages. You can imagine putting two vertical lines on the graph, and including all the points between them in a single "bin." And that data looks like fig. 5.4. Easier to see the trend, right? And look how much larger the R^2 is! And, as fig. 5.5 shows, the fewer bins we use (that's what's on the horizontal axis), the more the R^2 (on the vertical axis) goes up.

How come as we *throw away* information, things are getting *better* for us? That's because we've thrown away only the variation we couldn't do much with. This might seem puzzling if you remember, from chapter 4, that binning by our predictor could *decrease* our R^2. Ah, but then, we had *only* grouped by the independent variable. We might go from 1000 distinct values of x to only one hundred, but we left all 1000 values of y. Here we actually reduced the number of effective cases—because we got rid of variation in our dependent variable, not only in the predictor. And that's why our R^2 went up.

Tables, Cells, and Individuals

We've seen that aggregation that effectively obscures individual-level variation can increase our apparent success. In essence, we increased our percentage of variance explained not by boosting the numerator, but by decreasing the denominator. There's nothing intrinsically wrong with that, but it's important to bear this in mind when knowing how to interpret different results. If most of our data is data on the groups, it's natural

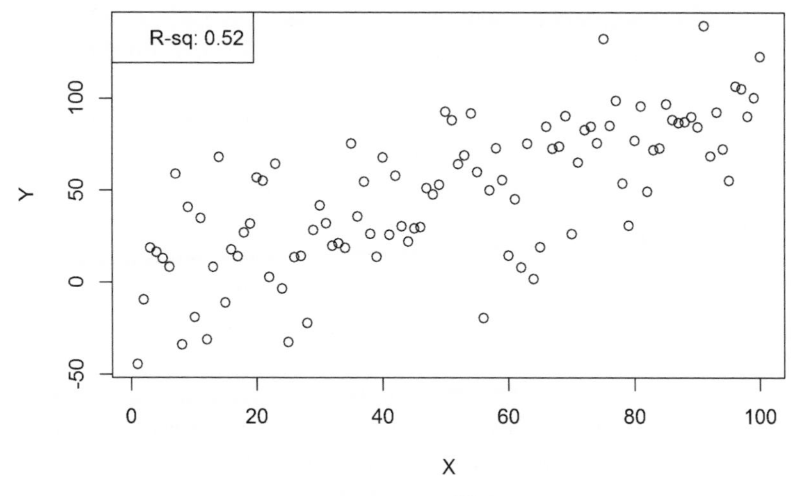

FIGURE 5.4. 100 Bins

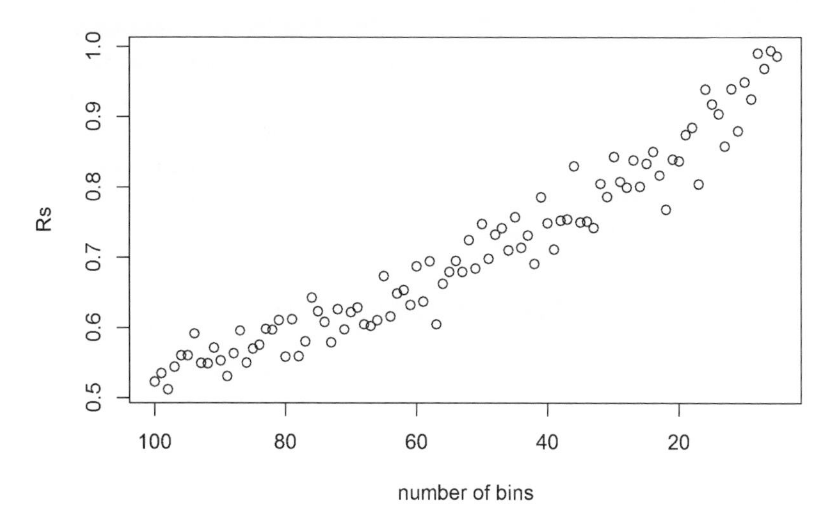

FIGURE 5.5. Fit and Binning

that we'll seem to have more success when we fit a model for the groups than when we fit a model for individuals.

For example, imagine that we have data on fights between schoolchildren, and we use the Bradley-Terry model that I described in chapter 3 to fit the data. It figures out a status score for each kid, and proposes that when kids of equal status fight, it's a toss-up as to who will win, and as the status difference increases, so does the chance of the higher status kid triumphing. There are two ways of seeing these data. One way is, if we have N kids, each fighting with each other M times, we make a table with

N rows and N columns. In each cell (say, the one in the 3rd row and the 5th column), we have the proportion of times that the row kid (kid 3) beat the column kid (kid 5). The other way is to look at all the fights, of which there are $MN(N-1)/2$, and predict which direction it goes in, to the first kid (A) or the second (B), depending on which kid is Kid A and which Kid B.

If the two kids 3 and 5 are of equal status, it might be a toss-up, and the model might correctly predict that, in the first case, the value in cell (3,5) is .5. The model would look like it was fitting very well. But if we treat the data differently, the model has to fit the M bouts between kid 3 and kid 5, and it can't do any better than a coin flip would. The model seems to be doing very badly! Which is right?

In this case, we should be pretty happy to go with the first, because all our measurements are at the kid level—none at the bout level. If we have information at the lower level, we don't want to throw it away, do we? Well, we can be tempted to, if we can't get much out of it. Sometimes most of our meaningful variation is only at the group level.

I think the only real way of handling these puzzles is to move toward a hierarchical way of thinking about data, and recognizing that we want to explain data on more than one level at once. And we'll get there soon. But we're not ready to solve problems yet; we're still working on identifying them. We've seen the importance of seeing where the mass of our variation is in a nested structure; let me illustrate briefly for a simple case of a cross-nested structure.

How Is My Data Shaped?

We've been looking at cases of nested data, like students in schools, where the variance can be partitioned as that between and that within schools. Now let's generalize to the case of multiple nestings—multiple ways of grouping our cases. That's a very common data structure—we see it in a conventional table (with rows and columns as groupings). Once again, we're going to see the Funkadelic principle—"You can't miss what you don't measure"—in action. If you have an analysis that turns on comparing the variance explained by different variables, anything that affects their overall range is likely to affect your conclusion. To illustrate, let's take a simple case where we have two independent variables and a dependent variable. We can summarize our data as a table, in which the rows are the values of the first independent variable, the columns the values of the second, and the cell values the mean (and standard derivation, if we want) of the dependent variable. A good example here would be the age-period-cohort tables that we examined in chapter 2; turn back to that for a visual representation. The key thing is that not all such tables are

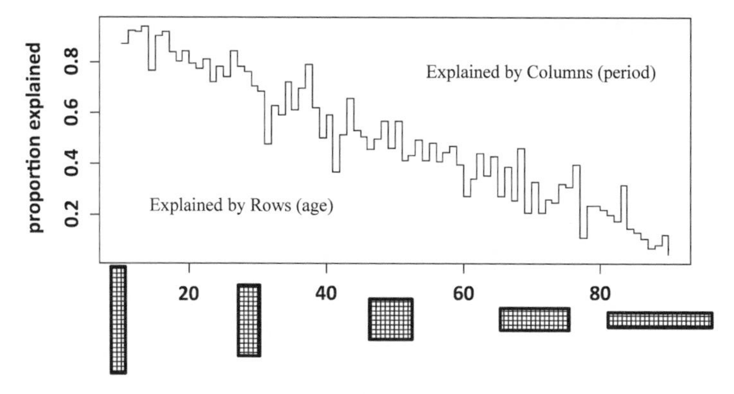

FIGURE 5.6. Rows and Columns

"square"—their shape depends on the range of the different dimensions of the independent variables.

The human age span is pretty limited. Depending on whether you are using census data or survey data, you're going to have an age range of only around 75 or 55 years. The period for census data could be more on the order of centuries, but survey data is usually limited to forty to sixty years, depending on the source. Thus different data-gathering approaches lead to tables that are non-square in particular ways.

That means that depending on what data source you use, you can find most of your variation explained by period or by age—even if the underlying processes are identical. For example, here I construct a table with two independent variables, let's say age and period (R 5.3). In each table, the overall strength of row and columns is identical—for each row (or column), I draw a random number from –2 to 2, to represent its contribution to the dependent variable (say, income). And I slowly change the shape of this table from sky-scraper to ranch-house—from a 90 × 10 to an 89 × 11 all the way to a 10 × 90 table. Fig. 5.6 below shows the percentage of the total variance explained by the rows (age) and columns (period) (ignoring unexplained variance). Along the axis, I give a cute visual representation of the table being analyzed.

If we're trying to explain variance, and we *use* variance to explain variance, then wherever that variance is, is where our explanation will be. Part of the attraction of the regression model is that, if our variables are continuous and have linear effects, we can try to compensate for this by standardizing by the variation in our independent variables. And yes, this *can* make it possible to compare across data sets that have different amounts of variance in predictors, but are otherwise similar. Still, this can lead us to be complacent and forget about the Funkadelic principle, which will be crucial wherever we rely on a partitioning of variance to make conclusions. And sometimes we are doing this without knowing that we are. We'll turn next to that problem.

Looking along the Lines

Partitioning by a Predictor

Now that we're used to thinking about the interplay of variables with different degrees of variance, we can handle a problem that creative analysts often stumble into without realizing it—coming to incorrect conclusions when partitioning their data sets. In chapter 3 we saw that attempts to look for causal heterogeneity could be misleading given imperfect specification—we *might* be finding that the treatment has more effect on those with a low propensity to *get* the treatment in the first place. Or it could be that we didn't specify the propensity well enough. We can understand that issue as one case of a larger category of analytic approaches, one in which we attempt to *partition* the variance according to a dimension from our data. (Remember that a propensity score is nothing but a projection from a larger-dimensional subspace to a single dimension.)

Imagine that we have a *strong* predictor x_1 of some y, and let's imagine that y is held within bounds (for example, it is produced by a scale, or a checklist of symptoms). For example, y might be depression, and x_1 might be health problems. That means that when x_1 is high (someone has a lot of health problems), y is likely to be up against its ceiling (they are pretty depressed). In that case, it's going to be hard for some x_2 (for example, income) to do a lot of predicting of y, right? And the same if x_1 gets really low. In chapter 4, we had examined this when analyses were symmetric—and so the bounding appeared as an interaction between x_1 and x_2.[2] Now let's look at what happens if we take an asymmetric analytic approach—binning our data by their value on the strong predictor.

Let me illustrate with a simulation (R 5.4). Fig. 5.7 below shows simulated data; each point is a case with a value on two variables, the first (our health scale; x_1) on the horizontal axis, and a second (income; x_2) on the vertical. The first is a strong predictor of depression, the second, a weak one. We regress depression (a dichotomy) on income within bins formed

2. It's worth emphasizing that this isn't *necessarily* true for an unbounded dependent variable (like income—*if* we gather it by looking at tax forms, as opposed to asking people to choose a category in a bounded scale)—and it also isn't true for loglinear type models (such as the logistic regression) for *bounded* variables (since such models are technically variance-indifferent). However, the finitude of samples makes this a familiar pattern even for logistic regressions.

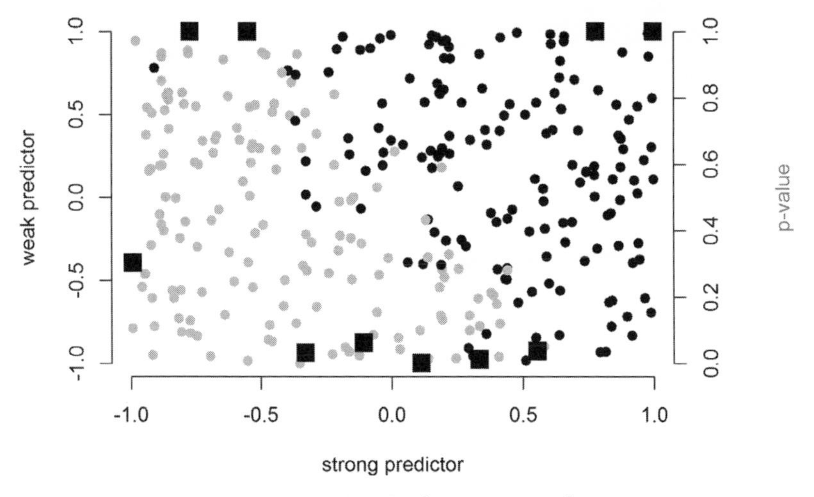

FIGURE 5.7. Partitioning by a Strong Predictor

by the cases' values on health.[3] Each dot is a data point, and the dot is dark if the dichotomous dependent variable is 0, and light if it is 1. I've divided the data into ten bins, and put a black square to indicate the *p*-value for the *income* coefficient (x_2) in each bin. (To avoid clutter, bin lines are not drawn, but you can visually group data points horizontally near to the square.) You can see that in the bins toward the extreme values of health (x_1), there isn't enough variation in depression left over for income (x_2) to predict anything. Doing this for 100 simulations, and averaging the *p*-values, produces a clear pattern, as seen in fig. 5.8. There's a tendency for the weak predictor to have explanatory power only when the strong predictor is in the middle of its range. So it can look like there's an interaction between the two predictors, when in fact, their contribution to the dependent variable is independent.

And that paper by Guo et al. (2015), which I used in the last chapter to show how we can mis-parameterize interactions, thought that it was investigating gene-environment interactions, when all it was really doing was this: trying to see how much residual variance is available for a second predictor given that there's also (or so they hypothesized) a strong predictor. In their case, they thought that the strong predictor was genetic propensity to binge drink, and the weaker ones were environmental (having

3. The equation is $y = b_1x_1 + b_2x_2 + \varepsilon$, where $b_1 = 7$; $b_2 = 1$, and $\varepsilon = N(0,3)$, where both x_1 and x_2 are draws of 300 points from a uniform distribution between −1 and 1. A dichotomous variable is then constructed $z = 1$ if $y > 0$. A logistic regression of z on x_2 is carried out in each of 10 bins formed by breaking up x_1 into 10 equally sized bins.

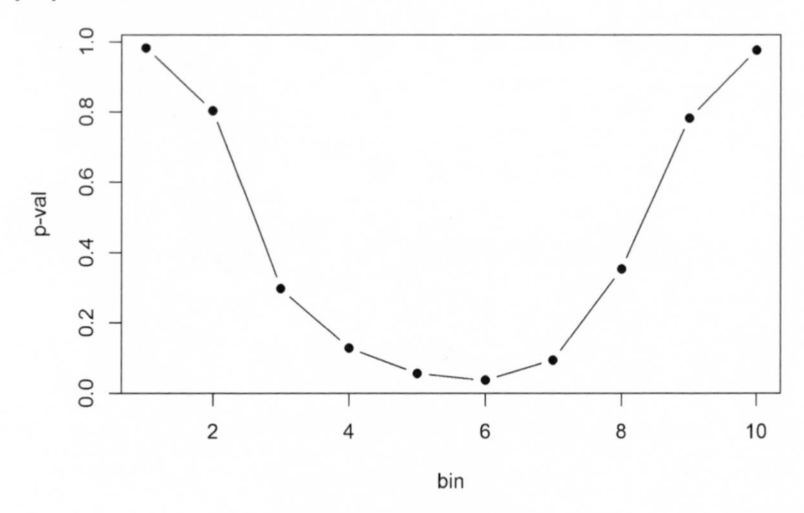

FIGURE 5.8. Average *p*-values

to do with students' roommates). But Guo et al. didn't even need to have their environmental variable measured to do their analyses! Presumably *something* led people to drink, and if it isn't genes, well, then, it's environment. They could have just looked at the dispersion over their genetic propensity measure. But then, of course, they'd have instantly seen that they were wrong . . . because, contrary to hypothesis and their attempt to maximize its power via stepwise regression, their x_1 turned out to be a *weak* predictor. Know thy data!

Constraint

So we can't always make fair comparisons across parts of our data if they differ in variance by construction. But there are some cases in which we attempt to do exactly that. The most vexing example here is the attempt to understand cognitive "constraint," understood as the covariance of beliefs within some population.

Now I've argued elsewhere that this whole notion is misguided (Martin 1999b, 2000), but let's put that to the side. Instead, let's see how even if the basic conception made sense, attempts to study it easily lead to problems. The assumptions of the conventional approach are, first, that people "have" beliefs, and that these beliefs are interconnected with one another by webs of implication, whether logical, psychological, or socio-logical. Second, we assume that the degree of this interconnection can be examined by looking at the aggregate distribution of many believers in terms of these two beliefs. Where association or covariation between two beliefs

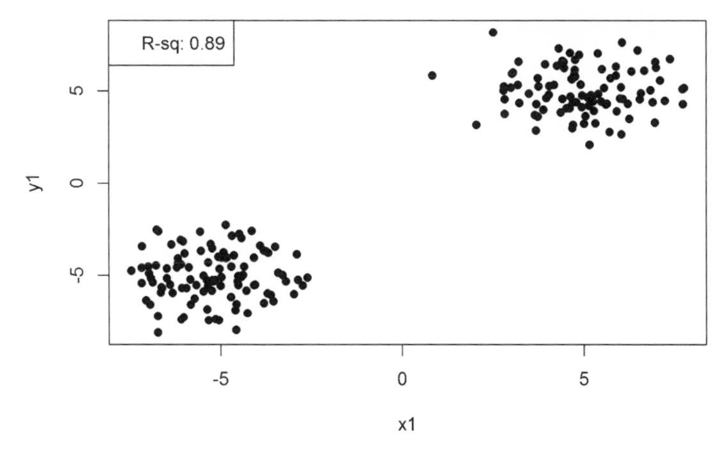

FIGURE 5.9. Two Clumps of Believers

is high—where lots of people hold both or hold neither—then the beliefs are judged to be tightly connected in each individual's mind. Where they are not—where holding one doesn't seem to increase the odds or degree of holding or not holding another—then the beliefs are judged not to be closely connected in respondents' minds. In other words, call it what you will, this is all about variance and covariance.

But what do we make of a situation like that graphed in fig. 5.9—an imaginary set of relations between, say, degree of support for gun control and degree of support for legalization of marijuana? Do we have one group, with high constraint (since those who are "high" on one belief are "high" on the other)? Or do we have two groups, in both of which there is *no* constraint (because the beliefs are independent)? Certainly, if we found that the top group was all women, say, and the bottom all men, we might feel confident going with the second interpretation, or that the former were all Catholics and the latter Protestants, but what if we had data only on these (and perhaps other) beliefs to go on, as opposed to data on group membership? Though then again, isn't the difference between Catholic and Protestant about beliefs? . . .

Well, there's no answer, so I'll just let that sit, and draw the lesson: how we partition our group is going to have big implications for our conclusion. There's an intrinsic paradox when we use a cross-individual statistic to tap an intra-individual characteristic, because, by definition, our statistic becomes non-existent as we approach the individual level. That's not true for most of our aggregate measures, like the average.

Ready for it to get worse? Imagine now we want to compare groups that we deliberately construct on the basis of their beliefs. Let's stick with politics: It's basically the case that in the US, there are two types of conserva-

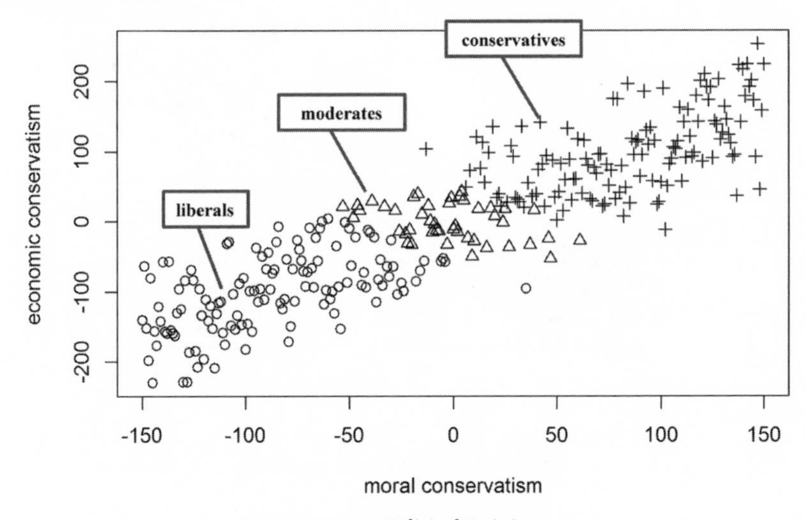

FIGURE 5.10. Political Opinions

tism, moral and economic. Sure, people who are one "tend to" be the other, but there's a separation that's worth keeping in mind, especially because rich people tend to be more morally liberal and economically conservative, while workers tend to be more morally conservative and economically liberal. It's been that way for a long time, though it's also always changing in interesting ways. Anyway, let's imagine that people are distributed in this two-dimensional space. And that they also choose a label for themselves, as liberal, moderate, or conservative, depending on their sum on these two dimensions. (That's more or less what most sociologists assume that respondents are doing when we ask them a question about overall political ideology.) Fig. 5.10 portrays imaginary data (R 5.5), with circles representing liberals, triangles moderates, and plus-signs conservatives.

Now let's look at the correlation *within* each of our "identities" (see fig. 5.11). The short dashed line is for liberals, the long dashed line for conservatives, and the solid line for moderates. Holy smokes! What's wrong with the moderates?! For them, the two dimensions are *inversely* correlated. How dumb are they? Don't they understand that the two things go together? But that sort of negative relation is *necessarily* the case given our truncation of the overall relationship! That's because the moderates occupy a thinner slice of the cloud. We can't (1) ask people to line themselves up by degree of liberalism-conservatism; then (2) ask them to assign labels to themselves based on where they sit; and then (3) insult those in the middle for a mathematical truism, can we? But we might.

Would you like to change your conclusions? Just change the bins! Whoever has the *largest* bin is likely to have a slope coefficient that is in line

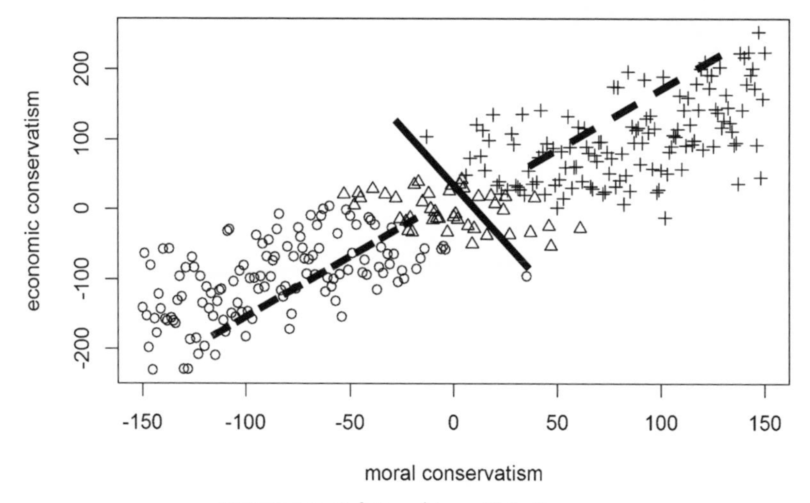

FIGURE 5.11. Relationships within Groups

with the *overall* relation. Whoever has the *smallest* bin is likely to have a relation that is *orthogonal* to the overall direction, because it's going to be picking up mostly on the residual variance! Thinking this through requires that we approach these data more rigorously as multilevel data, and so we next turn to that.

Meet the Simpsons

Part of the answer to our puzzle is that we need to decide if we have multilevel data, or not. If we don't, then partitioning it as *if* we did (as in fig. 5.11) can be perverse; if we do, then we need to analyze it as such. From that perspective, we've stumbled into an issue often called "Simpson's paradox." While this so-called paradox is named after a statistician, we'll remember it as Homer Simpson's "D'oh!" Because you come up with one understanding from looking at your data one way, and, by another view, it's totally wrong. D'oh! The key here is that when we have a multilevel structure, with people (say) clustered within larger units, intervariable relations *between* units aren't the same as relations *within* units.

For example, usually, those with more education are more trusting of science as an institution than are those with less. Evangelicals tend to be of lower education than non-Evangelicals, and to be less trusting of science. So you'd think the individual-level relation explains everything here. But Gordon Gauchat (2012) shows that *within* the set of Evangelicals, those with more education are *less* trusting of science, not more (something very similar was found by Regnerus and Smith [1998]; the reason,

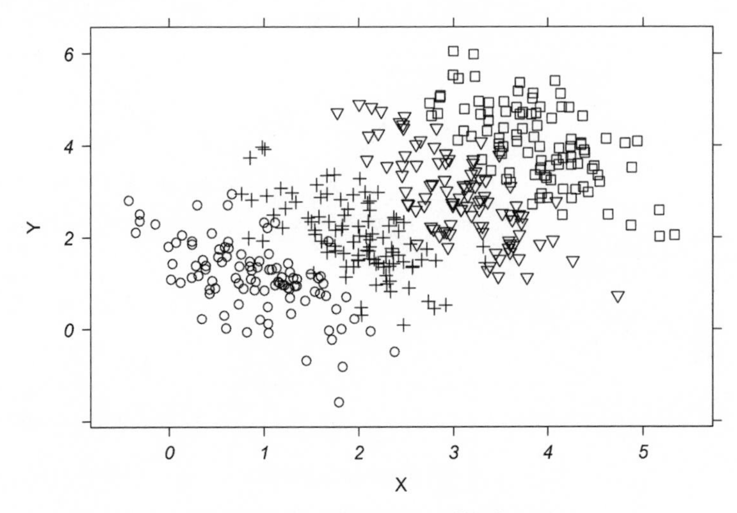

FIGURE 5.12. A Strong Overall Relation

if you're wondering, is that the more educated have a better sense of who their enemies are). Let me give a simulated example (see fig. 5.12; R 5.6). Imagine that we have a world in which high school kids can take as many after-school enrichment classes as they can stand. Most people think that if you specialize in taking math classes, you'll have less time for soccer, and vice-versa, so there should be an inverse relation between the two. Yet when we sample 400 high schoolers, we find a strong *positive* relation (here $r = .574$).

How can that be? Well, in this world, there are four classes: poor, okay, middle, and rich. The more money your parents have, the more classes you can take. The four classes are denoted with four different plot symbols. Look closely: *within* any group, the relation is clearly *negative*.

How should we think about these data? As having a positive or a negative relation? There are three ways to proceed. First, we can examine the relations *within* each group separately. Secondly, we can overlay these four groups on one another by re-centering all the observations as deviations from the *group* mean, and turn this overall positive relation to a negative one. That isn't wrong, and—as we'll see in a minute—is related to what is known as the "fixed effects" approach, but it isn't always what you want. The third is to try to examine both levels simultaneously. Sounds best, doesn't it? It's for these reasons that hierarchical linear models are increasingly popular—as I think they should be. These allow you to examine both within and between variation, and, in effect, to do five different lines, as portrayed in fig. 5.13.

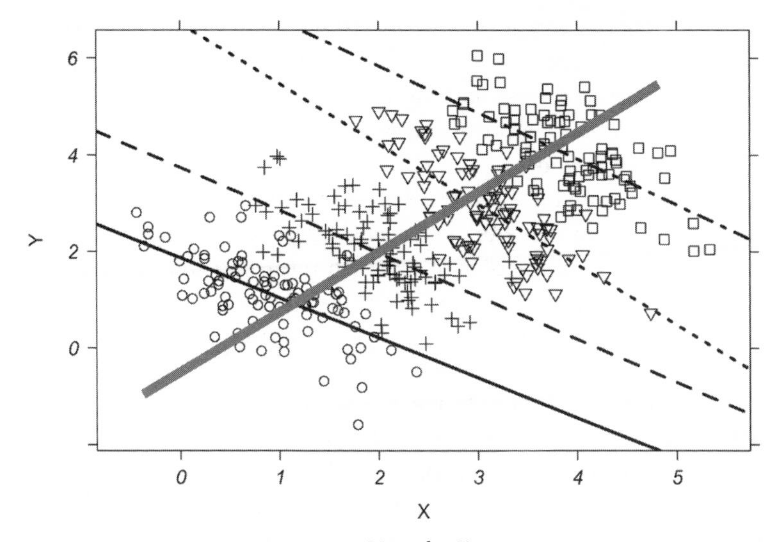

FIGURE 5.13. Meet the Simpsons

Such multilevel models are a very general and flexible way of thinking about data; they are also related to what are called *mixed* models (though not all multilevel models are mixed, and not all mixed models are multilevel). Mixed models contain both what are called fixed effects and random effects. In the next section, I give a simple introduction to the logic of this difference, and walk through the simplest ways to think about a hierarchical model.

Fixed versus Random Effects

Stay on the Cab

Social scientists are often interested in how individual-level processes are shaped by "contexts." For example, we might be studying political behavior, and interested in looking at the interaction of individual respondents' values on some independent variable x (say, their income), and some aggregate of x in the context that they are in (say, the average income in their neighborhood, $\bar{x}$). We are tempted to just go ahead and merge the context for the ith individual into her record, and then fit:

$$y_i = b_1 x_i + b_2 \bar{x}_i + b_3 x_i \bar{x}_i + c \qquad (5.1)$$

However, it turns out that these coefficients actually are all tangled up with the variance explained by the contextual level (Angrist and Pischke

2009: 194ff). There's no way to proceed without really taking seriously that you have two-level data (otherwise, as Angrist and Pischke point out, we ascribe all contextual-level variation to $\bar{x}$, and treat it without justification as peer effects). And the best way to do that involves some form of a hierarchical linear model.

The classic hierarchical linear model is one that treats some aspects of the contexts as "random effects." Because there are some prejudices and myths about such approaches, it is worth returning to our "analysis of variance" way of thinking we started in the chapter.

I was in graduate school, and I think I had just been reading a really interesting article by a methodologist that I wanted to build on. But one part didn't quite make sense to me, so I went to Jim Wiley for help. "The reason it doesn't make sense," he said smilingly, "is that it's all wrong." I'd already learned that some people use the phrase "all wrong" to mean "not my approach" so I was suspicious and cautious. "The reason," he explained, "is that these rows are a sample, not a population, so this is really an ANOVA problem and so the right answer would . . ." and he deftly sketched some formulae for me that I did not understand. I *had* understood this part of the original formulation in the paper, and I had liked it. There was nothing about sampling, only elegant formulae and structural parameters. In that sort of world—and mind—things were what they were. If you had a proportion of .35 in one cell, it stayed .35. What Jim was doing was making things weird and ambiguous.

But statistics is not a cab that one can have stopped at one's pleasure. Jim's point was that the approach I liked wasn't the right one if we wanted to make *inferences*. It used a fixed effects approach to a random effects problem. That felt better for my rigid mind, which wants a .35 to be .35, but in some cases, a .35 sample value suggests a .30 population value.

Here's the way of demonstrating the most fundamental issue to you—a classical problem in analysis of variance (ANOVA). Let's say that we have a population of $N = 1000$ with a single variable $x = N(0,10)$, which we randomly divide into 50 groups. If we do an analysis of variance, we find that around 5% of the total variation is *between* group means. But the ANOVA test—correctly—tells us to ignore these ($F = 1.014$, $p = .448$). Because we would expect *some* variation in group means even if all were a draw from the same population.

The same is true for when we are sampling schools. Let's say that we are back to our example, and regressing students' achievement (y) on, say, number of mathematics classes that they have taken (x). We're concerned that we might mistake something bound up with overall school quality—which we haven't measured—with both our dependent and our independent variables. A "fixed effects" model for this is one that basically lets each school have its own baseline average achievement. And that can be fit in the following equation, where we predict the achievement of the ith student in the jth school:

$$y_{ij} = bx_{ij} + c_j + \varepsilon_{ij} \tag{5.2}$$

We come up with an estimate for the *c* parameter for each school. And a standard error for that parameter. That works if we have a lot more students than schools. But suppose we have, say, 700 students, from 114 different schools? Are we really going to try to estimate 114 different parameters? Hmmmm . . . (yes, you can rescale, and we'll get to that below).

But we could instead treat these differences as coming from some *distribution* that we assume to be normal, and hence is characterized by a mean (which we can arbitrarily set at zero) and a variance, which we can call τ_c. In that case, we aren't actually estimating the values of the intercepts—the only parameter we're estimating is τ_c. For this reason, random effects often appear as a "poor man's fixed effects"—something you do when you can only afford one degree of freedom. And it's also often described as "weaker" than fixed effects.

Now it is true that with random effects, we make an assumption as to their distribution (and that they are uncorrelated with *y*). But aside from that, random effects are actually *better* when our cases are in fact a random sample, and what we want to do is make inferences to a population (so long as we correctly specify the distribution from which they are drawn, of course). Fixed effects are good when you really think the effects are . . . well, fixed, and that you want to be able to compare two groups to one another. Fixed effects aren't always the right way to go, and the difference between the two isn't always as clear as you might think. If we are trying to make inferences to a population—and that's usually what we *are* trying to do—fixed effects models are likely to give results that are too extreme, just like we might find empirical differences in means among a set of groups when we conduct a sample (see above). Yet it's this very tendency of random effects to reduce these estimates that has led a lot of people to jump off the train!

Shrinkage

The use of hierarchical models to fit the distribution of random intercepts isn't that controversial. In fact, the idea of "tossing in" random effects for any categorical attribute of our data has become quite common; it's often not really thought of as a hierarchical model (rather, it's a "mixed" model, as we mix random effects with fixed effects). So let me go to an aspect of hierarchical models that is more contested. The hierarchial linear model can be used to allow the groups to differ in not only their overall levels of the dependent variable (c), but also the slope connecting the predictor to the dependent variable (b), or:

$$y_{ij} = b_j x_{ij} + c_j + \varepsilon_{ij} \tag{5.3}$$

This often bugs people when it is applied to a type of data set where that hasn't been conventional. For example, imagine that most sociologists are used to, say, regressing mortality on BMI (the Body Mass Index), ignoring the fact that people come from different cities. A researcher now proposes allowing the slopes to vary randomly by city. Here I'm not going to work through the math, but this is the central issue: in the "global" model, we assume a single slope (b) that is the same across cities (j) for all individuals (i):

$$y_{ij} = bx_{ij} + c + \varepsilon_{ij} \tag{5.4}$$

In fact, we usually ignore the j subscript altogether, although in some cases, we'll note it to adjust the standard errors of the estimates.

In contrast, the random slopes model (eq 5.3) assumes that the slopes are normally distributed with $\text{var}(b_j) = \tau_b$. (Note that it also includes random intercepts [c_j]; we'll talk about that below.) Just as with the random intercepts, the HLM model estimates τ_b without actually having estimated all the b_j parameters, and it generally seems to do this rather well. But it can't necessarily give very good estimates of the b_j parameters themselves. The most common HLM way of making such estimates is to come up with two guesses. One is based on regressing y on x within any particular j (call this b_j^*). The other is the grand weighted mean of all of these, which is the global estimate b. We're going to compromise between these two values depending on how much solid information there is in j. If there are many cases within some j, we're going to lean more toward b_j^*, and if there are few, we'll lean toward b.

So again, we're seeing that shrinkage of the estimates of the particulars toward the average. Many people find this so counter-intuitive, that they insist that the HLM approach makes no sense. Yet the same people will have no problem in treating some wholly local slope b_j^*, one arising from simply fitting the model within the jth unit, if it were "the" value within j, while at the same time recognizing that its standard error is so high that it can't be seen as significantly different from some other $b^*_{j'}$. What do they imagine the true values of these estimated quantities are, anyway?[4]

Further, there's something a bit disingenuous about preferring the global model, which shrinks the estimates of the local parameters (b_j) to the global mean (b) *entirely*, to the HLM, because the HLM shrinks them *somewhat*. But even more fundamentally, we have to decide if we really believe the model, which says that (as we recall), the b^*_j slopes are *draws* from a distribution with a certain mean and standard deviation. If so, presumably, what we're trying to do is to get information on that distribution—it's an inference problem, right?

Now the thing that gets people worked up about the shrinkage is that there is a *bias* in the estimates of b_j—they're biased toward b. But, as Greenland (2000) points out, if we're trying to hit a bullseye on a target, there's no reason to prefer a gun that misses the target equally on every side as opposed to one that gets on the target, but pulls a bit to the left. Those shrunk estimates—if we believe what we are doing—are, as a whole, a lot better at telling us the actual distribution of slopes from which we draw than the unshrunk ones.

The critics who want to reject this will respond by arguing that the HLM mode is *making* all these assumptions. For example, the HLM model assumes that the slopes across units are normally distributed. That's certainly true, but by any reasonable point of view, the HLM is *relaxing* assumptions. In general, the "global" model—our conventional one—assumes pretty much the same thing as the HLM, but *adds* the strong constraint that the variance is zero.[5] Of course, we often need to make constraints to fit models. But that's something we are *forced* into; it doesn't accord one a position of epistemic privilege.

There's a second myth about hierarchical analysis we need to dispel. For a while, the rule of thumb of many sociologists was that fixed effects were a stronger and more defensible way of dealing with hierarchical data structures. We've already seen that this isn't quite so—if we have a sample of units, we *shouldn't* be treating them as if they're fixed (at least,

4. One reason I wanted to switch to random slopes to raise this issue of shrinkage is that, in a random intercept model, one can, if one wants, simply use the residuals as if they were fixed effects of group differences. (You average the residuals for all cases in group j to get the mean intercept of that group.) That might seem like cheating, as you get estimates of parameters without giving up a degree of freedom. Actually, the real problem here is that you *aren't* shrinking these estimates. They're probably too extreme.

5. Sure, it could be that the distributional assumptions of the HLM are wrong, and that the real distribution is, say, a triangle or a gamma distribution, and so we can see the OLS is just a degenerate form of that. But that's not what these folks mean.

not if we have a good guess as to the nature of the distribution they're drawn from).

Finally, we often hear that the random effects approach is less conservative than the fixed effects, because only the latter totally wipes out the group differences. But it's pretty easy, and, in many cases, advisable, to do this in the random effects framework (like that of HLM) by rescaling the independent variables so that they are in terms of their deviation from the group average.[6] The problem isn't that one *can't* get rid of this variation, but that how, and if, we do this, can be theoretically more consequential than we might want. So let's return to the sorts of graphs we started the section on the Simpsons with, and use this to visualize the problem at hand.

Random Slopes and Intercepts

I've been going over models that allow for random intercepts and random slopes. That all looks great on paper. But what about when our model doesn't fit all that well? Fig. 5.14 below displays two simulated groups (R 5.7) that might be fit by a hierarchical linear model. The top one has very little residual variation. For this reason, it's pretty clear what its slope and intercept are. But the second. . . . it's a lot harder to tell. Any of those three lines looks vaguely plausible. But they imply pretty different intercepts! And different slopes as well.

We can see that there is going to be a *necessary* tendency for estimates of slopes to be inversely related to the estimates of the intercepts. And that correlation increases with our imprecision—the bigger the slope, the smaller the intercept. So we're going to expect a negative correlation between these two values, and our uncertainty about one is linked to our uncertainty about the other. That's just saying that while the model is coming up with its best guess as to that line, a line that's a bit different would also fit the data pretty well. Similarly, if we had many such groups, we'd find that if there is a lot of variance in the intercepts, there will be a lot of variance in the slopes. So when we look at the estimates of these parameters, there's going to be a high standard error.

That isn't bad—that's the model working right. It's similar to cases of high multicollinearity in conventional regression. An OLS will have to admit it can't be sure about the results when relevant predictors are

6. It also might be pointed out that this technique of fixed-effects-via-turning-into-differences-from-the-mean can't necessarily be done for two cross-nested structures, like time and space.

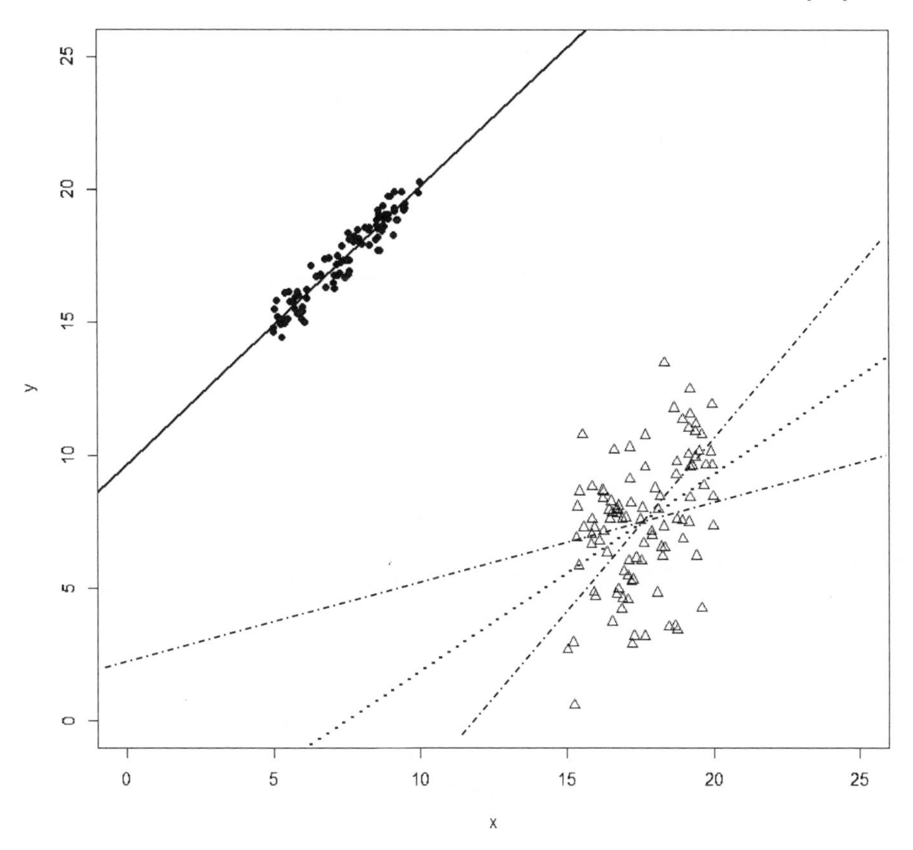

FIGURE 5.14. Slopes and Intercepts

too highly correlated. That's not wrong, that's right: it *can't* be sure. The idea that multicollinearity makes your data inapplicable for regression is silly—that's *why* we do multiple regression in the first place. We may indeed end up throwing out a variable, but we can't throw out the *interpretation* that hangs something on that variable. We have to admit that we don't know.

Same here. In many such cases, it's common to reach for a statistical fix: we center all the groups on their means. There are indeed good reasons to do this, both computational and interpretive, for the cases in which we are including cross-level interaction terms (we recall from the last chapter that how we center observations makes a big difference here). But we *can't* use it to handle a theoretical puzzle, namely whether these groups differ in their *net possession* of the dependent variable, or whether they differ in their *structural parameters* (the extent to which an independent variable predicts a dependent variable). We don't want to have our conclusions be uniquely determined by our assumptions! There's usually nothing to be done here but to closely examine the correlation of these parameters, and see how much the data can tell you.

Things can be even more complex when it comes to logistic regressions because—at least for some cases, and in some interpretations—it may not really make any sense to even *try* to distinguish between random intercepts and random slopes. And this gets to an issue that's well known by statisticians, but not really appreciated by most students I talk to, which has to do with the comparison of slopes from logistic models. It's a side issue, and so I've made it an appendix to this chapter.

To summarize, multilevel approaches have four important advantages. First, they let us explore heterogeneity within our data, which should *always* be our goal. Second, in contrast to models with *latent* groups (which we will explore in chapter 9), they are rather robust. Third, the HLM models are compatible with the global models as special cases, not just in the abstract, but there's a pretty smooth transition from one to the other. And in some cases (not all, unfortunately), it can be pretty clear whether we should move from a global model to a local one. Fourth, they do this with true statistics, as opposed to ad-hocery, like many other approaches to exploring complexity in data . . . which means they highlight our uncertainty. To shy away from them in favor of methods that *don't* let you know when they are guessing isn't the right thing to do. So I think that whenever we think about difficult problems involving multilevel structures, it can help to start by trying to formalize it as a hierarchical model, and then see what parts of it we can actually examine. I'm going to use that way of thinking to approach the difficult topic of cross-level inference.

Cross Level Inference

Double-Cross

Sometimes, unfortunately, instead of having true hierarchical data, we have our data at only one level—but we're interested in a different level. This is most problematic when we have data at an aggregate level and are interested in individual-level models. For example, when we use historical data like census data, we find that they are accessible to us only in aggregated form.

Social scientists are familiar with what is sometimes called the "ecological fallacy," by which people mean comparing *aggregates* of individuals to make claims about individuals. That's not the best ways of thinking about it, because it makes it sound like there is an inherent taboo on using some sorts of data. There isn't anything inherently wrong about analyzing aggregate data—indeed, individual-level data is often best understood as an "aggregate" of person-time-level data. I side with those who would prefer to speak of "cross-level inference." The point is that sometimes, we are interested in making a comparison at one level (say, individuals), when all our data comes from another level (e.g., towns, counties). We've learned the hard way that this doesn't always work. Although in many cases, we can use information at a more aggregated level to set *bounds* on possible relations at lower levels, it's a rare day that aggregate information identifies lower-level structural relations uniquely.

Here's a nice illustration of the problem. Looking at the nineteenth-century United States, we might assume that Catholics are likely to vote against prohibition candidates, and therefore that the proportion Catholic in a county should be correlated with the county's support for anti-prohibition candidates. What's wrong with this, even if our assumption about Catholics' voting propensities is true? The answer is that it might be that in fact, as the proportion of Catholics goes up, the prohibition vote goes *up*, not down. The Catholics aren't responsible; it's that *non*-Catholics become more pro-prohibition the more Catholics there are.

So what do you do? Here I would point you to Achen and Shively's (1995) *Cross-Level Inference* for one part of what you need to know, but to your own common sense for another. In almost every case, you're going to want to really think through the nature of the variation you are *trying* to explain, and then make sure your methods are appropriate for *that* variation. To focusing on the "ecological fallacy" is to assume that there is a privileged level of analysis, when there are some times when we are forced to use individual-level data to make statements about higher order entities! But let's be frank: most of the time, the lower the level, the better.

Of course, if your data is aggregated, it's aggregated. There's nothing you can do about that but treat it with caution and respect. When you can, it's worth going the extra mile to get finer data. Because, theory be damned, almost all the time, if something interesting is happening, *people* are doing it, and they come in ones. Yes, there are contextual factors that may be truly at a different level, but sooner or later, for some of your data, you're going to want to get as close to the individual as you can. (And it turns out that if you can add a wee bit of individual-level data to an aggregate data set, you can get a lot more out of your aggregate data [Glynn and Wakefield 2014].)

That's because in a face-off between analyses testing the same ideas using data from different levels—even if all the data are aggregated—the lower level usually wins. We know it isn't *mathematically* guaranteed to better approximate the individual model, but all our common sense pushes us in that direction. For example, pluralist critiques of populism in the 1950s and 1960s had gotten a lot of mileage from the fact that Senator McCarthy, pseudo-populist butthead extraordinaire, came from the same state as Bob La Follette, the darling of progressives everywhere, namely the great state of Wisconsin. In a stunning piece of work, Michael Rogin (1969) dropped down to the county level, not the state level. And we find that the counties that supported La Follette weren't those that supported McCarthy. Rogin won.

Aggregate and Individual-Level Data

Sometimes, however, we think we are interested *only* in group-level phenomena. In that case, we generally feel quite free to ignore all this hoopla about cross-level inference. And that can be true, but it's not quite as true as we often think. This makes sense if you *really truly* have only ecological data—and only need such data, that is, those that are intrinsically tied to the unit, as opposed to being aggregates of individual-level data. An example might be if you are interested in whether cities with managers have less debt than those with mayors. But if you are regressing, say, average income on average education in communities, it's a bit of a stretch to see this as inherently a community-level phenomenon—even if you have only the community level data. Further, unless there really is *no* relevant individual-level process that can vary across community, an ecological regression, even if it includes all the correct ecological predictors, won't even correctly identify the *ecological* relations (see Greenland 2001).

Of course, it's just as true that a misspecified *individual*-level model that leaves out confounding *ecological* variables will also be wrong. But if you have information on these ignored ecological units, you could take them into account in a way that you can't take individual-level variation into account where you don't got it. The Funkadelic principle again: You can't miss what you can't measure.

In some cases we might actually have only individual-level predictors, and an ecological outcome. This might be the case where there is a process that "aggregates up." For example, we might be trying to predict which states have passed a law or a resolution, using data on individuals nested within states. In other cases, however, we might actually have only ecological predictors, and individual-level outcomes. For example, we might be interested in how individual longevity is predicted by community characteristics—the degree to which an area is connected to trade routes, or its tax base, or its air quality. So there is individual-level variation in our dependent variable, but not in our predictors. If we lack any individual-level measures of possible confounders, we might want to collapse the individuals to the ecological units, and then do a straightfor-

Table 5.1. Achievement and Faculty Involvement, Pretend Data

	Model 1	*Model 2*	*Model 3*
	Naïve	RE	FE
Faculty Involvement	.980*** (.042)	.304** (.099)	.027 (.134)
Intercept	6.986	36.926	50.568

ward ecological regression, weighting our cases. (On issues of weighting grouped data in general, take a look at Angrist and Pischke 2009: 92f).[7]

But let's remember how we started—when it *appeared* that we might have individual-level variation in our predictors, but we weren't sure if this could be anything but noise. Remember Arum and Roksa wondering whether faculty involvement predicts student success? The item about faculty involvement asked about a *school*-level quality, but it asked *individuals*. Let me simulate (R 5.8) a world with similar data—in this case, 3694 individuals from 50 different schools report the involvement of faculty in their school, and we have their test scores. And we find (model 1, table 5.1) that there is a significant positive relation between the two. But one of our methodologically sophisticated readers points out that our cases aren't really independent—we've sampled them from a subset of schools. Our statistics are wrong.

So we try a random effects model (model 2). The effect is reduced by a third, but it is still significant. So we're relieved. But another friend (some friend!) tells us that fixed effects models are "better," because they are "more conservative." So we sigh, and run the fixed effects model in model 3. Darn! The coefficient is a fraction of what it originally was, and it isn't significant. There goes *that* theory!

Except that, in these data, by construction, faculty involvement *does* predict achievement, with a true slope of 1.0, pretty much what the naïve model found. How can this be? Because faculty involvement is a school-level, environmental variable. It's just that students report on it with a lot of error—they aren't "calibrated" to all use the same scale when they say "how much" involvement is "a lot," for example. In other words, if we had gotten the correct measures of the school-level environmental variables, we'd have a relation like that graphed in fig. 5.15. There's a real relation

7. I don't know where else to put this, but there is a common misconception that we should routinely weight cases by the inverse of their probability of getting into a sample. But results using weighted least squares aren't necessarily better than those from ordinary least squares, and it is very hard to explain when and why weights should be used to change coefficient estimates (Gelman 2007).

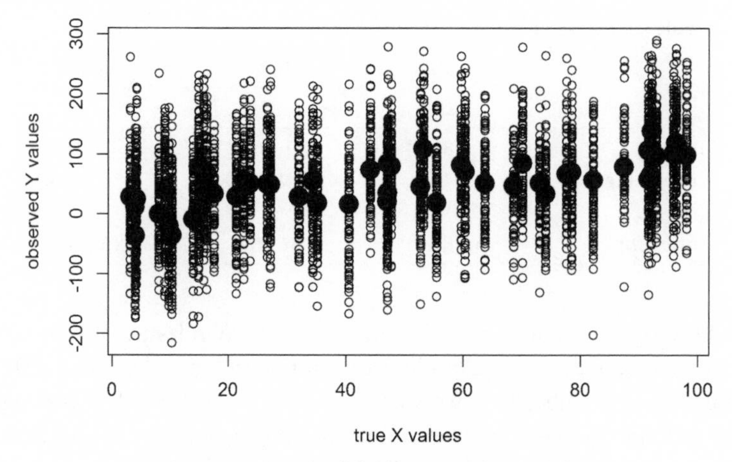

FIGURE 5.15. Meaningful Between-Units Variation

between the two variables, but it only appears at the mean level of the school. There's a lot of variation within schools—this isn't the *only* factor leading to success—but a clear pattern. It's harder to see, though, with the noise in the individual reports on *x*.

What the fixed effects model is doing is to *change the question*. The question the fixed-effects model asks is wholly a *within-unit* question. Sometimes that's the question that we want to ask. But if it isn't, then the fixed-effects model isn't a stronger answer to the right question, but the right answer to the wrong question. More important than knowing about different models is knowing *when you can change an across-units question to a within-units question*. When you *can*, your life is almost always a lot easier. But just wishing so doesn't make it so.

Conclusion

In social statistics, we use variance to explain variance. We need to know where our variance is before plunging forward, and very often, it is oddly allocated between different ways of grouping our cases. These sorts of groupings shouldn't be considered some sort of "problem" that you want to make go away—to reduce all social life to atomized individuals isn't a step forward in a social science! It's something to study, and to examine in terms of *substantive* understandings of what is going on. That means that there isn't an all-purpose way of "correcting" for such structures.

Ecological data are tricky data. In many cases, the only way forward is to consider each set of cases a set of *opportunities* for the observation of some phenomenon that will go into the computation of our dependent variable. Sadly, we have conventions for how to work with these sorts of data that are stunningly unreliable. And that's what we look at in the next chapter.

Appendix 5-A

What does it mean to compare coefficients from non-linear models? Because our general mindset is molded by the case of OLS, we don't have the right instincts. I want to focus on the case of the logistic model, for which we don't have "properly specified misspecification"—the parameters aren't unbiased if we omit predictors. What does that mean for model comparisons?

To start, I think I need to make clear the approach that I am taking to probability in this book. We often hear of two main approaches. The first is the frequentist (or "classical") or Pearsonian, in which probability is understood as a limit of a ratio of outcomes to trials as the number of trials stretches to infinity. The second is the Bayesian (or "subjectivist") theory, in which probability is a numerical expression of the magnitude of our subjective confidence in some belief. But there is a third, which comes from Fisher (1956), and it is, in a nutshell, the ratio of our knowledge to our ignorance (Anderson [2012: 102] also sees this as a different, third, tradition). Here, probability turns on the degree of *indistinguishability* of certain cases. I recognize that Fisher's approach works very well for contingency tables, and becomes more obscure as we move to continuous distributions, but I still think it's the right way to start. And yes, it's going to matter for us.

So the logistic regression is now often derived as a *collapsing* of an underlying continuous trait. For space reasons, I'll omit walking through the math; an accessible treatment will be found in Mood (2010). In this set up, we can imagine the underlying continuous variable being related to our predictors in our usual OLS-type way, where we have some "error" term at the end. Allison (1999) emphasizes that from this, we can derive that the coefficients we get from fitting a logistic model are a rescaling of those in the underlying continuous model—all proportional to the degree of error variance in the continuous model. This implies that—in contrast to OLS regressions—we can't really compare coefficients across different models for the same sample, or across the same model for different samples, unless the residual variation is the same.[8]

8. It is possible to think about this in terms of overdispersion and make certain types of clever work-arounds so that the variance can be decomposed without assuming the latent variable (see Browne, Subramanian, Jones, and Goldstein 2005). And the problem

Okay, that makes sense. But it doesn't necessarily make sense that we always have to interpret the logistic regression as a stochastic collapsing of an underlying continuous trait. People have tried to make that argument before, and they've never convinced anyone.[9]

So what if we treat the logit as involving a stochastic response process for inherently dichotomous data? Well, because of the fact that coefficients aren't unbiased with the omission of other predictors, it's also true that we can't compare coefficients across groups if there are relevant predictors we haven't included. So then we might have to say that the level of assumption for comparing coefficients across groups is a bit higher for logit models than for OLS. For OLS, we have to assume no omitted relevant predictors *correlated* with our included predictors; for logits (and other non-linear models) we have to assume *no* omitted relevant predictors, whether correlated with our included predictors or not. Note that since the *first* assumption is nearly as likely to be false as the second, it's a bit silly to allow the one and not the other. But let's push that to the side for now.

But Allison (1999) has urged us to see logit models for stochastic dichotomies as *also* having an unobserved, residual variance. That is, on top of the *ontological* probabilism of the stochastic process (every person has some probability of being positive on the dependent variable, a probability that is a function of the covariates), he slaps an *epistemological* probabilism (we aren't exactly sure what this is). This is a "definitely maybe" interpretation of the logit function (DeMaris 2002: 32).

Now there is one way in which we can think about this definitely-maybe, and it's well established in the statistical literature. We think that people have these probabilities that are linked to predictors, but because we don't have all the relevant predictors, our resulting data are overdispersed. In earlier work, Allison (1987) had discussed this as an "external" model of the error, as the error was added *in addition to* the non-linear function, as opposed to being incorporated inside it. He noted that one approach

---

can also be tackled, as J. Scott Long writes in an unpublished paper (2009), by comparing predicted probabilities, not coefficients, across groups. J. S. Cramer (2005) suggests using the average sample effect, which is a rescaling of the slope by the predicted probabilities, for comparisons. Finally, Breen, Holm, and Karlson (2014) demonstrate that in some cases, correlations of the underlying probability function with the independent variables allow for meaningful comparisons.

9. The best exchange was Pearson's argument with Yule as to the superiority of continuous versus discrete representations; Pearson ended up (if my memory fails me not) denying Yule's seemingly unshakable example of people being alive or dead. Experience with one's colleagues, Pearson pointed out, demonstrated that there were in fact many people who were somewhere in-between.

that has been used to deal with overdispersed data is to give the resulting probability function a beta-binomial distribution.[10] But he preferred the "internal model" because it "seems to capture better the notion of omitted explanatory variables." That is, rather than being about *variance*, it fit the vision of causal determination that I've been arguing against.

This vision of total determination convinces you that of course, you've *always* left out something—indeed, if you knew everything, you wouldn't be able to fit your model, because you'd have perfect prediction—and so you can *never* compare logit coefficients. Well, there's another way of deriving a logit regression. In general, in the Goodman system, a loglinear model links a cell frequency from a multiway table on the left side with a set of parameters for the marginals and all the subtables that can be formed (given an assumption of a multiway multinomial distribution). Leo Goodman (e.g., 1972) showed that we also can take continuous covariates into account in such a model.

Now note that if we have one dimension (corresponding to variable z, say) in our table that has a dimension of two, we can simply turn the cell frequencies into sets of ratios, by splitting the entire table into two portions, one for $z = 0$ and the other for $z = 1$. A loglinear model for this set of data is equivalent to a logistic regression. And so a logistic regression is really just that special case of loglinear model, just as a binomial distribution is a special case of a multinomial.

Imagine (and I simulate all this in R 5.10) that we have four predictors (x_1, x_2, x_3, x_4) of y, which I'll make dichotomies for the sake of simplicity, and we have two groups (we'll use the variable G to indicate grouping). According to Allison's way of thinking, if we do two logits, and get, say, b_{1A} for the effect of the first variable in group A, and b_{1B} for the effect of the first variable in group B, we can't say something like $b_{1B} > b_{1A}$. But now consider the table formed by $y \times x_1 \times x_2 \times x_3 \times x_4 \times G$, which has $2^6 = 64$ cells. Of the many models that we can fit to this data, one of them is equivalent to fitting the two logit models side by side for the two groups, and allowing the b's to vary. The parameter for the difference in the b_1's will be exactly $b_{1B} - b_{1A}$. Allison's logic requires that (assuming that the table is in fact multinomial) we treat this parameter as invalid.[11] But are the other parameters valid? And if so,why?

You might say, well, there's an asymmetry here between the y and the

10. The beta-binomial is a lovely and flexible distribution (see, e.g., recently Wiley, Martin, Herschkorn, and Bond 2015), and would allow for the comparison of parameters across groups.

11. I'm not sure if a comparison across two or more different sets coming from different sampling frames really should be joined in a single table like this. So we'll concentrate on the times when our groups are formed by subsetting a single data set. Once again, I

x's that you are forgetting about. No, there actually isn't. If we do a logit of y on (x_1, x_2, x_3, x_4) and get some b_1, guess what happens if we then do a logit of x_1 on (y, x_2, x_3, x_4) to get b_y? I can tell you what b_y will be. It's b_1. This is as much to say that these coefficients are referring to a 2×2 table formed by cross-classifying x_1 and y, once we adjust the subtable for the other variables.[12] The reason why one could claim that this interaction is wrong is the same fundamental one we hit upon before—that most non-linear models aren't indifferent to your exclusion of predictors, even those that aren't correlated with included predictors. That means even a saturated model for an M-way table, which we often treat as the "ground truth," can be wrong, if we really should be using an $M + 1$-way table. Your perfect fit of *this* table doesn't mean that your parameters correctly fit a different table made from the same data.

You might say, okay, well, since the coefficient fits the table we have before us, it is right, but we just can't interpret it as . . . what? A causal parameter? I would never wish such a fate to my worst enemy! So there's a fork in the road. One way, one that many methodologists will hold out to you, basically only ends in the ∞-way table that God only knows. That's because *all* our parameters are wrong unless we have the exact correct specification, with *no* omitted predictors! There are clever tricks that might get you a bit closer to that (the way the beta-binomial model does), but none of them can really be checked against the ground truth. So we spend our time attempting to better estimate numbers to correspond to processes that almost certainly don't exist in the world, and then try to judge them in terms of their deviation from total determination, which we don't believe in. The other way involves us admitting that we are fitting tables, making complex multiple comparisons, so that we can reject some theories that deserve to be rejected.

---

find that Richard Breen has a tighter exposition of this problem in a recent paper; see Breen and Karlson (2013: 172).

12. And in fact, with only three variables, even forbidding interactions between predictors, all the logit models from "regressing" one on the other two are actually the same model! In the Goodman notation, if we had three variables A, B, and C from two groups indexed by G, and attempted to fit the differential group models of $A = f(B,C)$, all would be the same: $(GAB)(GAC)(GBC)$.

* 6 *
Opportunity Knocks

The Problem of Dimensions and Dividing

In the previous chapter, we used notions of variance to explore hierarchical data structures, ending with the puzzle of ecological regressions. I want to return to that issue but from a slightly different perspective. Here we consider ecological data that consists of counts within the units in question—for example, the number of crimes committed in a city—and that leads us to try to adjust our analyses for the differential population of the units. We will see that our conventional approaches to this often fail abysmally.

Map: I start by showing the problems that arise when we compare outcomes in units like cities that vary in their size. To think this through properly, it turns out that we must determine which individuals in our ecological units are "at risk" for generating whatever outcome we are counting. This same notion is also expressed as "opportunity" or "exposure" in different fields. We see that since we rarely know the risk, our general approaches using ratio variables are quite likely to generate spurious findings. We then extend this way of thinking from exposure to the more general case of *conditionality*. Rather than statistically adjust our outcomes, we try to understand enough of the process that links outcomes to exposure that we can condition on exposure. It's another way of turning to decomposition to break up our problem, as opposed to relying on a tendentious model. We then show that, given data that comes from such processes, control strategies don't do for us what we often think they do.

Baby Stories

Why do cities have so many babies? I thought that urbanism was associated with declining fertility, compared to rural life? But if you look in the US, you see pretty clearly that there are more babies in cites than rural areas! Indeed, some pretend data (R 6.1) below will show just this. Of

Table 6.1. Babies and Towns

	Model 1	*Model 2*	*Model 3*
Dependent Variable	Count	Count	Proportion
Coefficients:			
Urban	1043.3***	-1.8219	-.0001
Number of Women		.1002***	
Intercept	89.3	-.7525	.1001
R^2	.257	.999	<.001

N = 50 areas

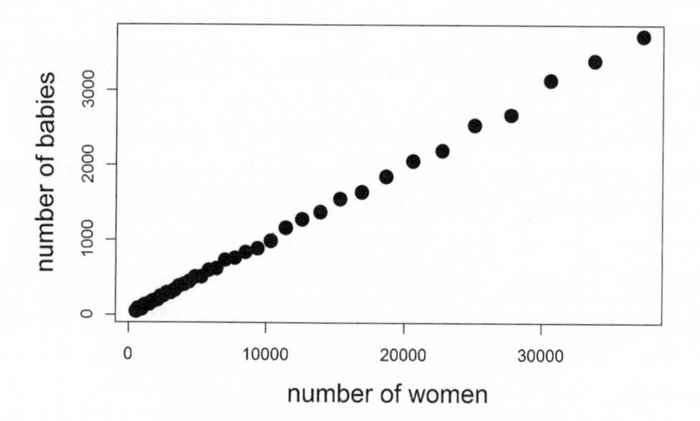

FIGURE 6.1. Population and Babies

course, as you probably know, you clever rascal, cities don't have babies. People have babies. And if there are more people in cities than in rural areas—as there are—we should probably take this into account. So, for this simple simulated data, a naïve model at the level of the city/town shows that urban areas have more babies (see model 1, table 6.1). But when we take the number of women into account, this disappears (model 2).

Don't get impatient! There's a reason to go nice and slow here. Model 2 fixes our problem and leads to a near-1 R^2. That's because, as we see in fig. 6.1, there's a linear relation between the number of women and the number of babies.

For this reason, we don't really need to include the number of women as a covariate. Instead, we can examine the *ratio* of babies to women. (This is informally often called a "rate," but I'll follow the general usage in statistics where we reserve that term for temporal processes.) When we make a new variable, dividing the number of babies by the number of women (n), we see (model 3) that this corrects our mistake in model 1.

Why does this work? Cities don't have babies, women do. In this simple simulation, all women were treated as being in the risk set "fertile." That means that any one could, at any year, pop out a baby. When we have such a risk set, within which we can make no distinctions, then it makes sense to divide an outcome by the cardinality of that set. But what if we *can* make distinctions? For example, between women of different ages? And what if when we do, we find that the risk *differs* across the sets? Well, unless our units all have the same distribution across the various subsets, we clearly aren't going to be able to adjust things by simply dividing by the number of women altogether, are we? We can only use this sort of division to *reject* the hypothesis that subsets are identical (for example, that 30-year-olds are the same as 20-year-olds, or Catholics the same as Protestants), not to *compare* across groups that differ in composition. It's for this reason that serious demographers will do a more detailed decomposition, and we'll get back to that toward the end of the chapter.

But most of us aren't that serious. We look for some "*n*" to divide our numbers by, simply because we know we can't proceed *without* dividing by *something*, and all we care about is that we go forward, even if it's straight to hell.

Lines and Curves

The sad truth is that most of the papers that I see that present "findings" for ratio variables have turned their misspecification into a seeming result. In a nutshell, dividing-by-*n* works only if your data *naturally* scale linearly with *n*, like we saw above. Otherwise, dividing-by-*n* makes $0/n$ sense. Let's think of a student who is impressed by the arguments about "civil society" and "voluntary organizations." He's always liked zoos, and thinks that maybe zoos are important for people to feel good about their community (let me make up some data; R 6.2). And that might decrease crime! He gets information on crime rates (violent crime per 10,000 per year, say), and on the number of zoos, and constructs model 1 in table 6.2. He is heart-struck to find that the number of zoos is associated with *increased* crime!

He shows it to his advisor, who reassures him. "Don't worry, you might still be right. The error you made was that you forgot to standardize by population. Larger cities have more crime, and they probably have more

Table 6.2. City-Level Models for Crime

	Model 1	*Model 2*
Zoos	1.270***	
Zoos-per-capita (in 10,000s)		−3.955***
Intercept	−2.890	11.150

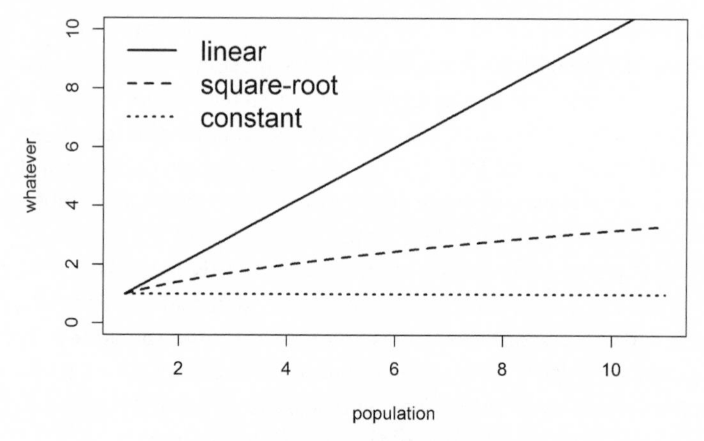

FIGURE 6.2. Different Scalings with n

zoos." So our student dutifully turns the zoos measure into zoos-per-capita and then redoes the analysis (model 2) and . . . the anticipated finding is there! He's so delighted . . . and so deeply wrong.

Why? Because (in this made-up world), there's no reason to think that zoos should scale with population linearly. We understand that 5 movie theaters in a town of 8000 is different from 5 movie theaters in a city of 8,000,000. So we might plan to divide the number of amenities by the population. But how many zoos do you need? Maybe a big city needs a bigger one, but they don't need proportionally more zoos. In many cases, you'll find that the count of some variable does increase with n, but not linearly. Instead, the relation is going to sit somewhere between total indifference to population (the horizontal dotted line in fig. 6.2 above) and a linear relation (the diagonal solid line). In the case of my simulated zoo data, the correct function is that square-root (the dashed line) between the two. By dividing by n (overcontrolling), we force a negative relation between our ratio and anything that scales linearly with population.

Of course, there are some cases in which your actual line will be *above* the diagonal linear ("divide by n") curve. For example, the number of Starbucks coffeeshops probably increases supralinearly with population, at least up to a point.[1] In that case, dividing-by-n undercontrols, and so we'd see *positive* relations between Starbucks-per-capita and something that is correlated with population.

Maybe you think I'm making too much of this. So I've simulated (R 6.3) four variables in 500 cities (say), one of which scales linearly with the

1. Greenhouse gas emissions scale supra-linearly with national population (Rosa and Dietz 2012).

Table 6.3. Can We Control for n?

	Model 1	*Model 2*	*Model 3*
	Naïve	Ratio	Control for n
x_1 (~n^1)	.383***	.067**	–.046
	(.054)	(.023)	(.065)
x_2 (~n^3)	.026	–1.378***	–.224***
	(.043)	(.024)	(.047)
x_3 (~$n^{.25}$)	.486***	.895***	.236***
	(.040)	(.018)	(.044)
n			.887***
			(.089)
Constant	.000	.000	.000
R^2	.713	.866	.761

city's population (n), one of which scales supralinearly (it correlates with n^3), and another two that scale sublinearly (they correlate with the fourth root of n). One of these will be our dependent variable. (Yes, I've made these relations very different so the pattern comes out—but this will be a *conservative* choice for what I'm going to show you.) Then all the variables are standardized to have the same scale.

All are interdependent only because of their interrelation with n. So what we want is a method that correctly tells us that one isn't the cause of another, and that any association is spurious. The first model (model 1, table 6.3) is a naïve regression. We have two spurious findings, but you will note that the second variable's coefficient is near zero. That's because with one variable scaling so supralinearly (x_2) and the other so sublinearly (y), there isn't much of a linear relation between the two.

Can we get rid of the spurious significance by dividing all the variables by n? Model 2 has the results and the answer is pretty clearly no. Indeed, it introduces a strong, spurious, negative effect with the variable that scales supralinearly. What if we instead put in a control for n? Model 3 has the results here. And what we see is that although this wipes out the effect for the variable that scales linearly, it doesn't wipe out the spurious effects for the others. Of course, if we had instead put in n^3, it would have wiped out the second variable.[2]

What can you do to deal with this problem? There are nonparameteric and semiparametric approaches that are worth mastering if you want to

2. Firebaugh and Gibbs (1985) derived the notion that we should add $1/n$ as a variable; that also doesn't fix things here.

work with this kind of data. But if that's a bit outside your comfort zone, there are still ways you can see whether or not this confounding with *n* might be leading you to spurious conclusions. One is to replace an attempt to enter a single covariate for *n* with a different research design: slice your data into, say, fifths by *n*, and see if your relation is reasonably strong within every bin. You can also plot the residuals of your model against *n*, and see if there is a worrisome pattern. But perhaps the nicest check is to sort your data by *n*, and take a succession of subsamples: first the 10 smallest cases, then cases 2–11, then 3–12, and so on. In other words, use a moving window. Within each moving window, you run your analysis not controlling for *n* in any way. Fig. 6.3 graphs the results for the simulated data in terms of three lines, one for each coefficient (plain, dash, and dot in order), here using a window of 7. The legend gives, for each variable in order, the percentage of windows in which the coefficient is positive as opposed to negative. You see that they spend as much time above 0 as below. That suggests that there isn't really any effect independent of *n*.

Now of course, you might say that a window of 7 is too small to lead to anything meaningful. But look at fig. 6.4: this charts the *average* across all those windows . . . showing how this changes from when we start with almost all the data (a window of 495 on the left) to the window of seven used above. Yow. Kinda says it all, doesn't it? This is what you do if you want to make across-case comparisons of units with different *n*s. That is, assuming you don't want to lie.

There's one other complication here, which returns us to the problem of skewed data that we raised in chapter 2. Often most of your data is crowded in the lower-left corner—the cases with, say, few zoos and few people. That means that while your brain is off in the upper right—imagining what goes on in large populations—most of your variance is somewhere else, and the misspecification at *this* end is what drives your results. If that's the case, don't just try to do a log-log plot (you'll see why in a second!)—figure out whether your results are really being pulled by the smudge in that corner: take successive samples in which the probability of inclusion is a function of *n*, and that function becomes steeper and steeper. If your results change, that suggests that they're really being driven by you getting the shape of the curve down in that corner a bit wrong (we'll return to this issue when it comes to estimating relations between variables and populations).

So—as I'll say again—before dividing-by-*n*, you want to visualize your data. In general, if you are thinking about using ratio variables, you should have a macro or chunk of code that, for any data set, will draw a contour map with the horizontal axis being the number in the denominator, and the vertical being the ratio. The height of the contours is your predicted

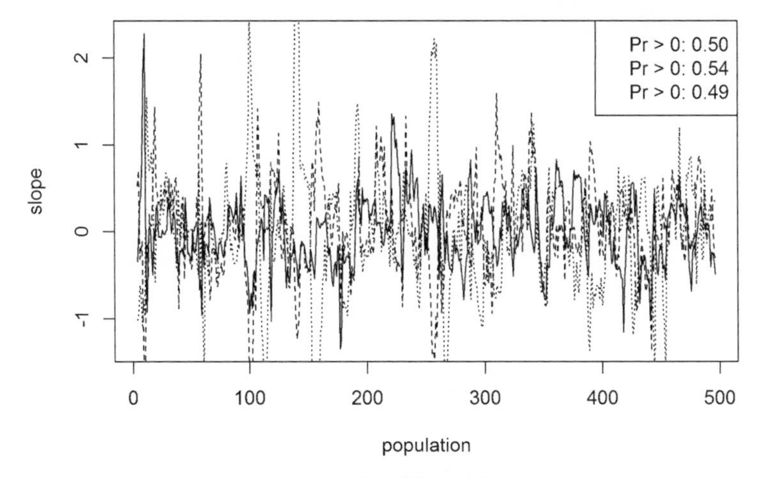

FIGURE 6.3. Local Regressions

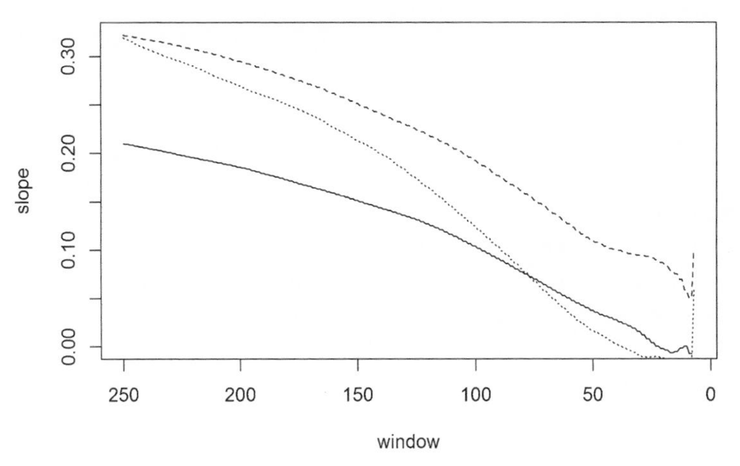

FIGURE 6.4. Size of Window

dependent variable. Make this macro flexible, so you can give it different inputs, so it can take coefficients that come from either a ratio or a count, or that take the numerator and the log of the denominator, and so on. Dividing by n without doing this isn't "normalizing." It's *abnormalizing*, as you create a monster. To avoid being the next Dr. Frankenstein, think it through, starting with the issue of risk sets.[3]

3. Is there a way to move past controlling-for-n? Benjamin Rohr and I are investigating certain nonparametric techniques as a way of flexibly dealing with many different functional relations between n and y. But before we advocate this, we want to make sure there aren't any traps we didn't think about.

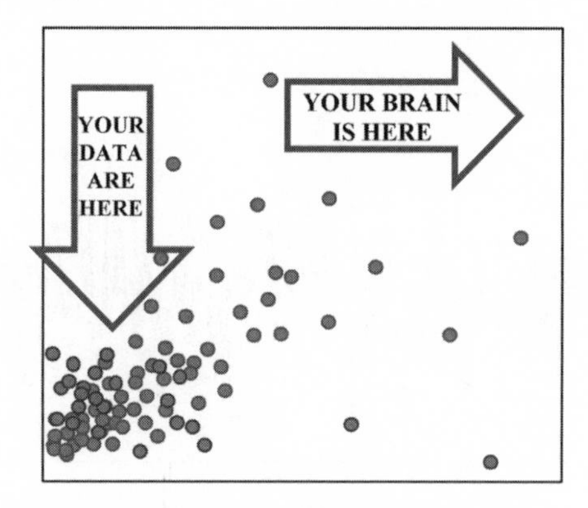

FIGURE 6.5. How It Goes

Who Is at Risk?

When it came to babies, we thought we could have at least a reasonable first guess as to who was at risk: women, especially between 15 and 40, say. But often we really don't know. Let me take an interesting and challenging empirical question, one that will serve as a running example for this chapter. We are interested, let's say, in the predictors, perhaps even causes, of hate crimes—for example, housing set aside for immigrants is burned in Germany; blacks are assaulted by whites in New York City (and not as part of another crime). We only have counts of the crimes in some ecological unit. Here I'm going to use data previously assembled and analyzed by Donald Green and colleagues (1998); he nicely made the data available (R 6.4). They got data on the prevalence of hate crimes against blacks, Asians, and Latinos from 1987 to 1995 in 51 New York City districts; the idea was to use it creatively to determine *why* some whites commit these crimes. And yes, later I'm going to show something wrong with his approach. But for now, we'll use this just to think through some fundamental issues.

So our basic data comes in the form y_j–that is, the number of such crimes in the jth unit. We have other covariates of each unit j, and we would like to use these to make our predictions. However, we understand that places with different population sizes may have different numbers of crimes. We don't want to make the same mistake that we did above in our model 1 for zoos!

It certainly seems reasonable, at least as a first approximation, that if two places have the same population breakdown, but one has twice as many people, we shouldn't be surprised to find twice as many hate crimes there. We aren't going to try to explain that result theoretically, since it is expected under what we might take as our null model. But suppose that

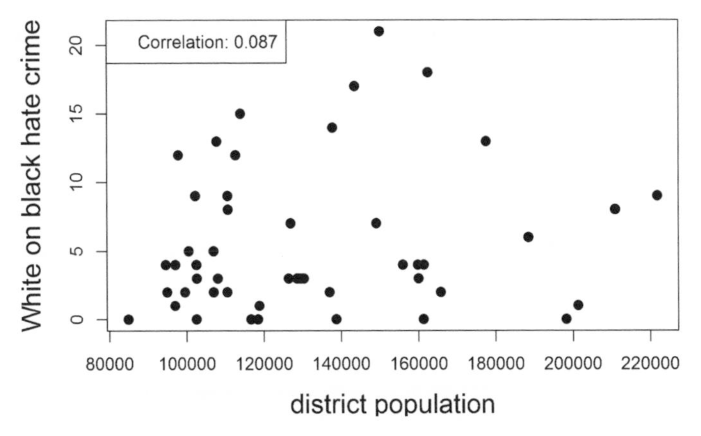

FIGURE 6.6. Events by Population

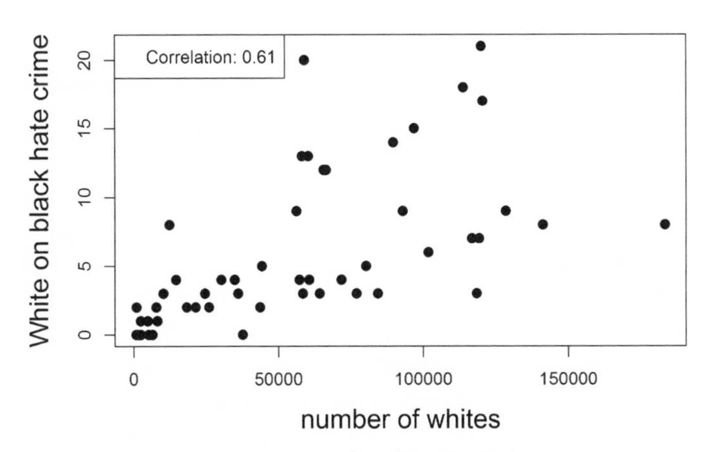

FIGURE 6.7. Events by White Population

the two places have different population distributions—different proportions of blacks and whites, say. Then it's hardly clear that we would normalize by the total population. Of course, it *might* be reasonable—if our data formed a nice line, like we saw with births. But as fig. 6.6 shows, they sure don't. Indeed, a simple correlation isn't even statistically significant (r = .087).

Well, perhaps we haven't seriously thought through what is our risk set. Only whites can commit white-on-black hate crime. And so we might imagine standardizing by the number of whites in any unit, as opposed to the total population. What do things look like then? Fig. 6.7 below shows that there is in fact a quite strong relation; the correlation is .61.

But it seems odd to argue that whites could commit the same number of crimes against blacks even if there are no blacks. We might de-

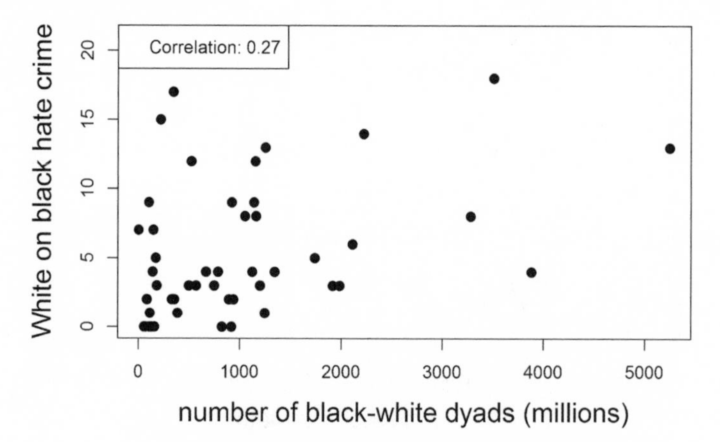

FIGURE 6.8. Events by Cross-Race Dyads

cide that the real unit at risk of becoming a hate crime is actually the interaction, and therefore standardize by the number of all dyads that are mixed race. (This is in accord with some great work by Peter Blau [1977].) This means, among other things, given any total population composed of whites and blacks, we expect the number of hate crimes to be highest when population is evenly divided between the two. But as fig. 6.8 suggests, this doesn't seem obviously correct. (And even less correct would be a hypothesis that the number of hate crimes would go up with the number of potential victims—it doesn't.)

Does this mean that we "know" that the proper denominator "is" the number of whites, and not the number of interactions? Not really. It's a null model (see chapter 1), and so far, this null model doesn't fit all that well (it explains 36% of the variance). It *could* be that the dyad is still the right unit at risk, but we don't see it here because we don't understand the heterogeneity within dyads: most simply, it could be the case that many whites and blacks are not really encountering one another regularly. What matters isn't the number of *potential* dyads, but the number of actual encounters. Perhaps we could narrow it, and look at the number of dyads of mixed race that live on the same block. (The problem with this, unfortunately, is that there is good reason to think that motivated whites will commute in order to find victims; see, for example, Pinderhughes [1997].)

Or there could be other heterogeneity that affects our risk set. For example, we might realize that the processes of ethnic transitions may tend to produce mixed race neighborhoods that have disproportionately elderly whites, who are not (we here presume) at risk for committing a hate crime. And so we might also want to take the age structure into account.

There's a problem with the way we're headed. And this is that we can imagine going farther and farther until our new ratio always approaches 1—because as we include more and more "predictors," we basically define to be "at risk" of committing the crime only those who *have*! This is not an attempt at a *reductio ad absurdam*, where I show how impossible any attempt to standardize is. Rather, I want to draw your attention to the fact that we often unjustifiably treat our attempts to control for the number of at-risk units in two different ways. On the one hand, we sometimes carelessly simply toss in a control for population or a division-by-*N*. This comes because we often think—though as we've seen, we usually don't think enough—that we are hereby purging our data of a well-understood problem. In this case, we are really basing our action on an assumption that there is a null model that's true but uninteresting, sort of like correcting for prices for inflation.

The other way of working is when we are making a theoretical claim that we are really interested in investigating. These enter as covariates in a model for the standardized value. The problem is that if our first conception is incorrect, our results for the second are wrong too. In other words, there are two ways that we can fix the non-homogeneity of risk across cases: we can try to divide by the number of those at risk, and also model the ratio as a function of covariates. There's nothing wrong with *first* dividing by the cardinality of a set that *isn't* homogenous in its risk for the event in question, so long as our additional covariates correctly soak up this heterogeneity.

But . . . if our model is wrong, then it won't fix an imperfection in our assessment of the risk set. And, as we've seen, dividing by the cardinality of the wrong risk set can introduce all sorts of biases. So our model has to be right *given the choice of the risk set*. It isn't even just about having all the right variables in—it's about correctly specifying the functional form.

The implication, then, is that we should try to normalize via division only if one of two things is true. The first is we have a *very* good a priori reason to make a statement about a risk set. And the second is if we see a near-line between our putative numerator and denominator. Just figuring that we can divide by *n* and then adjust by tossing in covariates isn't a safe strategy. But guess what—the problems of using ratios are bigger and worse.

When Does This Fail?

So if you've been paying attention, you'll be making a major mental note: always graph my relations, and use that as my guide. And that often *does* help. But it's not failsafe. And one way in which it blows up is related to one of the most marvelous failures of "quanties" in recent years. This had to do with the amazing proliferation of "power law" findings. It is true—and by no means without theoretical significance (see, especially, Zipf 1949)—that certain processes that follow Merton's "Matthew effect," namely that to them that have, more shall be given, will produce a distribution that is called a "power law." The idea here is that we can rank our units by some quantity x (such as population), and that the *natural logarithm* of the number of units with any value of x is proportional to the natural logarithm of x.

There's lots of work triumphantly pointing to the presence of this or that power law. The only problem is, it's a bit like saying, "according to my theory, that thing will look like a smudge if you look at it from a mile away." Most distributions of positive numbers, if they cover a wide scale, are going to look a bit like power laws.[4] And many of the ones we use have a big splot in the lower left corner, just like in fig. 6.5. So it's often impossible to empirically figure out the "true" relation of some variable with n by fitting a line to the log-log scatterplot.

And this brings us to a different aspect of this problem, one having to do with our understanding of ratios (and here I follow Nee, Colegrave, West, and Grafen 2005). This is key when we have a set of proportion-ratios (which can vary between 0 and 1), that are applied to cases of very different scale. This would be the case, say, if we looked at the proportion of GNP spent on defense, studying all countries with seats in the UN. The GNP of the United States of America is around 2000 times that of Togo. Fig. 6.9 shows the logarithm of a numerator (say, defense spending) and the logarithm of a denominator (say, GNP) for simulated data (R 6.5). The fact that our data forms just that sort of line that I've discussed above seems to suggest that we're okay to divide our data by the denominator to make comparisons across cases. Indeed, we might say even more; the line is so narrow that it seems that we have some sort of constant law here, doesn't it?

But actually, all the data are random. First, we choose a value on the x axis (and exponentiate it to get our denominator). Then we randomly choose a proportion (from 0 to 1), multiply it by the denominator to get the numerator, and log it. The reason that we have this line appearing is basically that the data are on very different scales (which is why we took

4. This is similar to the problem in diffusion research that I discuss in *Social Structures* (2009), in which eyeballing distributions that *looked* logistic led people to make claims about the presence of a diffusion process. But pretty much anything that goes from low to high looks logistic. Lots of distributions look like a power law if you ignore the very tails . . . and most people do.

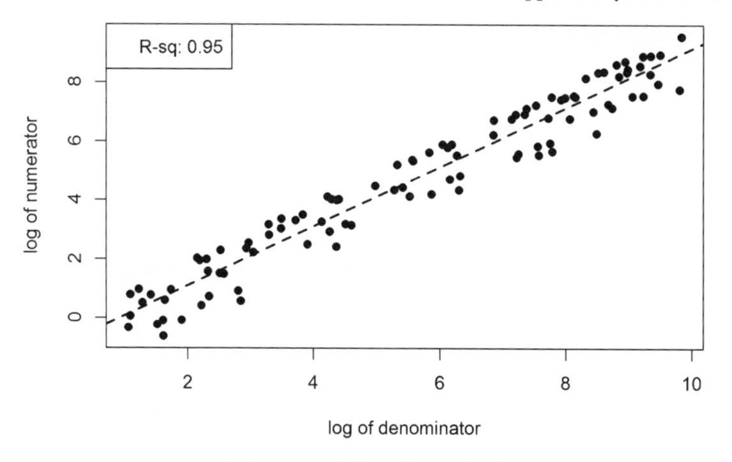

FIGURE 6.9. A True Power Law!

the logarithm). Really, the only regularity is that the numerator can't be larger than the denominator.

What's the lesson? Any time you have things of very different scale thrown together, log-log plots can look incredibly clear and linear. Should you divide by the log of *n*, then? It's hard to make that claim, because we've been talking about ratios in the absolute metric. Really, we should be dividing by *n*; dividing the log-numerator by the log-denominator isn't the same thing as dividing the numerator by the denominator. That's true, even though the numerator/denominator plot *lacks* a crisp line like this! A visual check is the first word . . . not the last.

And it's important to bear in mind that we can't use goodness of fit criteria to decide whether or not we "should" use a ratio or count measures in any analysis. That's what we've seen here in the (common) case of the log-log plots that have huge R^2s (we'll see an example in the next chapter). The same is often true, we might remember, with models that simply ignore the size differences altogether, and use raw counts. They'll have great explained variance, since they in effect have *n* on both sides of the equation.[5] On the other hand, we've also seen that we'll get great ex-

5. Such possible inflation was pointed to by Kasarda and Nolan (1979: 217) in the debate over the use of ratio variables with common components. This debate (e.g., Fuguitt and Lieberson 1974; MacMillan and Daft 1980) has demonstrated that there is no general answer to the question of whether common components are intrinsically problematic; in each case, we must consider the theoretical entities (to answer the question regarding the proper specification) and the data set (to answer the question regarding the results of improper specification). And that means that we can expect a lot of findings that are just b.s. . . .

plained variance if the proper model *should* link counts of x to counts of y, but instead, we divide both by n. Ouch.

Further Complications of Ratio Variables

Degrees of Freedom

There are inherent *statistical* problems with using ratios which we'll get to in a second. But there's another one—it can tie our data together in knots. Most of us understand that we can't have a linear dependence in a set of independent variables. If you divide people up into male and female, you can't have a coefficient for both, because the categories are exhaustive and mutually exclusive. So, too, if you have ratios for your cases coming from a single denominator, you can't include a set that adds up to one. So, for example, if you are studying firms, you can't include both the portion of employees who are men and the proportion who are women, since this adds up to 1.0.

But even if you know this, it's easy for us to come far too close to this to comfort. And we can see this by returning to that paper by Donald Green and colleagues I discussed at the beginning of this chapter.[6] They had a hypothesis, which was that white-on-black hate crime was a result of some whites acting to defend their neighborhood against the inmigration of minorities through the terrorizing of random blacks. Their approach was to take the data introduced above, the number of hate-crimes in New York City districts, and then adjust them for differences across districts. But once the adjustments were made, they wanted to test the theory that these crimes would be most likely to be observed where a majority-white district was having an influx of blacks.

And indeed, it does seem that way: here's a scatterplot of the actual data they used (fig. 6.10). The vertical axis is the count of white-on-black hate crimes (y), the horizontal axis is the proportion white in 1980 (p_{white80}), and each district appears as a point; the points are separately labeled according to whether or not there has been an increase in the proportion black between 1980 and 1990 (i.e. $\Delta_{\text{black}} = p_{\text{black90}} - p_{\text{black80}} > 0$, where p_{black90} is the proportion black in 1990, and p_{black80} that in 1980). As we see, the

6. I choose this because when it came out, I saw the problem and wrote a comment for the *AJS* that was rejected after review because it had ballooned and a reviewer didn't understand some of the statistics involved. I would like to thank Joshua Goldstein, Megan Sweeny, and Julie Philips for their comments on these matters; even more, I would like to thank Donald Green not only for his thoughtful comments on that MS, but for his exemplary practice of making these data as well as all the analyses easily accessible to the public.

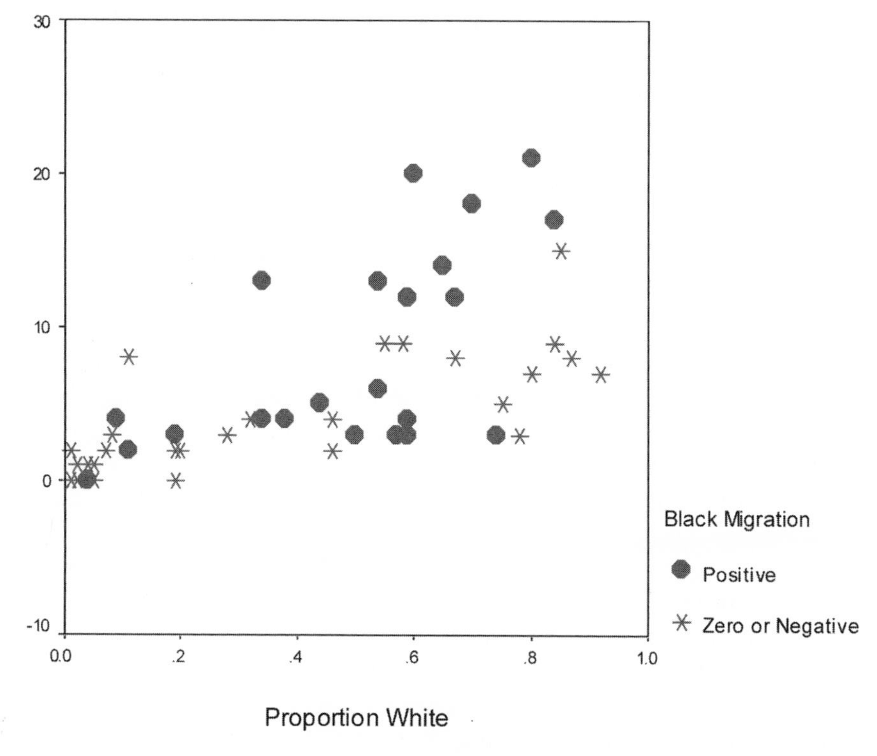

FIGURE 6.10. Original Data

number of hate crimes rises with the percentage white (the correlation is .677, $p < .001$). Furthermore, this rise is indeed sharper where there has been an in-migration of blacks. So eyeballing the data doesn't obviously lead us to reject their interpretation. But they wanted to go further, and show that this pattern survived controls.

Most obviously, there's good reason to think that this might simply be because there are more whites around to commit the hate crimes. Let's instead consider our dependent variable the chance that any white will victimize a black (or the number of white on black hate crimes divided by the number of whites, i.e. $R = y/N_{white90}$, where $N_{white90}$ is the number of whites in the district in 1990).[7] Then we'd find the correlation of hate crimes and percentage white to be mildly *negative* ($r = -.222$ when we weight the cases

7. To make the pattern easier to see, I eliminated one district with an extremely high rate and a small proportion of whites (i.e. a *worst* case for Green's hypothesis). While Green et al. use the proportion white in 1980 as an independent variable, this correlates quite highly with the proportion white in 1990 [$r = .979$], allowing us to speak more generally about the "proportion of whites."

by the total population). Yet Green et al. *did* control for these population differences . . . and still found support for their claims.

How could that happen? It's because there is a *positive* correlation between the in-migration of blacks (Δ_{black}) and the number of whites ($N_{white90}$) ($r = .386$, $p < .01$), which leads to an increased *count* of hate crimes where black migration is high, given that $y = N_{white90}R$ (and that $N_{white90}$ is of course highly correlated with $p_{white80}$ [$r = .913$, $p < .001$]). Why is there such a counter-intuitive association of the number of whites and the increase in the proportion black?

Think back to the chapter on knowing your data, and thinking through floor and ceiling effects. Where the proportion of blacks is already high, there can't be a big increase in the number of blacks, right? And (bracketing other groups, which we can for these data), the greater the proportion black, the lower the proportion white, right? So the largest *absolute* percentage increases in black-migration must tend to occur where there are more, and not fewer, whites. Since Green et al. didn't divide the percentage increase in blacks by the percentage of blacks already there, they found a positive association between the in-migration of blacks and the count of white on black hate crimes.

There's a deeper, and simpler, issue. Their key theoretical proposition is that hate crimes are increased by the *interaction* of the proportion white ($p_{white80}$) and the change in the proportion black (Δ_{black}). But the proportion of blacks in a city is pretty close to 1 – the proportion whites! That means that our variables are tangled in ways that aren't necessarily apparent. Appendix 6-A works through what happens if we use this relation to make a reduced form equation, where we can regress the dependent variable on two more easily interpretable variables: the proportion black in 1980 and 1990. Let's now visualize the results as a two-dimensional "birds-eye" contour view (fig. 6.11). The lighter color indicates higher predicted rates of hate crime.

Green's idea was that where whites originally had a strong majority, but there has been large black immigration (the upper left corner of the figure), the rate of hate crimes will be high. But the model also implies the *opposite*—that where the proportion of whites was originally small, but there has been a recent white influx, the rate of hate crimes will be far greater! This, then, is an empirical implication of the crucial interaction term, but this was not, of course, part of Green's theory. Remember what we learned in chapter 4—interactions are tricky things. We tend to focus on the "marked" sides, but most of the interactions we use are inherently symmetric.

There are, not surprisingly, no districts that fall into the extreme cases

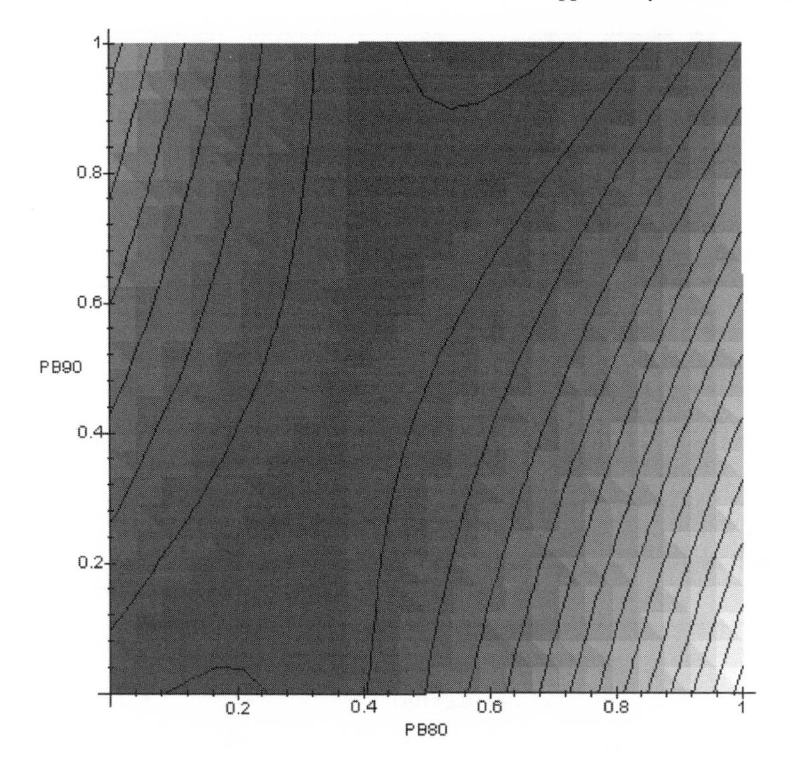

$$\text{PB80} = p_{\text{black80}};\ \text{PB90} = p_{\text{black90}}$$

FIGURE 6.11. Predicted Hate Crime Rate by Proportion Black in 1980 and 1990, Contour Map

corresponding to the upper left and lower right quadrants: indeed, the proportion black in 1980 is rather closely correlated ($r = .980$) with the proportion black in 1990. So, let's return to chapter 2, and ask ourselves, where *are* the data? Fig. 6.12 below arranges the case in a way corresponding to the predictions in the previous figure; the cases have been ranked by their observed rate of white-on-black hate crimes, with the larger circles corresponding to the larger rates; beside each case is the number of hate crimes observed. Green's hypothesis implies that we will tend to see large numbers above the diagonal, at least toward the left side of the page. Clearly, it does not imply that districts with white minorities where blacks *leave* should have an increase of white on black hate crimes. But consider the case labeled (#47)—this district was 2% white in 1980, and 57% black (most of the rest were Hispanic) (see R 6.6). By 1990, the percentage black had fallen to 39%, and the proportion white had increased

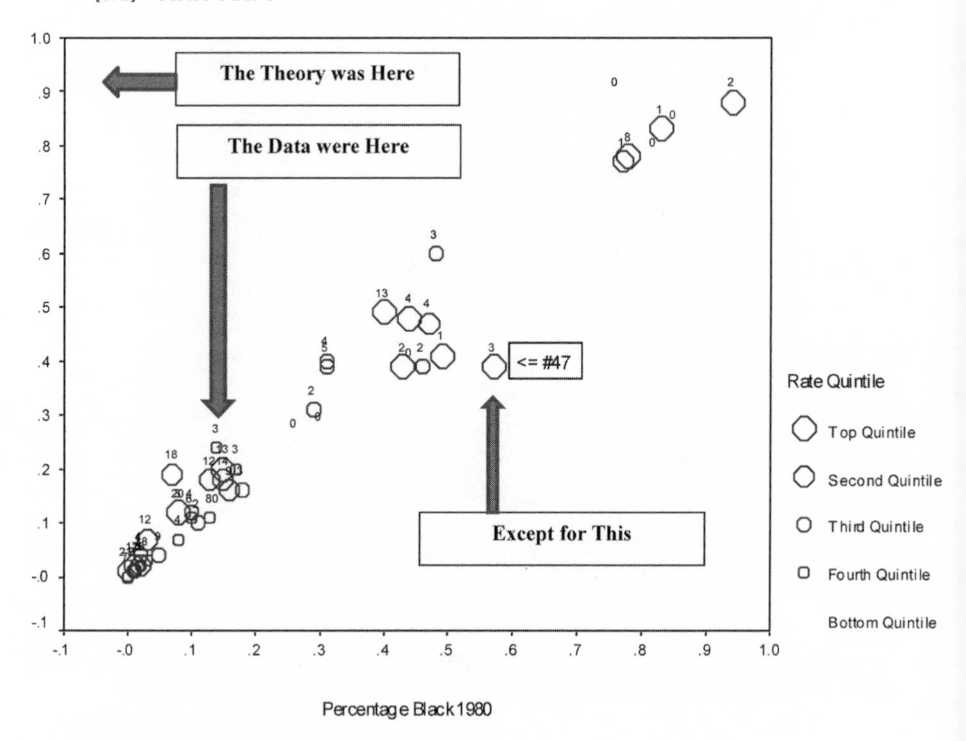

Cases weighted by number of whites

Total number of white on black hate crimes near each case

FIGURE 6.12. Observed Hate Crime Rate by Proportion Black in 1980 and 1990

to 8%. The three hate crimes committed by whites against blacks led to a rather high rate, given the small number of whites. This case, then, fits the counter-intuitive prediction demonstrated by graphing predicted probabilities, that cases in the bottom right of this joint distribution are likely to have high rates of hate crimes. And it is upon this case which the significance of the interaction rested!

I wish that I could say that this was an unusual case. But I frequently see papers with complex claims involving intertwined ratio variables, and sorting them out into a simpler formulation leads to a radically different view of the matter. That's what turned out to be going on in the data on religious participation I discussed in chapter 2. The lesson is that in such cases, the variables are too closely interrelated to allow for a rigorous adjudication between complex explanations. If you're *right* about how to construct your ratios, all the data originally will lie along a thin mountain range. When you divide by the right denominator, you should turn a mountain range into something more like a nice little hill in Teletubbyland. But if you're wrong, by dividing all your cases by something, you'll probably now *force* your data to sit on a line, one constructed by your division. Your theory is likely to be trying to get the variation *across* this range.

That's bad enough, but if you try to establish interactions between parts of these ratios, you can take the same basic finding, and make it appear in five or six different ways. It doesn't matter how much your theory guides you here—this is thin ice. Back up, and get away quickly.

Tangles

We've seen that ratios can lead our variables to be tangled in ways that we don't always understand. When that's the case, it's important that we don't attempt to interpret the ratios. So what do we do? A nice example is seen in a recent, very ambitious, paper by Elizabeth Bruch (2014). She's trying to understand the factors that predict a family moving from one place to another. She figures, quite reasonably, people can be priced out of certain rental markets, so that one predictor might be the ratio of housing costs to household income, or RENT/INCOME. But movers might also want to be around people whose income is like their own, and so she also includes INCOME/MEDINC, where MEDINC is the median income of the target area. But all other things being equal, families might prefer to move to richer neighborhoods, and so she enters MEDINC as a zero-order control. As a result, there are three terms using three different variables in different combinations.

These are pretty tangled, aren't they? Bruch's approach is to estimate these parameters, which she thinks are important, but she doesn't focus on the parameters themselves; they're probably too intertwined to be interpretable. Instead, she looks at the model predictions across values of these parameters—these can be interpretable even if the parameters aren't. I noted above that if you want to use ratios, you should have code to turn each one into a scatterplot. In this case, where we have multiple terms, it would make sense to go further, and try to visualize a four-dimensional space (perhaps coloring points by the dependent variable and rotating a three-dimensional cloud of points corresponding to RENT, INCOME, and MEDINC).

In sum, if we have ratio variables that share terms, it's usually a good idea to un-ratio them, just as we did with Green et al., and to see if the predictions across ranges of the components, taken two or three at a time, make sense to us. If not, we should walk away. But even if you can't quite untangle them, you can use their predictions to shed light on questions you have, so long as they aren't specifically about these dynamics. And there's one last case where we should walk away, namely when following the principles of chapter 2—know thy data—should lead us to distrust their compilation in ratios. Let me close this section with attention to this issue.

Final Puzzles of Ratios

We need to go through something that may seem quite simple, but we still need to do it slowly. Ratios are a numerator n divided by a denominator d. In many cases, n is a count of a certain type of event, and d the number of units believed to be at risk for the event. The idea of dividing n by d is that we hope that it will allow us to compare units of different sizes. That logic is fine so long as we really have our risk sets exactly right. But before we start analyzing these ratios, we need to pay attention to the components and, in many cases, to focus our attention *on* the denominator. Recall that in chapter 2 we found that while many of our statistics are inherently symmetric, our corresponding imagination of what is going on is often heavily asymmetric; we focus on the "marked" half, and it can be difficult for us to take seriously the implications of the unmarked.

Often our numerator is a relatively rare phenomenon, and our denominator is mostly the unmarked category. For example, we might be considering forces that lead religious festivals in different countries to turn into riots. Our numerator n_{it} is the number of riots in country i at time t, and our denominator d_{it} is the total number of religious festivals. Now here's the thing: *very often our data on* n_{it} *and* d_{it} *come from independent sources.* That isn't a problem if both sources are *one hundred percent perfect*. But if they aren't—and they almost always aren't—then you may have to worry that the variation that in *your* mind is about n_{it} is actually, in your data, being driven by $m_{it} = d_{it} - n_{it}$. That is, changes across cases, or differences over time, in the unmarked cases are what are driving the differences in your ratios. So what do you do? *Before* you start mucking about with your ratios, you examine the variation in your denominator. Is there variation there that doesn't make sense to you? If so, check on this. Are you letting totally different agencies in different countries give you the "same" number? Is there a good chance that these agencies have different procedures, or are of different qualities? If so, you do not want to divide n_{it} by anything.

This is especially true for time-series data. If your denominator behaves as you would expect (say, slowly creeping up every year), that's wonderful. But if it's going up and down, you need to understand why before just "dividing by" it. For a number of interesting processes, there can be reasons why the numerator is going to be relatively *constant* over time. Perhaps in some place, police officers can only arrest so many people, and they have a strong incentive to arrest as many as they can process. Let's say we are dividing the number of arrests by the number of offenses, to see how likely it is that a wrongdoer is apprehended (ignoring the fact that not all arrested are guilty). The estimates for "number of offenses," on the other hand, might be extremely error prone, and jump up and down for all sorts of reasons. (We'll see another example of this process later in this chapter, when we examine attempts to quantify the force of attraction by comparing a numerator of friendships to a denominator of potential friends of different types.) If you are dividing some good data on rare events by low-quality data based on estimates or summaries of various data streams, you should be fed to the sharks.

How to Deal with Ratios When Things Go Bad

So there's no one-size-fits-all solution for these problems. It often has to do with the particular pattern of relations in your data. But let me work through some fixes. As we saw with the Green paper, one problem can be that our ratios add up *nearly* to one, even when we omit a category. In that case, we're doing that thing of trying to balance on a very narrow ridge. Our estimates are still the best that can be gotten, but that doesn't mean they're worth doing. We might be better off actually throwing away some of our ratio variables, so that the ones that remain in the model don't add up to something near 1.0.

Other times we have problems because our ratios have extreme distributions. That is, we have a ratio variable with minimum 0, maximum 1, and a mean of .04. This is a problem in the same way that other skewed independent variables can be a problem: it isn't necessarily wrong, but often, we suspect that there is important information in the small variations close to 0 that are wiped out by the few outliers. But sometimes we *introduce* this sort of skew by computing ratios, because our denominators vary widely across cases. In such a case, it might indeed make sense to simply reshape the independent variable somewhat. Don't just reach for some "transformation" that you've heard is intrinsically appropriate for such skew—it can make things worse. There isn't any fix that is theoretically or methodologically appropriate for all cases. The first thing is to try to bin your skewed variables. You might try making four bins, with some sort of compromise between having enough cases in each bin and doing justice to the true distance between them. Do you see a clear trend in the mean of the dependent variable across your bins? If not, maybe toss the variable out of the model, rather than risk screwing everything else up.

Let's say you have reason to think there's a relationship between a ratio variable and some dependent variable, but it's hard for you to get the model to work right—perhaps the coefficient for the ratio has a huge standard error, or perhaps you can't simultaneously estimate it with other important variables in the model. One thing to try is to replace the ratio with one measure of the numerator, and a second of the denominator. Now when it comes to loglinear models, like a logistic regression, there's a way in which this is already a ratio, and some people will say you "should" do this for a logistic regression but not for an OLS. That's not true. There's nothing about a model that guarantees for you what the world is like. You can even use different metrics for the numerator and the denominator—for example, in some cases, you might use the absolute count of a numerator, yet the log of a denominator. This might work well in practice; even more, for some processes, it will make theoretical sense.

Whatever you do, make sure that you *graph the predictions* of all the different approaches you take on the same scale. See if all of your different ways of parameterizing lead to the same graph. If they do, that's great! If not, you probably have a head start in figuring out why.

In sum, often we do have data in which we have good reason to think that our units differ in the number of individual cases at risk for having some value on some variable, which would lead to bias in any naïve comparison of these units to one another. We've seen that the key in identifying likely bias was when we realized we weren't really sure what the risk set was. This way of identifying our problems in terms of risk, or exposure, or opportunity, isn't just important for identifying problematic assumptions about scaling. It also allows us to reduce some intractable problems to tractable problems, and, in some cases, to test more complex hypotheses that involve combinations of preference and opportunity. The next section works through this logic.

Conditioning

Conditionality

Instead of thinking about the aggregate results of many cases that might share some risk or opportunity, let's focus on a single case. Why did Marilee Mossman go out and buy a KitchenAid stand mixer? If we talked to her, we might hear about the large amount of baking she had to do for the local Shakespeare troupe fundraiser, her frustration with her previous mixer, a sale at the local store, and perhaps even more. It's very hard to find all such factors that might go into a decision, and implausible that these would be widely enough found among other purchasers to justify fielding a survey to gather data for a model that predicts why someone would buy a KitchenAid stand mixer.

But imagine instead that Mossman had *first* decided to buy "a" mixer, and *then* decided to choose the KitchenAid. Then we can model the probability that she buys a KitchenAid (BK) as follows:

$$\Pr[BK = 1] = \Pr[K = 1 \mid B = 1] \times \Pr[B = 1] \tag{6.1}$$

That is, it is the product of the probability that she buys *some* mixer, times the probability that, *given* that she's going to buy a mixer, she chooses the KitchenAid.

This is definitionally true. But it's only useful for us in situations in which these two terms on the right hand side can be treated as *substantively* independent. In the case of impulse buying, where the decision to buy isn't independent of the decision to buy this particular model, we aren't going to be able to profitably move from the left hand side to the right. But where the two terms actually tap substantively independent processes, then, if we are actually only interested in inter-brand comparisons, what we can do is ignore the $\Pr[B = 1]$ term, because it's brand-independent.

Are you jumping up and down in your seats with excitement? Well, maybe you should be. The reason is that the left hand side is something that has a gigantic, sprawling denominator. Every day of the year Mossman was "at risk" for buying a KitchenAid stand mixer. Do you want your model to include all those "didn'ts"? But only when she actually makes a purchase is she at risk of choosing a KitchenAid. The set of events is far more tractable.[8]

For dichotomous acts like a decision to do or forebear, this formulation links a conventional logistic regression to a conditional logit. The former has a function of $\Pr[K = 1]$ on the left side, and the latter has $\Pr[K = 1 \mid B = 1]$. It's worth perhaps re-emphasizing that you can't simply make this switch for reasons of convenience—there are assumptions that, unlike many statistical assumptions, you really need to think through. I've given the simplest, and most important one—that the factors that lead one to make a decision *to* choose aren't closely bound up with those that influence the choice. But it gets a bit more complicated—the relative proba-

8. There have been some very interesting approaches to dealing with the fact that not all choosers have the same effective choice sets (Swait 2001). However, as we'll see in chapter 9, such latent category models usually look better on paper than they do applied to most data sets. They're great where we have some information, or an experimental setup, but they don't solve our practical problems.

bilities of two options can't change when a third is added or subtracted (this is called the "independence of irrelevant alternatives"). Here's an example: It's 1980, and you are trying to decide between a Mercury and a Ford car. These are, in case you don't know, basically the same chassis, just with different fripperies. You are leaning toward the Mercury, which has more fripperies. Then the salesman shows you the Lincoln. (If you are guessing that it is just the same chassis with the *most* fripperies, you are right.)

It's much nicer than the Mercury. But you can't afford it, no way. So you think, "hell, it's all the same car; only a sucker would pay more for the power windows and the plush seats. I'll go for the Ford." Choice processes like that can't be well modeled by the conditional logit. Another way of saying it is that you might not like rational choice theory, but wherever someone who *does* like it would be happy, that's where you are safe using conditional logits.

Exposure and Opportunity

The beauty of this way of thinking is that rather than having to simply make an assumption about a single risk set, we can break down a complex process into stages of what in *TTM* I called "dependent arising." And then we can focus our attention on different parts of this chain, very often by using different subsamples. Merely thinking this through avoids a whole class of easy errors. That's because we (as human beings) have a tendency to interpret what people *do* as evidence of what they *want*. And as sociologists we translate this into thinking that data on *outcomes* tells us something about *preferences*. But more generally, we must say that

$$\textit{Outcome} = \textit{preference} \times \textit{opportunity}.$$

For example, when it comes to marriages, the fact that marriages are dyadic means that each marriage removes an opportunity for others. And that means that if we are looking for ways of attempting to find the "force of attraction" between different groups, we can't simply take the rates of intermarriage as indicating preferences. Because, for example, there might be many whites who would like to marry Native Americans, but as the latter are a relatively small proportion of the population, the opportunity is constrained.

In this case, we also might realize that the opportunity that is given to the actor in question (say, a white woman) involves a decision by a different actor, one who might not be the focus of our sampling (in this case, a

Native American man). For a real example, there has been a fair amount of work using aggregate data on white-on-black lynchings in the American South to test claims about the motivations of the white populace in different areas. However, Hagen, Makovi, and Bearman (2013) realized that the law could intervene between the formation of a lynch mob and the commission of the crime. A completed lynching *given* the formation of a mob assumes that there was no successful intervention by law officers to head off the lynching. And it turns out that a great deal of the variation that was being imputed to the white populace really had to do with the willingness of law to intervene.

As we attempt to think through the chain of dependent arising, we are likely to generate testable implications. For example, with the Green hate crimes data, we realized we weren't sure who was actually at risk for committing a hate crime. As we mull over different possibilities, we might decide to see if we could check whether crimes were committed on late weekend nights (compatible with the idea that they are more-or-less unplanned results of encounters of victims with drunk and belligerent men), or during the times when youth walk to or from school (compatible with the idea that they are planned, and planned to strike particular fear into black residents by targeting children). This is actually the strategy Durkheim used to good effect in *Suicide*.

Relationships and Exposure

When we first considered how to normalize the counts of hate crimes across districts, we realized that the answer wasn't obvious to us, because we were really looking at a dyadic relationship (between a white offender and a black victim). The issue of how to determine the exposure to such relationships has proven a very tricky one for sociologists; though here they are usually considering more benign relationships, the basic problem is the same. If there are 100,000 interracial marriages, is that a lot or a little? If there are 1 million blacks and 10 million whites (say), those marriages involve a small proportion of the whites, but a big proportion of the blacks. We see that as the population becomes more homogeneous (% black goes down), it will be less and less possible for there to be interracial relations . . . if only because there aren't that many blacks. Conversely, a rise in the percentage black could lead to more interracial relationships if only because there are more blacks.

There have been different solutions offered (e.g., Qian 1998, Huckfeldt 1986, and recently Skvoretz 2013) but the most common is to standardize by the number of possible opposite-sex relationships between blacks and whites (which is the number of black men times the number of white women, plus the number of black women times the number of white men). The idea makes sense. We might imagine that each relationship has a probability of transitioning to marriage. We want to get the average probability. But if we think about it this way, we realize that where blacks and whites are living in totally different areas, the dyadic "relations" we might compose are quite unlikely to have any real counterpart, and so are quite unlikely to transition to marriage. What do we do?

One way is to add covariates, and hope that they soak up the remaining heterogeneity. But as pointed out above, if we don't get this quite right—and we usually don't—we can introduce spurious correlations between our variables by dividing by the wrong risk set cardinality, and this divisor is correlated with the coefficients of interest—as it always is when we are interested in compositional factors such as "percentage white." The other way is to attempt to decompose the overall phenomenon into separate parts, using the conception of "dependent arising." This can help us understand how we might take this sort of exposure into account in our models.

Let's work through one of these classic problems of intergroup dyads a bit slowly. Network researchers have models for *ties* (we'll explore these in chapter 8)—thus, on the left-hand side is the probability $\Pr[x_{ij} = 1]$—the probability that a tie forms between i and j. Demographers have models for *counts*—thus, on the left side, they have f_{gh}—the number of pairs between groups g and h. Koehly, Goodreau, and Morris (2004) pointed out that these models were related to one another in ways similar to those that connect conditional and unconditional models to one another; in the network case, we have a model for a logit $\Pr[x_{ij} = 1 \mid i \in g, j \in h]$ whereas for the demographers, it is for $\Pr[i \in g, j \in h \mid x_{ij} = 1]$.[9] After they point this out, it's hard to understand how sociologists can often propose *the same right-hand side* approaches to the two expressions! And guess what—our most common ones don't work for *either*.

Here my example for considering models of interracial relationship formation is going to be Strully (2014), but I won't be saying that she did it wrong—only because I'm not even sure there *is* a right way to do this! If there is, I sure can't figure it out. The problem: we have two, equally compelling, and perhaps equally valid-for-some, ways of modeling this process. Here's the first way:

A. Focal actor runs into other person randomly.
B. Focal person may ask other out.

9. They propose nesting these in a more general unconditional model.

C. Other may accept.
D. Until one party breaks relation, it exists, though any other actor may "ask out" an already matched actor (this simplifies our math).

In this case, we take the *opportunity* into account in a simple way at stage (A), and preferences at (B–C), which we can't really separate. Thus we here say

$$\Pr[D] = \Pr[D - B \mid A] \times \Pr[A]. \tag{6.2}$$

One reason to write it like this is that we can't separate stages B and C. All we know when a relationship doesn't happen is that *someone* didn't want it to. But we don't know *who*.

Now let's say that, like Strully, we're interested in interracial dating. For simplicity, let's just assume heterosexual relations between whites and blacks, with the number of males and females in each being w_m, w_f, b_m, b_f. Then, if we use D_{gh} to mean that a boy from racial group *h* (*h* can be either black or white) is going out with a girl from racial group *g* (ditto), we might say

$$\Pr[D_{gh}] = \Pr[D_{gh} - B_{gh} \mid A_{gh}] \times \Pr[A_{gh}] = \Pr[D^{*}_{gh} \mid A_{gh}] \times \Pr[\mathrm{A}_{gh}] \tag{6.3}$$

where D* is simply that "they're going out," collapsing stages B, C, and D. Further, we usually make the following (potentially problematic) assumption:

$$\Pr[A_{gh}] = \Pr[A_{gh} \mid M] \times \Pr[M]. \tag{6.4}$$

That is, we say that the probability of a girl from *g* and a boy from *h* meeting is the probability that, *given* a chance meeting, the girl is from *g* and the boy is from *h*, times the probability of a chance meeting. (We never make this assumption explicit, because it seems so obvious.) The reason this is key is that we have no traction on $\Pr[M]$ and so we assume it is constant across all actors. If that's wrong—if this varies by person in a way that is related to other parts of our model, we're in trouble. But if we can assume it, and that encounters are truly random, we can get some traction on comparing the chance of a meeting between a girl from group *g* and a boy from group *h* meeting, to the chance for a *different* type of meeting, one between a girl from group *g′* and a boy from group *h′*.

As shown in Appendix 6-B, with this assumption, we can estimate the differential force of attraction between the two pairs of groups by including the *n*'s for our groups as offsets in a logistic regression. And that basically justifies the idea of doing a logistic regression of the *observed counts* of matches of (g, h) and (g', h'), regressing them on the number of persons in each category (which lets the coefficients for the size of each group deviate from 1, thereby allowing—whether reasonably or not—the sheer number of persons in a group to affect the force of attraction).

But now consider a different scenario. Here, the stages are as follows:

A. Focal actor constructs preference structure for other groups;
B. Focal actor finds opposite sex member of top ranked group;
C. Focal actor asks out.
D. Other actor accepts.

Here, it is extremely difficult to write a parsimonious model (though one important, and underutilized, approach was developed by Huckfeldt [1986]; Skvoretz [2013] has recently built upon this). There are three challenges that then arise:

1) We find it extremely a priori likely that it is not the case that either model is true for all persons;
2) It is extremely difficult to figure out any plausible way of testing the assumptions of either;
3) We often *use the first model to explore the second.*

Huh? What I mean is that we use ways of adjusting for composition that *assume* the null model of an *undirected* search, but only in order to try to estimate parameters tapping homogamy (or preference) that, if taken seriously, push us more toward the *second* model. That's because we've made an assumption, one that might be true, but certainly isn't obviously true, that the degree of deviations from the first (null) model can be used not only to *reject* that model, but to measure the force of attraction. That might be defensible as a "we don't have anything better" sort of way, but it just isn't *true*. Deviations from a *false* model don't automatically measure anything that corresponds to a quantity of interest in the *true* one.

Further, we often imagine that we can use this analytic approach, one involving deviations from a null model of random matching, regardless of scale. Suppose we are investigating whether children have a tendency to make friends with those of their own ethnicity. Do we have data from classrooms? Control for the proportion of same-ethnic children in their classroom! Do we have data from schools? Control for the proportion of same-ethnic children in their school! Do we have data from neighborhoods? Control for the proportion of same-ethnic children in their neighborhood! Do we have data from cities? Control for the proportion of same-ethnic children in their city! Do we have data from nations? Control for the proportion of same-ethnic children in their nation!

Okay, it seems that the idea gets progressively sillier—as if all the children in the world were at risk for becoming your friend. And I think that it *is* silly. But it isn't obviously the case that the procedure makes sense for classrooms either. I don't know about you, but I didn't enter each year's class with an empty slate of friends which I populated from those in my classroom. My friends in fourth grade were basically my friends in third grade, and most of my friends in third grade were the kids who lived on my block or thereabouts. If the odds of my friends being same-ethnic changed from year to year, it was largely just because of the different non-friends that might get thrown in the class. (Another nice example where fluctuations in a ratio have only to do with the non-marked denominator.)

In any case, with this under our belts, let's turn to the interesting and ambitious work of Strully (2014). She was trying to determine whether having a racially diverse school leads to interracial dating. The problem, as you might guess, is that there's more than one way to think about this, and what's a finding from one perspective isn't from another—because no one doubts that there will be more interracial dating in racially diverse schools if only because of the difference in *opportunity*. But she wanted to see whether there was also an effect on *preferences*, which requires having the right null model of random choices.

Let's say we are thinking about the probability that any dyad has of transitioning to a romantic relation (scenario 1), and whether interracial dyads are more or less likely to make this transition. In this case, if the total number of between-race romantic relations (r_b) increased linearly with the total number of interracial dyads ($d_b = w_m b_f + w_f b_m$), we wouldn't be surprised, and probably we wouldn't say that there was a school climate effect. Instead, we might compare the ratio r_b/d_b to the same one for within-race dyads (r_w/d_w). On that basis, we might start with the (definitionally true) expansion:

$$r_b = (r_b \mid d_b) \times d_b. \tag{6.5}$$

That is—we want to decompose the number of romantic relations between the races into two portions: the total opportunity, and then the preference given the opportunity. We could of course turn this into probabilities by switching from the aggregate to the level of the dyad.[10]

10. Strully actually had a more complex set of data, because she had relations both inside and outside of school. And while it's one thing to imagine that the "school" composition represents a decent bag of marbles from which a prospective dater might "draw,"

Let me work through a simpler scenario. Imagine that we have a number of schools with different distributions of the races of the students, and we also know all the dating relationships therein. We cross-classify the race of girls and boys who are going out, thus making a 2 × 2 table, with girl's race as the rows and boy's race as the columns. The tendency toward homogamy in this school can conventionally be expressed as the odds ratio of this table. Now one nice thing about the odds ratio is that it has already knocked out the marginal—that is, if you believe the probabilistic assumptions that we start with, that the chance of meeting is proportional to the number of possible dyads, we don't have to take into account how many boys and girls of each race there are. That's the beauty of the logistic regression, which is based on that odds-ratio.

In any case, what Strully tried to do was to "control" for the racial composition of the school by entering, for any person, the proportion of the other students who were of a different racial/ethnic group, and thus *could* be chosen as a romantic partner. Presumably, if other predictors are significant even taking this into account, we have a sense that choice is nonrandom.

Unfortunately, that's not quite so. Strully's problem is that, like most of us, she's really thinking about the logic of *making* a choice, but analyzing the logic of *made choices*. These aren't the same thing; something that correctly predicts whether a partnered girl has a black or a white boyfriend might not affect whether or not she chooses to date a black boy! It's a very difficult problem, and let's simplify it, examining some simulated data (R 6.7). Let's imagine that there are only two races in the school, and every girl and every boy are in a relationship; further, they are randomly paired. They pay no attention to race whatsoever. Let's only look at the girls' choices here. Here I am making 50 schools, each with 400 boys and 400 girls, all of whom are dating someone. The proportion of boys who are black is pretty close to that of girls in any school, both based on random draws from the same school-level parameter, but a parameter that varies across schools.

We want to examine the role of race on the force of attraction, using logistic regressions. There are two ways we can do this; first, we can look at the odds of a cross-race relationship, as did Strully. Second, we can also just look at the odds of any girl having a black as opposed to a white boyfriend. If we were taking's Scully's approach, for the first question (the dependent variable would be CROSS-RACE) we'd control for the proportion of a different ethnicity (model 1 in table 6.4). Fitting this model,

---

it's a lot harder to claim that the racial composition of one's community as a whole has the same relation to the choice process. But it's all we can do.

Table 6.4. Analyses of Cross-Race Relations from Simulated Data

	Model 1	*Model 2*	*Model 3*
Dependent Variable	Cross-Race Relationship	Black Boyfriend	Black Boyfriend
Coefficients			
Girl's race	.032	-.008	-.012
Proportion boys of different race	4.760***		
Proportion boys black		4.760***	
Log-odds of black vs. white boys			1.002***
Intercept	-2.392***	-2.392***	.006

we find that indeed, the proportion of another race is highly significant, which seems to make sense—there are more "others" to randomly choose. But now let us consider the constant, which is what we would be looking at if, say, we wanted to compare schools in two different states to one another, to determine which had a greater openness to cross-race relationships. This is significantly negative, which seems to imply a strong tendency of girls not to choose boyfriends of a different race. Yet we know, from construction, that this isn't the case! If the control strategy was working correctly, and the term for the proportion of boys from another race was correctly making all our predictions from schools with different compositions equivalent, this should be near zero. Something seems to be amiss.

Perhaps we are confusing ourselves by combining two different types of relations into these "cross-race relationships." Let's see what happens if we ask our question phrased in the second way. Now our dependent variable will be BOY_IS_BLACK, and so we'd naturally control for the proportion of boys in the school who are black (model 2). Holy smokes, our coefficient for the exposure is the same as the effect of the proportion of boys who are of a different race on our *other* dependent variable! Let's take a moment to enjoy for a second the seeming puzzle that we get the same answer to our two, seemingly different, questions.[11] And the intercept is

11. How can this be? Make four cases coming from the cross-classification of girl's race = (0,1) and boy's race for any girl. Denote any case by these two numbers (e.g., [0,1]), where the first indicates the girl's race, and the second the boy's race (thus [0,1] is white girl, black boy). Make two versions of the dependent variable, one for "boy's race" and the other for "boy-of-different-race." Note that these are the same for [0,0] and [0,1], and different for the others. Now make two versions for the contextual variables, one for "percent black" and the other for "percent of different race." Notice that these are the same for [0,1] and [0,0], but for the other cases, the second independent variable is 1 – the first. Now do you see?

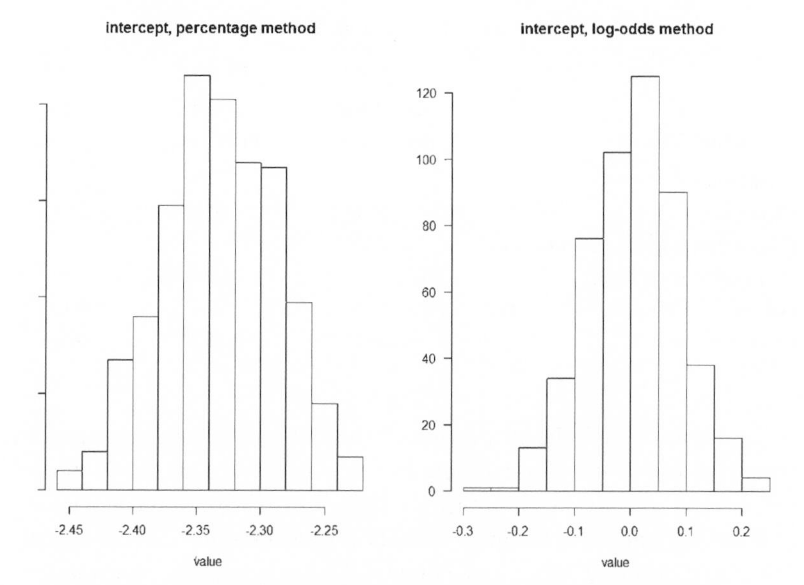

FIGURE 6.13. Distribution of Intercepts, Models 2 and 3 from Table 6.4

also the same—though now it has a substantively *different meaning*. Now we seem to find that all girls have a tendency to prefer white boys. And we know *that's* not true either. What is going on?

The error is that composition should affect the choice not via the *proportion* of black boys in the class, but their log-odds. Model 3 enters this as our control, and there are two things to note. The first is that its coefficient is very close to 1.000. That is, it's working well at turning into an "offset"—a way of taking exposure into account by entering it as a predictor where we fix the coefficient at 1. If this coefficient were different from 1, that might indicate some non-randomness. The second thing is that the intercept is now basically zero. There's no evidence of any general preference.

Of course, sometimes, by chance, we might get a significant coefficient for the girl's race, or even the intercept, from either model, but the key is that model 3 correctly tells us that there is no generic preference for girls in terms of the race of their boyfriend. Fig. 6.13 below shows the distribution of intercepts from model 2 (on the left) and model 3 (on the right) across 500 simulations each.

The implication is that if you are interested in, say, contextual factors that lead to something like intergroup friendships, unless you already know the process whereby friendships are made, you can't "control" for composition by adding in terms for the proportion in any category. That means that you can't make claims about subjectivity (for example, people being more or less likely to make intergroup friendships depending on context). And if you have several variables that are likely to be bound up with *size*, the chance that you can disentangle them through linear controls is near zero.

What do you do? First, visualize the data. Look at how your numbers change with size and by composition. Second, simulate how certain measures are likely to change with size given the linear model you are considering. Third, you might even want to switch to a simulation-based approach. But you need to incorporate *alternate* models of the fundamental dynamics. And be prepared for an inability to make any conclusions, because you might find yourself saying, "if the first model of friendship formation dynamics is correct, then I think that the force of attraction between these two groups is large and positive. But if the *second* model is correct, the force is probably large and *negative*." As we'll see in chapter 9, simulations are less robust than most of our other methods to the nature of our assumptions about the world.

Conclusion

We've gone from what might seem to be a very simple issue—how to think about ratio variables—to an exploration of some key principles for thinking through many statistical problems, namely the notion of "dependent arising." Whenever we can, we want to use the remarkable law of conditional probabilities to tease apart different aspects of a process, most importantly, those pertaining to conditions or opportunities, and those pertaining to choices or preferences.

Further, this way of thinking brings us back to one of the key points of chapter 3; rather than control via statistical adjustment, we can often decompose. Let's say we have aggregate data and are trying to figure out whether people have become more conservative over time. But we guess that people tend to become more conservative as they age. How do we compare times where there are different age structures? There are three ways we might do this, *bad*, *better*, and *best*. *Bad* is to do some sort of ratio: for example, we divide the average conservatism at any time by the average age, or we divide individual's conservatism by their age. *Better* is that we examine the excess conservatism of people at any time, once we take age into account. That is, at least as an exploratory tool, we residualize conservatism by age (and now we might have a quadratic as well as a linear term for age—which should be your standard operating procedure when it comes to age, by the way). *Best* is to do a full decomposition (as in chapter 3).

We use this same sort of conditional thinking to estimate the set of parameters—how likely is it that a person is conservative, *given* that she is between 30 and 39? What about between 40 and 49? And so on. That's the same basic approach that we were using to try to separate opportunity from preferences. Maybe we shouldn't want to separate these as badly as we do. But the worst thing is to want to talk about choice, and instead, really have data on constraints!

And this brings us to another, more technical, issue regarding this sort of separation. We sometimes have data on careers where we are trying to predict who gets a promotion to a certain position. But in some cases where we have a small number of positions, results are influenced by the fact that one can't get a job unless there is a vacancy. If there is a temporal structure to the data, this can lead to spurious results. Since this is a bit technical, I'm putting a discussion as appendix 6-C.

Anyway, sociology is hard enough when all our data are straightforward—measurements or counts. Things get tougher when we need to consider the difference between outcomes and risk sets, or between outcomes and opportunities. Ratios often seem like they're a fast solution, but as we've seen, that's rarely the case. It's worth taking the time to do it right. And this is necessary if we're trying to make comparisons of units as they move through time. That gives us the opportunity of changing our question—if this makes sense—and conducting only within-case comparisons. But we'd better hold off on that until we get to the next chapter.

Appendix 6-A

Green et al. proposed that hate crimes are increased by the *interaction* of the proportion white (p_{white80}) and the change in the proportion black (Δ_{black}). Thus their key model (e.g., their models 1, 6, 8) can be expressed

$$\hat{y} = b_1 p_{\text{white80}} + b_2 \Delta_{\text{black}} + b_3 p_{\text{white80}} \Delta_{\text{black}} + c \tag{6.6}$$

and their key claim is then that $b_3 > 0$ (although they did negative binomial regressions, I bracket functional form to highlight the logic; this can be turned into a non-linear model whenever we want). But the Δ_{black} may be expanded as $p_{\text{black90}} - p_{\text{black80}}$, giving

$$\hat{y} = b_1 p_{\text{white80}} + b_2 (p_{\text{black90}} - p_{\text{black80}}) + b_3 p_{\text{white80}} (p_{\text{black90}} - p_{\text{black80}}) + c. \tag{6.7}$$

Assume that the population is either black or white; then the proportion white in 1980 would be linearly related to the proportion black in

1980; that is, $p_{\text{white80}} = 1 - p_{\text{black80}}$. This assumption is workable; although around a quarter of the sample is Asian or Hispanic, the correlation between p_{white80} and $(1 - p_{\text{black80}})$ is .85 when the sample is not weighted and .81 when it is. (This assumption is even less problematic for Green et al.'s work, as they combine white offenders with offenders whose race was unknown to boost sample size.) Our model then becomes

$$\hat{y} = b_1(1 - p_{\text{black80}}) + b_2(p_{\text{black90}} - p_{\text{black80}})$$

$$+ b_3(1 - p_{\text{black80}})(p_{\text{black90}} - p_{\text{black80}}) + c = (c + b_1) - (b_1 + b_2 + b_3)p_{\text{black80}}$$

$$+ (b_2 + b_3)p_{\text{black90}} - b_3 p_{\text{black80}} p_{\text{black90}} + b_3 p_{\text{black80}}^2. \tag{6.8}$$

Were we to make an unconstrained model using these independent variables, we would retrieve

$$\hat{y} = \gamma + \delta_1 p_{\text{black80}} + \delta_2 p_{\text{black90}} + \delta_3 p_{\text{black80}} p_{\text{black90}} + \delta_4 p_{\text{black80}}^2 \tag{6.9}$$

where $b_1 = -\delta_1 - \delta_2$; $b_3 = -\delta_3 = \delta_4$ (a constraint is added here); $b_2 = \delta_2 + \delta_3$; and $c = \gamma - b_1$. Indeed, fitting this model to the data finds that $-\delta_3 \approx \delta_4$, hence the constraint imposed by this version of the model does not lead to loss of fit.

Appendix 6-B

Assuming a single chance of meeting and random encounters, we can say,

$$\Pr[A_{gh}]/\Pr[A_{g'h'}] = \Pr[A_{gh} \mid M]/\Pr[A_{g'h'} \mid M] = n_g n_h / n_{g'} n_{h'}. \tag{6.10}$$

This expression, then, is simply the ratio of the number of intergroup dyads of each category. Then

$$\Pr[D_{gh}]/\Pr[D_{g'h'}] = \{\Pr[D^*_{gh} \mid A_{gh}] \times \Pr[A_{gh}]\}/\{\Pr[D^*_{g'h'} \mid A_{g'h'}]$$

$$\times \Pr[A_{g'h'}]\} = \{\Pr[D^*_{gh} \mid A_{gh}]/\Pr[D^*_{g'h'} \mid A_{g'h'}]\} \times \{\Pr[A_{gh}]/\Pr[A_{g'h'}]\}$$

$$= \{\Pr[D^*_{gh} \mid A_{gh}]/\Pr[D^*_{g'h'} \mid A_{g'h'}]\} \times n_g n_h / n_{g'} n_{h'}. \tag{6.11}$$

Hence

$$ln(\Pr[D_{gh}]/\Pr[D_{g'h'}]) = \lambda + ln(n_g) + ln(n_{h'}) - ln(n_{g'}) - ln(n_{h'}), \tag{6.12}$$

where the first term (λ) is used to summarize all of our size-independent model terms that correspond to the brackets in eq (6.11). We isolate the size-independent terms by including offsets for the log of our group sizes.

Appendix 6-C

It's very common for us to have a set of data on, say, candidates or workers, and to try to explain their promotion or attainment on the basis of their individual covariates. That would be well and good if you could *force* a job to appear for you. As you probably know very well, it ain't that way. Other people have control over the jobs, and the key thing is whether or not they *give* you one.

When the number of candidates and jobs is large, this is unlikely to matter; we assume that there is "frictional" unemployment, and there are always *some* jobs. But if you are trying to become a circuit court judge, or a supreme court judge, you may have to wait for a vacancy to open up. If the number of vacancies changes over time, and covariates of candidates change *also* over time—as they almost certainly do for most cases—then a conventional approach is wrong.

Here's one of those happy cases in which math and being smart are on the same side. What if you could change your question from "will person i get hired for a job" to "*given* that someone is hired, will it be person i?" That is, we go from an unconditional to a conditional model. We're simplifying the math, and doing better at modeling the actual process that happens. And once we think in this way, we can reverse-engineer a proper expression for the individual process, conditional now on the presence of vacancies.

Let us consider a set of N persons observed over T time periods, where each person can have one of J ranked jobs, as you might have in an organization. (Note that we can use the same logic for simpler cases, in which we aren't looking at more than one time, or more than one job.) Let $y_{it} = j$ indicate the job held by person i at time t, with $j = 0$ indicating that person i is out of the organization. Let p_{it} be a variable indicating the probability of a particular promotion at time t, that is, that $y_{it} > y_{i(t-1)}$. For now let us assume that people get promoted only one step up at a time and never get demoted; that is, for any $y_{it}, y_{it} \leq y_{i(t+1)} \leq (y_{it} + 1)$.

Now, we recognize that the probability of getting *some* promotion isn't the same as being promoted to any *particular* position. And if someone is promoted to one position, they can't be promoted to a different one. So let's think about what we'll call P_{it}, namely, the probability that person i gets *some* promotion at time t. That's got to be equal to one minus the probability that he doesn't get *any* promotions, right? If there are, at any time t, $V_{j(i)t}$ vacancies for the slot above the one that i occupies (denote this V_t for short), and we assume that the probability of promotion into any of these for individual i is the same p_{it} then the probability of being promoted P_{it} is the probability of being promoted to the first vacancy ($= p_{it}$), plus the probability of not being promoted to the first vacancy but promoted into the second ($= [1 - p_{it}]p_{it}$), plus the probability of not being promoted to the first vacancy nor the second vacancy but promoted into the third ($= [1 - p_{it}][1 - p_{it}]p_{it}$), and so on, or

$$P_{it} = \sum_{v=1}^{V_t} p_{it} \prod_{g=1}^{v-1} (1 - p_{it}) = \sum_{v=1}^{V_t} p_{it} (1 - p_{it})^{v-1} = 1 - (1 - p_{it})^{V_t}. \tag{6.13}$$

(This is akin to a traditional hazard approach, as used to study internal labor market promotion by, e.g., DiPrete and Soule 1986; see Allison 1982. But in this case the multiple exposures refer to vacancies, not times.)

Let's also assume that p_{it} is a function of a set of K covariates,

$$p_{it} = f(\mathbf{X\beta}) = f\left(\sum_{k=1}^{K} \beta_{ik} x_{ik} \right) \tag{6.14}$$

Now if we assume that this function f is one of our most common ones, like a logistic or a probit, things get a little awkward. But let's instead assume that f is the inverse of the "complementary log-log function," namely

$$f(\mathbf{X\beta}) = 1 - \exp[-\exp(\mathbf{X\beta})]. \tag{6.15}$$

Then

$$P_{it} = \sum_{v=1}^{V_t} p_{it} \prod_{g=1}^{v-1} (1 - p_{it}) = 1 - (1 - p_{it})^{V_t}$$

$$= 1 - (1 - \{1 - \exp[-\exp(\mathbf{X\beta})]\})^{V_t} = 1 - (-\exp[-\exp(\mathbf{X\beta})])^{V_t}$$

$$= 1 - (-\exp[-V_t \exp(\mathbf{X\beta})]). \tag{6.16}$$

This implies

$$1 - P_{it} = -\exp[-V_t \exp(\mathbf{X\beta})]. \tag{6.17}$$

So with minor manipulation,

$$-\ln[-\ln(1 - P_{it})] = \ln(V_t) + \mathbf{X\beta}, \tag{6.18}$$

which means that the complementary log-log allows us to estimate the covariates, so long as we put the log of the number of vacancies in the equation as an offset.

* 7 *
Time and Space

Introduction

In the past two chapters, we considered what changes we might need to make to our general control strategy when we embed our data in structures—so far, only in aggregates. In the next two chapters, we consider when these embeddings are themselves structured. In this chapter, we're going to look at when these embeddings are *times* or *places*. The central problem we will focus on is not an issue of statistical interdependence and the correction of standard errors. It's a much bigger one: there is information about similarity between our cases that is not gathered by us in our data regime, but is held in the world. *When* observations were made probably tells us something about the cases . . . but we don't necessarily know *what*. Ditto *where*.

Most of our problems, we recall, come from the fact that we don't have the right model. Something is happening, and you don't know what it is.[1] And usually, some of our omitted predictors are, contrary to our optimistic wishes, correlated with our included ones. The fact that unobserved predictors are likely to be more similar among cases close in time and/or space than those farther away leads to statistical problems. But it can also provide ways of uncovering, and perhaps correcting, problems we wouldn't otherwise know about! That is, if we assume that events closer in space-time are more likely to share unmeasured predictors, we can push on the robustness of our analyses.

Interestingly, we've had very different orientations when it comes to time and to space. That's in large part due to the fact that we assume that time and space are intrinsically different. Time—"monarchical time," in Bergson's words—is inflexible and possesses its own principles whereby it forces an order upon our experiences. The past is ever inaccessible to us. Space—"democratic space"—is the opposite. While time is inflexible,

1. Do you, Mr. Jones?

space is generous and yielding. Unlike with time, when it comes to space, we *choose* how we proceed through it to order our experiences. For this reason, we've tended to trust time to do things for us that it doesn't, and so analyses turning on time have tended to be weaker and worser than most conventional analyses, while those using space have been stronger and more conservative.

But for most of our analytic purposes, the way in which time and space affect our analyses is more similar than it is different. The *directionality* of time has given many sociologists false assurances; directionality rarely solves problems of omitted predictors. Let me give a brief example: you have an awesome set of data—all the babies who were born in Sweden from 1900 to 2000. You want to use this to estimate, say, the degree to which "family" (whether genetic or social heritage) seems to explain some behavior—by comparing brothers within families to people *across* families.

But, given this time span, brothers are more likely to be born close in time, and to experience similar social conditions, than non-brothers. That's going to increase the amount of the variation that seems to be within *family*, but is really in *time*. Notice that it doesn't matter which way time flows, or that we find out that there is some particular confounder we can identify. It's just about *closeness*, just as with space. That's the sort of problem we're going to bump into again and again.

Map: First for time, then for space, we're going to start with how certain techniques that might be used ritualistically don't help us with all our problems. We're going to think about how we can use such techniques to get a better sense of our data, and then we're going to note some of the problems that we might actually *create* by trying to simplify our data. In particular, smoothing can sometimes help, but sometimes lead to false findings. More specifically, we'll start by reviewing common forms of longitudinal (temporal) analyses, "first differences" and "fixed effects" models. We'll then examine the sort of periodicity in data that these approaches rarely help with. We'll then look at space, again, focusing on how cases may be more alike than our predictors can account for. We'll work through some different ways of representing space, and think through the logic of smoothing data.

Longitudinal Analyses

The Problem with Time

Here's the basic problem with time data—lots of things go up and down together. Most statistics textbooks have a cute chart like the number of chimneys and the number of babies in some city—the more chimneys, the more babies. This is *compatible* with the hypothesis that storks make babies, and they will leave as many as there are chimneys to receive them. But it's also compatible with the hypothesis that with population growth, there are more dwellings. As Mr. Rogers used to sing, everything grows together.[2]

You might think, well, then all I have to do is to control for time in my models! That's our weird orientation to time. Because it seems to flow on and on, we don't think it at all strange to assume that its effect is linear—yet you probably wouldn't assume that the effect of space is two variables, latitude and longitude, each as linear predictors! Even more, controlling for time returns us to a problem that we examined in the previous chapter, when trying to "control for size." Recall that we found that in many cases, controlling for size could make things worse. That's because if our variables scale with the size, but in different ways, entering size as a linear control can actually *increase* spurious correlations. The same thing is true for time. (So I won't bother redoing those arguments here; you get the idea.)

And as you might imagine, the fact that everything grows together, but not at the same rate, can lead to real headaches. That's especially true when the *range* of variation in our dependent variable is large, and doubly especially when that's true for an independent variable as well. Such data structures can lead to that extremely high correlation that we saw characterizing the log-log plots in chapter 6. (An R^2 of .97 for such an analysis is hardly newsworthy; see, for example, Massey, Durand, and Pren [2016: 1569].) There's nothing intrinsically wrong with the high R^2 from such models—so long as the variation that you are left with after adjusting for range is meaningful. But as you might have guessed from the results of the last chapter, when there are relations this strong, small misspecification can lead to big problems.

Time and Trust

I noted above that one reason we've seen so many errors in temporal analyses is that we assume that time *solves* problems, not creates them: at least if we are concerned about the direction of a causal process, we can rely on temporal ordering to sort things out. We know, say, that x and y are robustly correlated. But it could be that x causes y or y causes x. So

2. In fact, that's where our statistics originally come from—measuring how your hand grows as your arm grows as the rest of you grows, because you're all one piece.

if we find that x comes before y, we figure, well, that proves it! It's that x causes y.

That's actually pretty weak reasoning, even if, for the sake of argument, we accept that all effects are later than their causes. Why? Because there is a lot of slippage between the temporal ordering in the world, and the temporal ordering in our data. For one thing, often the gaps between cause and effect are far too short, or far too long, for us to gain traction on. Temporality works only for things that are synced to the general rhythm of social research. And as we saw in chapter 3, sometimes, causes "appear" in our data later than they first started. A person who drops out of high school at 18 and develops prodromal schizophrenia at 22 may have suffered from mental disease since 16, but it just didn't leave a record.

So I hope you don't think that looking at temporal ordering can solve your problems of identification. Because an incredibly common, though obvious, mistake is for researchers to think that by comparing a lagged value of one variable (x) to another unlagged variable (y), they have isolated the causal contribution of x to y. Indeed, weird though it is, would you believe that reputations are made by analyzing time series data on a *single data point with only 50 or so observations*? If you are laughing, you are not a sociologist. But it's become completely unremarkable for the *ASR* to publish articles that make a claim that some x causes y and to demonstrate this by taking, say, yearly data from the United States, and showing that a change in x precedes (by one year, say) a change in y.

Let me present some simulated data of a single unit (say, a nation) observed at 100 times (R 7.1), where we are interested in predicting economic growth (y) from, say, average political liberalism (x). The hypothesis is that liberalism is actually *good* for economic growth. We might first start with the basic equation $y = bx + c$, assuming that x is a cause of y (model 1, table 7.1). However, someone points out that it could be that y causes x; maybe liberalism is a luxury that people can afford when their basic needs are met. Thinking that, well, y couldn't cause a *past* x, we regress y on the value of the previous observation of x (model 2). Thus we are fitting $y_t = b'x_{t-1} + c'$. Our coefficient is basically the same, and so it looks like we've proven that x is a cause of y!

But this confidence assumes that, say, if *time had gone the other way*, we *wouldn't* see liberalism still appearing as a cause of growth.[3] (We're going to refer to this technique as "running time backwards.") We can do this (in this case) by lagging y instead of x and repeating the model (model 3; $y_{t-1} =$

3. Ken Frank and Tom Dietz inform me that this practice is done in the policy field; we should have been doing it long ago!

Table 7.1. One Case, Multiple Times

	Model 1	*Model 2*	*Model 3*
Dependent Variable (Growth)	Simultaneous y	Simultaneous y	Lagged y (1)
Independent Variable (Liberalism)	Simultaneous x	Lagged x (1)	Simultaneous x
Coefficient	.679***	.631***	.664***
Intercept	.120	.147	.128
R^2	.358	.314	.341

$bx_t + c$). Uh-oh! It's basically the same as model 2! In this case, the relation between x and y is spurious—both are functions of time.

What have we learned? That temporal *closeness* can be more important than temporal *ordering*. Fig. 7.1 (R 7.2) shows 39 replications of model 2, but changing the lag from −19, to −18, to . . . , to 0, to 1, to 2, and so on. But we're always fitting the same data. You'll note that the coefficient gets smaller the farther we are from a 0 lag—in either direction! In other words, there's some sort of commonality to y and x such that observations that are closer in time are more similar. This sort of temporality is one that *can't* be corrected by relying on the fact that time flows one way. It's more like a spatial ordering (in which closeness matters). Temporal ordering works if we have processes that happen in bursts, so that if you saw them over time, there'd be, say, a burst in x, and then, shortly after, a burst in y. (See, for example, fig. 7.2; I'll call this the "gamma burst" picture because it reminds me of data on gamma radiation from a *Planet of the Apes* sequel I once saw, but technically, we can call it "stationarity," because we don't see the means *going* anywhere.) But anything that tends to ebb and swell is going to pose a much bigger problem. And unfortunately, most of our temporal questions are about the ebb-and-swell processes. So what do we do about them?

First Differencing

We've realized that the problem of temporality isn't one that can be solved by just lagging our observations. But some sociologists argue that there is still a solution, which is to take the "first difference." If the dependent variable of case i at any time t (denoted $y_{i,t}$) is a linear function of a set of

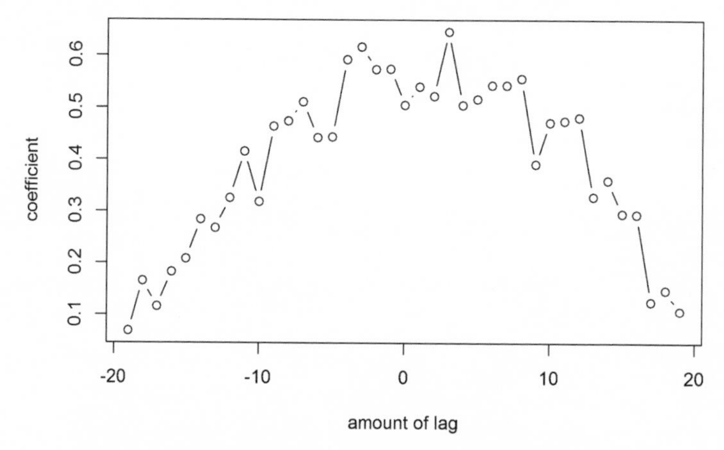

FIGURE 7.1. Signs of Data Not Amenable to Causal Analysis via Lag

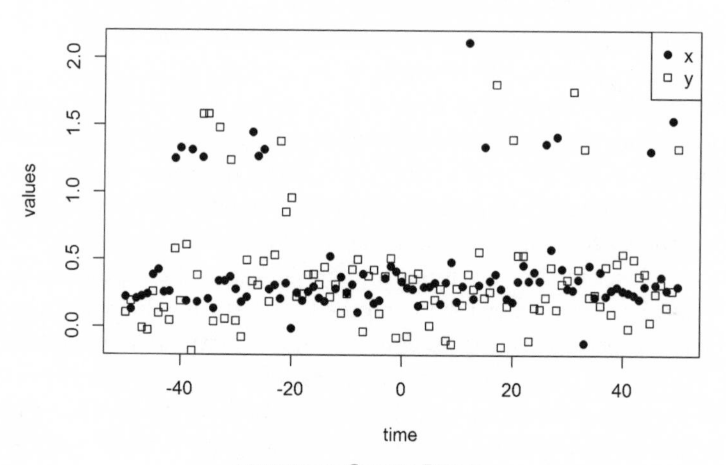

FIGURE 7.2. Gamma Bursts

other time invariant variables Z_j $(j = 1, 2, \ldots, J)$ and time variant variables X_k $(k = 1, 2, \ldots, K)$,

$$y_{i,t} = c + \sum_j b_j^A Z_{j,i} + \sum_k b_k^B X_{k,i,t} + \varepsilon_{i,t}, \tag{7.1}$$

then we can subtract this from the same equation for the next time $(t + 1)$ and get

$$y_{i,t+1} - y_{i,t} = c + \sum_j b_j^A Z_{j,i} + \sum_k b_k^B X_{k,i,t+1} + \varepsilon_{i,t+1}$$
$$- c - \sum_j b_j^A Z_{j,i} - \sum_k b_k^B X_{k,i,t} - \varepsilon_{i,t} = \sum_k b_k^B (X_{k,i,t+1} - X_{k,i,t}) + (\varepsilon_{i,t+1}^{*}). \tag{7.2}$$

Table 7.2. Results from First Difference Model

	Coefficient
First Difference of x	-1.00****
	$(5.97 \times e^{-16})$
Constant	2.00
N	49 (we lose one because of the differencing)
Multiple R-squared:	1.00
Adjusted R-squared:	1.00

That is, all the time invariant disturbances cancel out—whether or not they are measured and included in the model, and the new errors can be assumed to behave nicely, leading us to get the original parameters via this change score model. This method, Jacobs and Myers (2014: 758) assure us, is "conservative" because it "helps eliminate spurious relationships by removing mutual trends in variables." That means that even though we really have only a single case, and are looking at that case in an unbroken stretch of time, we should be able to trust results such as those seen in table 7.2, from a simulated data set with x and y observed over 50 times (R 7.3).

That's a pretty damn good fit! And yet this is totally spurious. There's a non-linear temporal effect (both x and y are quadratic functions of time). First differencing doesn't do anything for this except get it nice and ready for a spurious linear relation. With only 50 data points, this conservative test assures us that we have a t value of . . . 957400000000000. That's around as conservative as Che Guevara.

Do you think that you can make an even more conservative test by combining the two strategies we just saw not working individually—that is, by lagging the difference scores? And therefore by fitting the model below?

$$y_t - y_{t-1} = b(x_{t-1} - x_{t-2}) + c \tag{7.3}$$

You might think, if there's something wrong, at least we bias our coefficients to zero (which is what it means to be conservative). Vaisey and Miles (2014) demonstrate that this can lead to results very different from what we might have thought. That's because if we actually have a simultaneous effect of x on y, the lagged coefficient will not be biased to zero, but will actually be *negative* one-half the actual slope![4]

4. Vaisey and Miles point out that a piece on which I was the methodologist uses exactly this method as a robustness check; they tactfully don't point out that our lagged results

Further, if there *isn't* a temporal trend, but in fact you have "gamma bursts," taking the first difference probably isn't going to hurt, but it usually won't help either . . . unless there actually is some sort of change process to your bursts, and you know the proper lag (if there is one). When *will* first differencing help? Only when you have a gamma burst process *on top of* a reasonably linear "creep uphill" type trend.

There is a more general way of handling these problems, the "AR[i]MA" approach. This stands for "auto-regressive [integrated] moving average." The idea here is that we strip away anything like an overall movement from a time series, and then predict the resulting gamma bursts. This *might* be something for you to do. But if you do, you can't assume that any particular specification "works." You can have an ARIMA model that still fails the "run time backwards" test. So you should always visually confirm that the data, after being ARIMAd, look right, and make sure that forward lags aren't picking anything up. You *can* make this work, in the sense that if you put in enough terms, you'll get rid of any temporal organization.

While you might think that purging the temporal trends from your dependent variable is enough to allow you to estimate your model correctly, in some cases, the existence of time trends in the independent variables makes it hard to disentangle their effects. Of course, it's possible to ARIMA them as well. However . . . if you've ever stripped paint off of woodwork, you know that there is a bit of a dilemma in choice of methods. The most effective paint strippers don't just take away the paint—they take away the wood or ruin it. That's the problem you may have with an ARIMA model. Let's say you are interested in protest events, and want to predict them using time-varying covariates. But you do just what I've said here, and realize that the covariates and the events ebb and swell in a way that means your results are just as strong whether you run time backward or forward. So you ARIMA the f*** out of it. But what's left after you strip away the moving averages and auto-regressive effects doesn't necessarily have much to do with what we normally understand as the protest movements you were originally interested in. You may indeed have gamma bursts, but they might turn out to be about the scheduling of Presidents' Day, about cloud cover, about all sorts of things very remote from your theoretical interests.

But this brings us to a deeper puzzle regarding the use of fixed effects for temporal processes.

---

do in fact seem to suggest that the lagged model is wrong! Luckily, our finding holds up either way. And even luckier is having folks like Vaisey and Miles figuring out what we are *really* doing.

More Bad Times Ahead

So far, we've been exploring methods for dealing with a single series. The lesson is basically that if you visualize your data, and it looks like bursts of radiation, that's good. But if it's something that wanders up and down, *and so do your main predictors*, the chances of being able to identify a causal process via conventional statistics is too small to be worth your while. But what if we have more than one unit being observed? Here it's not so crazy to try to fit a model. But we still have the same problems.

Let's assume that we have N cases (say, countries), which we observe over T times. We want to use this to see if some x (we'll stick with "liberalism") is a reliable independent predictor of y ("economic growth"), or whether the relation might be due to something else. Often the way a statistician will see this problem is that our different observations, coming from the same cases, aren't statistically independent. And for a while, because of Stata's great "cluster" command, sociologists thought that they were able to solve their problems by "taking this non-independence into account." But that assumes that we have the right model. If we don't, our problems have to do with omitted predictors. The insufficiency of this approach for most cases is now pretty well understood, and so I won't even bother presenting simulations demonstrating this.

So if we're not really worried about getting the significance test (for the *right* model) correct, but rather, are worried about the effects of misspecification, we'll find that the techniques we've already found to be inadequate—entering a control for time, lagging the predictor, and first differencing—all remain inadequate. Or, to put it more simply, they are inadequate for all but the simplest possible versions of confounding. If the true model is simply that there is a linear time effect, then all of the methods will work just fine. But if there is some sort of ebb-and-swell pattern, almost none do anything.

Let me demonstrate this with a set of simulations (R 7.4); all the results are in table 7.3. I want to compare four simulated data sets. In all of them, there is no real direct relation between liberalism and growth, but both are functions of time. We want to make sure that we correctly reject the hypothesis of a true x-y relation. We'll look at ten cases, and ten time periods. Each row of table 7.3 represents one type of relation between these variables and time, and each column represents a model. Each cell has the regression coefficient we get for the effect of liberalism on growth.

The first row is for a world in which both liberalism and economic growth are linear functions of time—a kind of Whig history—though with some error. If we just look at the two in a bivariate regression (the first column), we'll of course incorrectly see x (liberalism) as a potential

Table 7.3. Which Methods Detect a Spurious Relation?

	Naïve	*Control for Time*	*First Difference*	*Case Fixed Effects*	*Time Fixed Effects*	*Both Fixed Effects*
Linear Time	.698***	.010	.019	.739*	.027	−.015
Linear Change	.746***	.147*	.227*	.902***	.037	.007
Uncorrelated Two Cycles	.262**	.235*	.963***	.926***	.007	.085
Correlated Two Cycles	.464***	.431***	1.016***	.995***	.238*	−.059

cause of y (growth). But throwing a control for time (second column) correctly eliminates this spurious relation. So does switching to a first difference model (third column), which we've seen in equation (7.2).

Difference models are closely related to the fixed effects models we discussed in chapter 5—by subtracting one equation from the other, all the time invariant predictors drop out, which is also the result of adding fixed effects for each case. Adding fixed effects for cases leads us to reformulate our question as "does the degree to which this case's current liberalism deviates from its overall average predict its growth?"

The results in the fourth column come from a model that includes *case* fixed effects. Here we simply include a dummy variable (D_i) for each case. A general formula for this approach with a number of independent variables would be as follows[5]:

$$y_{i,t} = \sum_k b_k^B X_{k,i,t} + \sum_i b_i^C D_i + \varepsilon_{i,t}. \tag{7.4}$$

So now we can see that while we can imagine that we are modeling "change" with case fixed effects specifications, we're really examining the deviations from each case's average value. And so this does not help with

5. Note that if our true model includes any unmeasured time-invariant variables Z (as in eq. 7.1 on page 199), the D dummies will fit the variation due to Z, and hence our estimates of the b_k^B will be the same as if we had included those effects (assuming proper specification). (Minor differences may exist depending on the estimation routine and the number of fixed effects and their magnitude, as well as whether there are linear dependencies in the complete set of equations.) Also note that we no longer include a constant, as the fixed effects may be thought of a set of constants, one for each and every group.

our confounding due to both our variables being functions of time, because it is precisely within-case variation that is driving our results.

But we can also add fixed effects for *time periods*. That means we reformulate our question as "does the degree to which this case's current liberalism deviates from the average for this time predict its growth?" And adding such fixed effects using dummy variables, just as in eq. 7.4, works fine at reaching the null result (column 5).

In other words, this approach makes sense if we are concerned not with something that affects each *case* across all *times*, but something about certain *times* that characterizes all *cases*. Some years happen to be unusually "high" on our dependent variable, and some years unusually low.

The second row presents the result from data where time (linearly) affects the *change* in the values of liberalism and economic growth, and not the values themselves. (That fits a world of exponential growth.) In this case, the control for time (column 2) doesn't work to remove the spurious relation. That's because the rate of increase grows with time. Even further, the first difference model (column 3) doesn't work—because in fact it's the differences between adjacent years that are spuriously correlated in time. Unless we adjusted the functional relation of x and y, our simplest fixes don't work. But entering fixed effects (column 5) for time works because it is equivalent to a wholly flexible specification of that functional relation.

The third row presents results from data in which there is a more complex relation of liberalism and economic growth to time. There are two sinusoidal cycles of the effects of time on both of these variables (that is, both variables go up and down and up and down). That fits a world more like the one most of us live in. The wheel turns—good times, bad times, good times; short hair, long hair, short hair. For these data, not surprisingly, a linear control strategy (column 1) fails to remove the spurious relation, and going to first differences actually *increases* this relation, because the generating process is one that works in terms of change as opposed to y and x being a simple function of time. But once again, fixed effects for time (column 5) successfully elimates the false relation.

But for the fourth row, we make one change: at the first period, liberalism and economic growth are correlated. Their *changes* aren't correlated, and so, if we were looking for a causal effect taking advantage of the longitudinal data, we'd want to conclude that this relation was zero. The initial correlation is a "correlation without causation." That's usually the way things go in our data; it's rare that we can get data "all the way back," before there was any evidence of the correlation of our variables. Now notice that in this case, adding time fixed effects (column 5) *doesn't* work to reveal the negative relationship. Why? Because these fixed effects basically turn the question into a cross-case, within-time question. But we already know that, by construction, the cases vary at time 1. Even though adding case-fixed effects (column 4) only increased the spurious relation, there is serious cross-case variation that we think of as being non-causal. It's only the specification that adds *both* case *and* time fixed effects (column 6) that recovers the correct absence of a direct relation.

Now it's important to emphasize that I'm not saying that this approach is the "right" one or even that it's "most conservative." It's the answer to one specific question, and that might not be your question, and it's not always the best answer either (multiple sets of fixed effects mean that they no longer have an interpretation as "deviations-from-a-center," and there can be all sorts of estimation issues when we have lots of fixed effects). If you have a well-defined research question, you might not need to scrape away at the data in these different ways. But if you're like most of us, and you have some interesting non-experimental data from multiple cases over time, you probably want to try these approaches to see whether any of them makes your findings go away and, if they do go away, you want to think about that seriously. But this is only the beginning, not the end, of puzzling through how to deal with such longitudinal data. Let's consider the relation between change scores and lagged scores more closely.

Fixed Effects and Change Scores

We saw that case fixed effects didn't help eliminate temporally induced correlation between our variables; as you probably learned in your stats class, this approach should be used to get rid of the effects of time-*invariant* unobserved variables. You probably also learned that this wouldn't help when we have unobserved heterogeneity that is time-*variant*. But there is another class of questions for which this is a poor choice.

I often see talks in which someone is doing a fixed effects model and someone in the audience asks about a certain form of time-invariant heterogeneity that seems like it could confound the results. For example, we're looking at growth and liberalism, and we might think that countries that are part of the EU would have a different relation than would other countries. The speaker then admits that she's thought about this, but since she is doing a fixed effect model, she *can't* look at this. Because all time-invariant characteristics *have* to drop out.

Well, the choice of model can't answer the key substantive question—and you can't ignore a plausible hypothesis because it requires a change in model. When do we need this sort of change? When we suspect that a *time-invariant* characteristic (membership in EU) is related to *change* in your dependent variable. If that's the case, you have to confront the fact that you need to do a model *for change*, and not a *change model for the non-change dependent variable*. This will *not* have a fixed effect interpretation.[6] But it's definitely the right thing to do. That's especially the case if you want to interact your time-invariant with your time-varying predictors (like liberalism).

You might think that what you should do is switch to a lagged dependent variable. Here's the thinking. Let's say you believe that the true model is

$$y_t - y_{t-1} = b_1 x_t + c. \tag{7.5}$$

That is, x predicts *change* in y, not y itself; in this case, liberalism doesn't predict growth, but the increase in the rate of growth. However, if your dependent variable was *de facto* bounded, you might anticipate that at higher values of y, it will be difficult for the change to be great. For this reason you might, quite reasonably, decide to add y_{t-1} to both sides, as so

$$y_t - y_{t-1} = b_1 x_t + b_2 y_{t-1} + c. \tag{7.6}$$

This looks a great deal like a lagged dependent variable model, or

$$y_t = b_1 x_t + b_2^* y_{t-1} + c \tag{7.7}$$

That's because we can rearrange eq. (7.6) as

$$y_t = b_1 x_t + b_2 y_{t-1} + y_{t-1} + c = b_1 x_t + (b_2 + 1)\, y_{t-1} + c \tag{7.8}$$

which means that the lagged dependent variable model (7.7) produces a coefficient that adds 1.0 to the coefficient of the lagged *change* model (7.6). The two models will fit the same, but differ a bit in interpretation. Estimation issues aside, the values of the *other* coefficients won't change. That's because a model for *y given* previous *y* is *itself* a model for change, isn't it? It might be worth starting with the first change model (7.5) and comparing the results to a cross-sectional model to help get a feel for how to think about your predictor's relation to the dependent variable.

So the main question we have to ask ourselves is whether or not we actually have a theory that predicts the values of change in the depen-

6. In some cases, you might have a model that doesn't really make sense, but a model that doesn't give the right estimates for certain parameters can still be right in terms of rejecting the null hypothesis, and that might be all you really want to do.

dent variable, or whether we are only looking at change as a way of trying to purge unobserved heterogeneity. If the latter is our aim, a better way of combining lags and differencing might be the Arellano-Bond approach. This includes both a lagged dependent variable and a fixed effects for the cases, and can be seen as more flexible than either of these alone (Halaby 2004). There are, however, certain technical issues of estimation that you'll need to give serious attention to.[7] For very clear research questions, such "cocktail" methods might work, and, if you can identify real substantive concerns that lead you to think that you need to be conservative in the face of trends and unobserved individual heterogeneity, they might indeed work. And the mightiest cocktail of all—the Long Island Iced Tea of panel methods—is the ARIMA method I discussed above, which can also be applied to multiple-case series of data.

But I'd like to emphasize that most of these techniques are designed to help you better identify a model that is for the value of y and not the change in y, and that these are different. The sorts of models that are convenient for one aren't always convenient for the other. And finally, it's not even quite true that with information about the change dynamics, we can switch to a model for statics. As Wolfgang Köhler liked to point out, the force of gravity—and the changes it induces in a free particle—are the same for a pendulum swinging back and forth and one hanging still. The difference is in the initial conditions. And as we saw in row 4, table 7.3, those initial conditions can be more complex than simply the value of an intercept.

For a very simple example, if you think that x can be transformed into y and vice-versa, this is compatible with any cross-sectional relation between the two. For example, wealth can be used to buy consumer goods, or goods sold to get wealth. That doesn't imply that those with the most wealth have the most consumer goods—because if they've used their wealth to get consumer goods, we no longer see their wealth! This can hold if we are looking at other, more vaguely defined "capitals." If they really *are* capitals, and not simply attributes, then we would expect that converting one into another lessens one's stock. Whether this is true or not won't be determined by examining cross-sectional relations. In this case, we're going to need models *for*, and data *on*, change.

Finally, there's a subtle problem with many of our approaches that as-

7. Given the lagged dependent variable, it's hard to estimate the fixed effects correctly; as you can imagine, it tends to bring in all sorts of error correlations. For this reason, researchers need an instrumental variable approach, and they usually use a past lagged value or lagged difference. Here see Angrist and Pischke (2009: 244ff), who suggest seeing fixed effects and lagged dependent variables as bounding the estimate of a causal effect.

sume that there is a single process linking some independent to some dependent variable, one which we can better estimate through, say, difference models or fixed effects. If you believe that there are any aspects of social life that are *systemic*, you'll have to entertain seriously the notion that the causes that explain the relation between x and y are different from those that explain the relation between *changes* in x and *changes* in y.

Think of the bowling game among the Norton gang famously analyzed by Whyte (1981 [1943]). Whyte was interested to find that social status (x) was positively related to bowling score (y). You might imagine that one caused the other: that bowling is a route to status, say ($y \rightarrow x$); or maybe the reverse, that the confidence of being high status means that you bowl better than others ($x \rightarrow y$). But what Whyte saw happening was different. It was definitely possible for a low-status member to start doing well at bowling. So he was a big positive residual: $y > \mathrm{E}[y \mid x]$. But this didn't lead his status to increase. Instead, this discrepancy was brought back into line by the group razzing the bowler. The sum total of these dynamics kept the x-y scatterplot looking as it should. But that scatterplot had no direct relation to the dynamics involved.

And even when we don't find systemic behavior that leads to the preservation of a static relation, there are plenty of times in which the things that best explain the *change* in a distribution just aren't important in giving us insight into the nature of the distribution itself. In a way, we're going back to Lieberson's point (1985), with a Funkadelic twist—if the only variation we have is on relatively insignificant rises and falls in, say, class position, we're going to end up attributing the class structure to irrelevant factors. Lords who get the plague in their manor may go broke and become peasants. But the presence of a feudal class structure isn't explained by health. Remember, fixed effects isn't a *stronger* answer to the same question (why are some people lords and others peasants?), but an answer to a *different* question (what explains within-person changes)? That sort of question fits a rigorous—often overly rigid—vision of causality better than it does our "casual causal" view . . . which might, precisely because of its vagueness, not be obviously wrong. Just manipulating equations doesn't guarantee that this latter answer is relevant for your question.

Dichotomous Data

Models for change become especially problematic when we have a dichotomous dependent variable. Many people advocate a fixed effects model here, which is basically going to take the cross-classification of time 1 and time 2, and focus on the ratio of the off-diagonal cells. That is,

Table 7.4. Change in a Dichotomy

		Time 2	
		Voted Republican	Voted Other
Time 1	Voted Republican	f_{11}	f_{12}
	Voted Other	f_{21}	f_{22}

if our dependent variable is voting Republican, and we are interested in change, we would look at table 7.4, where f_{12} is the number of persons in row 1 and column 2, and so on. To model change, we then analyze f_{12} and f_{21}. There are number of problems with this approach. First, it should be clear that it requires that you throw away a lot of your data. This can be very confusing when we have some predictors that are heavily associated with the outcome but simply don't vary. Beck and Katz (2001) give the example of genes and cancer: we might be interested in predicting the onset of cancer (which changes over time), but we might not want to ignore the (time-invariant) effect of genotype—which might also be related to time-*variant* predictors. But that is required by a fixed effects approach. We're back to the problem I raised with the example of needing to take EU membership into account—you don't want to let the statistics drive you to ask the wrong question.

Even if we don't bias the estimation of our time-variant effects when doing the fixed effects model, we can lose too much power when our dependent variable is highly skewed. For example, we throw out a tremendous amount of the information if we ignore the fact that many cases were, are, and will be at low risk for the dependent variable.[8]

Finally, this approach assumes a fundamental symmetry between the categories of the dependent variable. But there are many dependent variables that are asymmetric, and in many cases, we're better off thinking about a transition only one way. That isn't only for things like being dead, that really don't go backward, but for things that are *nearly* asymmetric. For example, with some types of cancer, there is a significant non-zero chance that a person who "had" the cancer at one time will "not have" the cancer five years later. Still, while hoping for the best, we might prefer to treat such cases as if there were change in only one direction, and instead, to do a hazard model.

However, there are other cases that are asymmetric, but transitions still

8. Further, omitting the effects of invariant predictors can also change our findings if we have a non-linear model such as a logistic, in which the "effect" of other variables on a case depends on *where* it is along the probability function. This is another way of restating the point about how non-linear models don't have any form of "properly specified misspecification."

happen in both directions. So what can we do? We don't need a single "all purpose" perfect answer and we don't need a single model for that 2 × 2 table. We might be better off examining two different 1 × 2 tables, as it were—transitions from 0 to 1 on the dependent variable, and, separately, transitions from 1 to 0. Of course, we lose statistical power, but, as Lieberson (1985) emphasized, allowing for asymmetry can be key to understanding the actual processes in question. In chapter 3, I gave the example of a paper (Habinek et al. 2015) where we found that the right way to examine change in friendship relations was to split the sample and look separately at those who were and those who weren't friends at first, because the factors that lead people to *make* friendships aren't "the same but upside down" as the factors that lead people to *break* them.

In sum, there are times when we may turn to change models, but it doesn't make sense to have a one-size-fits-all fix for generic problems. Rather, if we turn to models for change, it should be because we have good reason to think that our change is going to vary across units and that we need to be able to explain this. Some of this explanation may invoke time-invariant characteristics of our units. We need to return to Aristotle, who pointed out (in the *Physics*, Book IV, chapter 2) that some things have a kind of inherent causality that can lead themselves to change. You can be a convinced Hobbesian and try to deny that, but you're going to need to come up with an alternate explanation for facts like some seeds grow faster than others. You can't just pretend it isn't true in order to justify your approach.

Periodicity

We've seen that many of our approaches to dealing with time-series data don't help us when we have correlations due to the waves of simultaneous changes that rise and fall and drive correlations between variables. So it's important for us to develop the capacity to recognize such waves in our data. For an example, let's imagine we have data on the degree of attention paid by some newspaper to a single theme, in terms of the column inches devoted to it each day. For fig. 7.3, I simulate data (R 7.5) for 200 days, and then smooth them to see if there are patterns.

It's pretty clear that we're seeing periodicity on the order of every 10 days. Well, maybe it's not so clear. Fig. 7.4 reproduces the previous figure, but now with the raw data superimposed. Actually, the data are totally random, with a constant probability at any time. Why the periodicity? It's an artifact of the smoothing routine, which aggregates all the observations over some time as a moving average. We're basically pushing pink noise through a filter, and, like holding a shell to your ear, hearing a tone.

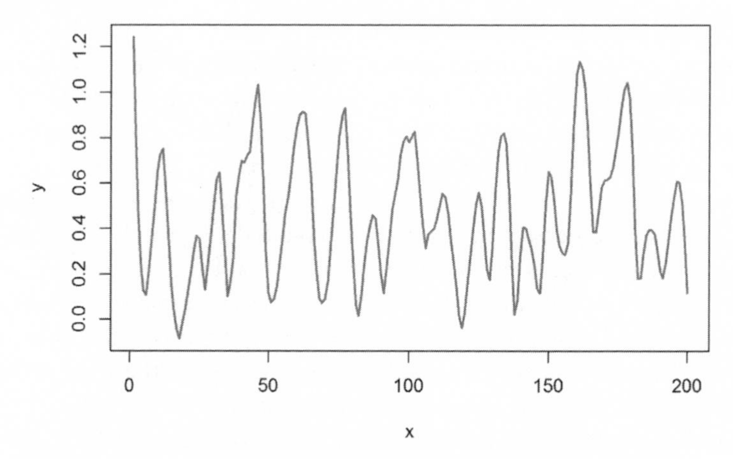

FIGURE 7.3. Periodicity in Temporal Data

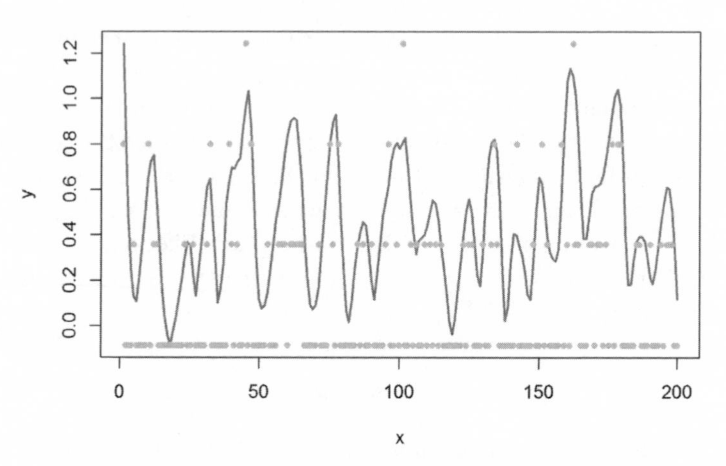

FIGURE 7.4. Periodicity in Temporal Data, Data Added

Okay, lesson taken, you think: beware of visual confirmation of periodicities based on some smoothing procedure. But it gets worse: this can lead to false relationships in your data. When we have a number of such data streams (like, how much attention is there to *this* news theme? What about to *that*?), smoothing tends to increase the risk of finding false positives—concluding that two streams are related, when they are actually independent (see simulation R 7.6; I'm not going to print this stuff here).

So beware of looking for patterns in smoothed data—many times, random data can give elusive hints of a pattern out there, like seeing a ghost ship in the fog. This is one of the cases where you shouldn't just believe your eyes. When in doubt, trust the math—if spectral decomposition methods like a periodgram (look it up!) don't show clear evidence of a disproportionate tendency for repetition at a certain frequency, don't pursue it further.

Conclusion to Time

In sum, there are some cases in which the temporal nature of time—the fact that it flows one way—really matters for your analysis. But more generally, temporal data structures produce relations of sameness that might be due to any number of things. One of them might be increasing selection bias over time (see chapter 3). Others come from omitted predictors. In that case, it makes sense to treat time more like space. What matters is that some observations are more similar to others than our covariates would have you believe. And that can be good or bad for us, depending on whether we're willing to think it through.

Space Is the Place

Spatial, not Special

There are two sorts of ways of thinking about spatial statistics, one for statisticians, and one for dataticians (if we can call them that). The first sees data that are located in space having certain very troublesome error properties that require very complex corrections. (The bogeyman word here is "autocorrelation.") It means that if you don't do this, your work can be dismissed. This autocorrelation is seen as a worrisome thing to be disposed of, if possible. As you might guess, while I completely respect this view and think that you should learn it and understand it, I don't think that's going be a useful way to think about these issues.

Instead, I side with what I called the "dataticians" here. They are incredibly excited about the potential of spatial data, largely because if you have spatial data, you can link it to anything *else* that is spatially located. I like that positive attitude. And we'll go with it. But the statisticians are right—all sorts of problems can creep in here. In a way, the problem isn't that the data are autocorrelated—it's that when we haven't properly specified our analyses, adding data with a strong spatial organization can increase our biases. But if we are aware of this, space is an asset, not a liability.

Let me begin by disposing of the bogeyman. (And we're going to see the exact same issue in the same terms in the next chapter, on networks.) Two sociologists can have very similar data sets. One has a conventional random sample of the 48 continental US states, say. We might have information on region, but we rarely use that. No one will object to a conventional OLS of, say, income on education. The other has complete data on everyone in every last county in California, say, down to what block each person lives in. The experts will often pounce on the second, declaring that *these* data are "spatial" and that the estimates are "biased" and the same analyses they accept in the first case are "wrong" in the second.

Huh? How is the other data *not* spatial? Where *are* all those Americans? Floating in the 0th dimension? Everyone is *somewhere*. End of story. Just because you don't *know* where they are doesn't mean you are exempted from spatiality. We need to be very suspicious of any approach that favors ignorance over knowledge.

So was I kidding when I put "biased" in quotation marks, as if I didn't acknowledge that the results *were* biased? Not really. Because *of course* they're biased. *All* of our results are biased, and they're biased for the same reason. We don't have the right model, and there's no reason to think that our excluded predictors are irrelevant, and no reason to think our measures of what we *do* have are perfect. So it's a bit silly to treat one model as if it were perfect and the other as if it were guilty until proven innocent.

But let's focus on one of the particular biases that spatial statisticians are interested in. It can be confusing for us when we speak in terms of correlated errors, or "autocorrelated" errors, because this tends to evoke in our mind the case of, say, a large number of people drawn from the same PSU ("primary sampling unit")—when some but not all PSUs are chosen, and those living in a non-chosen one have no probability of getting in the sample. Yet autocorrelation can even exist when we have only a single unit sampled from each place. What we mean is that our cases are *drawn* from distributions that are correlated. And that means that our estimates can be biased.

With spatially located statistics, we have a reason to suspect that there are such correlated errors. The first type of correlation I raised—that in which individuals are not independently *sampled*—leads to the problem of *clustering* of sub-samples, and there are well-known fixes for it. For many statistical analyses (though not all), assuming we have the correct model, this affects only the standard errors of our estimates, and not the estimates themselves. Since there are straightforward fixes for this, that's not what we're worried about in this book. When it comes to the second type of correlated error—that produced by shared omitted variables—there's no one right fix. We're going to have to go slowly and use our brains.

The key issue is that, for example, people who live near each other are likely to be exposed to many of the same conditions, and so they're likely to have similar values on a whole host of variables that we haven't measured. So they have the kinds of correlated errors that will bias our other variables. For example, I am a professor who works in the social sciences, voting left, with a family, white, interested in spatial statistics. I live on a block that has five or six other professors, at least one of whom works in the social sciences, with a family, white, interested in spatial statistics. I don't know how he votes. So there is something—probably the fact that we came here to work for the same employer—that is similar about us, and that isn't captured in a normal survey.

But guess what? There's someone else, who lives in another city, in another state, who *also* works in the social sciences, votes left, with a family, white, and is interested in spatial statistics. And he's a lot like me in many ways that aren't captured by a social survey. Why? Imagine that we can arrange pairs of people in terms of their "likeness" in some way. Then let's follow Sorokin (1927) and others and imagine a "social space" in which we arrange people so that those who are alike are closer. We're right next to PLUs ("People Like Us") and far from "those jerks." There are plenty of ways in which we are *more* like those close in *social* space than those close in *geographic* space. So why all the kvetching and moaning about spatial correlation of errors?

The answer is (or it should be) . . . space isn't the *problem*. Space is the *solution*. If we had information on PLUs the way we do on PSUs, we'd be in better shape than we are. We should be *happy* when we have spatial locations because it means we have a stand-in for dealing with some of the correlated errors that we *know* are always out there. Sometimes (though as we'll see, not always) PLUs tend to share PSUs. Corrections for the "clustered" nature of the data aren't relevant here (that's only about the non-independence of sampling, not omitted predictors). Hierarchical models may be able to help, but that depends on the particular nature of our most worrisome omitted predictors. We need to think about the structure of the most likely serious confounders for the questions we're trying to ask of the data. If they have to do with geographic organization, then yes, we shouldn't forget about space. In fact, knowing about space can help us solve problems that we otherwise might ignore.

For example, in a very nice recent paper, Gamoran, Barfels, and Collares (2016) attempted to see how racial isolation in high schools affects later labor market outcomes. When they thought about the major methodological problems, their attention was drawn to those that are typical of *school* research: the "non-independence of students within schools and nonrandom selection of students to segregated and integrated schools" (1134). But in this case, it's quite possible that a bigger problem comes from the fact that overall racial composition is related to both the local labor market structure and the pool from which students are drawn. Gamoran et al.'s approach was to use some measures of place (urbanicity and region), but to focus on selectivity into schools using propensity methods. That's the way things are done in education research. But in this case, as they undoubtedly knew, their conclusion—that "white students who attended schools with lower proportions of whites found themselves in less white-dominated workplaces" (1154)—might have to do with the ethnic composition of where they live. It might not matter whether they even *go* to high school! (It's still a great paper.)

Having data on spatial location means that we can try out some assumptions as to how spatial location could allow us to compensate for some of our ignorance about these omitted variables, and we can bring in contextual data from other sources to examine those ideas. Unfortunately, I'm going to go on to say that some of our usual assumptions about what these corrections do are easy to misinterpret. As we'll see, they don't do what many of us want them to do.

Our Big Problems Are the Same

One of the cool things about thinking about spatially located processes is that we have the chance to examine the *diffusion* of certain states, behaviors, or opinions. Let's think about crime rates. You can imagine that the crime rate in any block group, say, is, in part, a function of the crime rates in other blocks right around it. Why? Maybe the police are over-strained. Maybe law-abiding people, seeing all the crime around them, get discouraged and stay inside, decreasing the "eyes on the street." So even if crime is in part a function of something that we measure (say, percent living in poverty), we want to see whether there is this sort of diffusion of crime.

The most elegant way of specifying this is known as a "spatial lag" model. First, let's see if we can take a guess as to the degree to which each area is affected by all the others. We will put this in what we call a "weights" matrix. For example, we might say that $w_{ij} = 1$ if areas i and j are contiguous (that is, they touch one another) and 0 otherwise. Or we might way $w_{ij} = \alpha / d_{ij}$ where α is some constant we come up with and d_{ij} is the distance between the two areas. There are plenty of ways to do it; but somehow we'll take a stab at creating this **W** matrix. Then we might propose the model

$$\mathbf{y} = \rho\mathbf{W}\mathbf{y} + \mathbf{X}\boldsymbol{\beta} + \mathbf{u} \tag{7.9}$$

where the Greek letter rho is an estimated parameter indicating the degree of spatial correlation, and **u** the error. **X** is all our independent variables, say, in this case, proportion under the poverty line and population density, compiled as a matrix; **β** is all the corresponding slopes. So this is like a regular regression model, just with this first term. This first term is a way of taking the non-independence across cases in terms of y that re-

Table 7.5. Models for Crime Rate

	Model 1	Model 2	Model 3	Model 4	Model 5	Model 6
Model:	Lag	Error	OLS	Lag	Error	SDM
x_1 (Poverty)	.517***	1.561***	1.583***	1.002***	.993***	1.555***
x_2 (Density)				1.019***	1.010***	
Intercept	-.046	-.411	-.386	-.015	-.013	-.015
Rho/Lambda	.821***	.962***		-.012	.006	.962***
Gamma						.003
x_1 direct	.619			1.002		1.559
x_1 indirect	2.273			-.012		.050
x_2 direct				1.019		
x_2 indirect				-.012		
Resid Var (σ^2)	2.898	1.850		.977	.977	1.850
LL	-5045.66	-4605.63		-3517.92	-3518.18	-4605.62

mains unexplained even when we take **X** into account, and it does so by positing that cases are interdependent to the extent specified in the **W** matrix. When we are doing spatial analysis, that matrix is going to have certain properties (like, it has to be symmetric and non-negative; the diagonals have to be zero; we may also require that it satisfy the triangle inequality, namely that the distance between two places can't be greater than the sum of their distances to any third place). But the method can be used for non-spatially organized interdependencies as well.[9]

Well, let's say that we fit this model for our crime data, with a single exogenous predictor—say, poverty (x_1), and we find that there is indeed strong evidence of this sort of diffusion! (See model 1 in table 7.5, which uses simulated data [R 7.7].) Our exogenous variable is significant, but the spatial lag is also very, very strong. That seems like a very theoretically exciting result!

But there's a very different way of thinking about the correlation in the spatial data, known as the error model, which posits that the problem comes because the *errors* (and not the values of the dependent variable) are correlated. That is, we start with a regular regression

$$\mathbf{y} = \mathbf{X}\boldsymbol{\beta} + \boldsymbol{\varepsilon} \tag{7.10}$$

9. However, if the matrix isn't symmetric, there can be complexities—literally, in the form of complex number solutions. For a network-based version of the same approach, see Friedkin (1998). And there are some versions of **W** (like **W** = **1**) that lead to real problems.

but posit that

$$\boldsymbol{\varepsilon} = \lambda \mathbf{W} \boldsymbol{\varepsilon} + \mathbf{u}; \tag{7.11}$$

in other words, the errors are correlated to the degree held in the weights matrix. It's the same basic idea as in eq. (7.9), but now applied to the residuals. Now here's the thing: a little math will convince you that these models are very similar—indeed, if you recall that the error is just the difference between the true and the predicted values of the dependent variable, you'll see they'd *better* be closely related—but it isn't that one is nested in the other.[10] That means that if someone suggests that you redo the model, but as a spatial error model (see model 2, table 7.5), the chances are good that if model 1 fits well, model 2 fits well. As it does.

Now it isn't the case that there aren't diagnostics that you can use to determine which of these is best . . . if either. My point is that if you just jump into fitting a spatial model, you can get encouraging responses that aren't due to what you think they are. In this case, neither model is "right." The reason that there seems to be spatial autocorrelation is that there is an important predictor (say, population density, x_2), that is omitted. It's correlated with x_1 and with y, and, like x_1, it has a spatial organization.[11] Fig. 7.5 below graphs the distribution of x_1 on this grid. So near places are likely to be similar on each predictor, and those high on one predictor tend to be high on the other.

The problem is that we didn't observe x_2 (population density). Now if we were to do an OLS regression omitting the second variable (see model 3, table 7.5), we'd greatly bias the value of the poverty variable (the true $b_1 = b_2 = 1$). No surprises there. And, if we were to include x_2, both the spatial lag model (model 4) and the spatial error model (model 5) would correctly tell us there were no spatial effects.

However, we might expect that the spatial models would help us get the right parameter of the included variable. Isn't that the whole point of the spatial statistics? To fix things? Let's start with model 1. The parameter for x_1 is half what it should be. But this is deceptive. We can't interpret a coefficient in a spatial lag model so simply, because the effect of changing an independent variable on the dependent variable in any one place

10. For an overview of the relation of these different models to one another, see Anselin (1988: 34–37).

11. The slopes of both variables on y are 1.0, and the two are correlated ($r \sim .5$). What I do is to make three Gaussian random fields, z_1, z_2, and z_3, on a 50×50 grid, and then create the variables $x_1 = f(z_1, z_2)$; $x_2 = f(z_1, z_3)$. Because the two variables share a component, they are correlated with one another. The equation that adds them together also adds a random disturbance of $N(0,1)$.

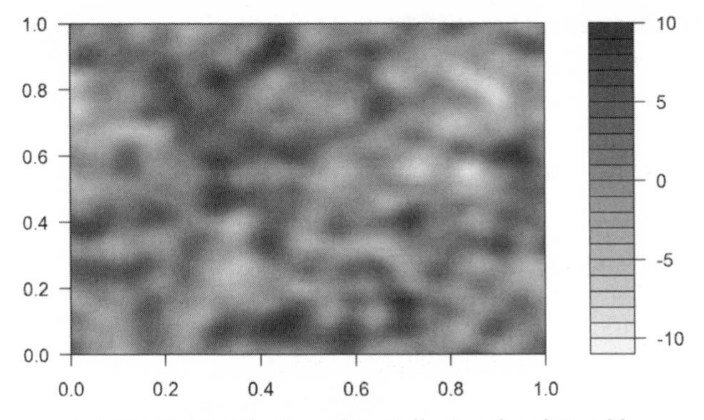

FIGURE 7.5. Distribution of Spatially Correlated Variable

is mediated by its effects in all the *other* places! We instead divide up the effects into those that are direct, and those that are mediated via other places. That's been put in the column for model 1.

It's hard to compare indirect effects to a normal OLS coefficient. But it's clearly telling us that most of the effects of a change in poverty would be mediated by the interdependence of places in terms of their crime rate. So the model is definitely thinking that the spatial correlation is pretty serious. What about the spatial error model? After all, this really fits the bill, for what we have are exactly spatially correlated errors due to an omitted predictor—though, of course, we didn't actually specify the spatial dependence correctly.[12] But although model 2 correctly identified the presence of spatial correlation (something we often first use diagnostic statistics to accomplish), it really doesn't change our estimate of the included parameter at all!

How can this be? It's because we forgot that the spatial correction is really only for OLS models that have the fortunate circumstance of "properly specified misspecification," in which uncorrelated variables can be ignored. When we use a normal OLS model, we don't worry about the absence of predictors that are uncorrelated with our included independent

12. Like many analysts, I simply had a dichotomous weights matrix that was equivalent to contiguity, while the Gaussian random field has a continuous function. However, when the scale of order in the field is short, this tends to work well enough in practice. And given that we rarely know the true nature of the spatial weights, any other guess can be worse, or better, we just don't know. It is worth emphasizing that even contiguity isn't always something we can ritualize, since we have to decide whether to consider two areas as contiguous if they share only a single point (also called "queen contiguity" for a raster) or if they must share more than one point ("rook contiguity"). Going with the first means that New Mexico and Utah, as well as Colorado and Arizona, are neighbors. Going with the second means they aren't.

variables. We assume that they "cancel out." But in a spatial regime, we do worry. And *that's* the problem the spatial error model is designed to solve. It will give you better standard errors of your parameters if you have the right model. It *doesn't* solve the problem—the one we should assume that we have—of the omission of *relevant correlated* predictors (LeSage and Fischer 2008).

The absence of *correlated* predictors implies a model that is like a combination of the spatial lag and spatial error models, known as the "Spatial Durbin Model." This model is written differently depending on how close one wants to keep it to other notations; in some ways of writing it, it just has *both* an autocorrelation term for the dependent variable y, *and* a lag term for the independent variables **X**. But LeSage and Pace (2009: 29f, 69) demonstrate that it can also be derived by imagining that there is an omitted variable that is correlated with the included variables. In that case, we can express it in this form:

$$\mathbf{y} = \lambda\mathbf{W}\mathbf{y} + \mathbf{X}(\boldsymbol{\beta} + \gamma) + \mathbf{W}\mathbf{X}(-\lambda\boldsymbol{\beta}) + \boldsymbol{\varepsilon} \tag{7.12}$$

Here λ is an autocorrelation parameter, and we see it not only determining the degree of autocorrelation in the dependent variable, but also the degree of lag of the independent variables. And γ is a parameter that corresponds to the degree that the included predictors are related to the excluded predictor![13] LeSage and Pace (2010) show that for this reason, the Spatial Durbin Model (SDM) can reduce the bias in an included parameter that is due to an excluded predictor.

But can it compensate for our ignorance of the true model (in this case, the importance of population density)? Not in this case. The Spatial Durbin Model (model 6) basically gives us the same estimate of x_1.[14] That's

13. The key is to take the spatial error model, and change the equation for the error term from $\boldsymbol{\varepsilon} = \lambda\mathbf{W}\boldsymbol{\varepsilon} + \mathbf{u}$ to $\boldsymbol{\varepsilon} = \lambda\mathbf{W}\boldsymbol{\varepsilon} + X\gamma + \mathbf{u}$. So it then taps how the included predictors would "predict" the omitted predictor, which ends up in the error term. There is another model, which Elhorst (2010) calls the Kelejian-Prucha model, that also includes the spatial error and spatial lag model as special cases, and it looks more like putting the two of them together. However, there are reasons to prefer the Spatial Durbin Model in practice. Both of these are nested in the "mother of all spatial models," what Elhorst calls the "Manski model," which includes not only the autoregressive term for the dependent variable, but also lagged terms for the independent variable, and autoregressive error terms. Why wouldn't you use this? Because of identification problems, the parameters are uninterpretable. Sigh. . . .

14. You will notice that the SDM isn't improving the fit of the SEM, and is basically recasting the results of the SEM as an autocorrelation among the dependent variables (the lamba here is the same as the lamba in the SEM). Also note that the estimation routine that I used returns not the values λ, β, and γ, but rather, λ, (β + γ) and (−λβ). I

because the omitted variables interpretation of the SDM is for omission of predictors that are folded into the error term.[15] What the model can do is return the estimate of the included predictor back to around what it would be if the omitted variable *weren't* folded into the error term.

I don't want to scare you away from these models; they're beautiful, and they make a lot of substantive sense. And there are, not surprisingly, many other, more complex and more elegant spatial models.[16] But we've once again encountered the same lesson we learned in chapter 4—nothing beats being right. The important changes don't have to do with what model we use—even for wonky data structures like spatial data. They have to do with having the correct specification. OLS will do well enough if you have the proper specification—it'll get the right parameters for the simulated data if we include both x_1 and x_2, even if it doesn't get the right standard errors of the estimates. And if we *don't* have the proper specification, we can use the spatial model to tell us what's going wrong, but we can't expect it to *fix* our estimates. Omitted variable bias can easily lead us to convert ignorance into knowledge: to go from "I'm not sure what all the relevant predictor of y are" to "I *know* that y has a diffusion-type process!"

So if we can't rely on the groovy models to fix our problems, what are we going to do? Rather than try to jump to *modeling* a spatial process, we should begin by *visualizing* it. We plot the residuals from an OLS regres-

---

then use simple manipulations to get the theoretically important values. This means that the tests of statistical significance I indicate here really apply not to the theoretical parameters, but to the conjoint ones.

15. I am very grateful to R. Kelly Pace for communication on these issues; the key is that their interpretation (2009: 29f, eq; 2.19; 2.23) of the SDM requires solving the autoregressive equation for the error term *including* the omitted predictor, as opposed to first solving it, and then adding all the other predictors, omitted or not, into the generating function of y.

16. Finally, it should be noted that it's also possible that we think that the values of independent variables *outside* our focal unit influence its value on the dependent variable, only they are attenuated. For example, the crime rate might be greatly affected by unemployment in any county, but also rather affected by unemployment in neighboring counties, and somewhat affected by the unemployment rate in neighbors' neighbors, and a teeny bit by the rate in *their* neighbors. This leads to a model that looks like the spatial lag model, only lagging x's instead of y:

$$\mathbf{y} = \lambda \mathbf{WX} + \mathbf{X}\boldsymbol{\beta} + \mathbf{u}.$$

If you can take a guess at the weights matrix, you can make this a pretty straightforward model to estimate (though you can, of course, take into account the correlated errors of the measures at the different places, which actually enter your individual observations because of the lags, but that's a real headache, and basically puts you back at the spatial error model. . . .).

sion on an actual map. This isn't "error" to be made to disappear—it's a clue.[17] Can you come up with a provisional guess as to what an important omitted predictor might be? Does it seem that you are under-predicting rural areas? Perhaps population isn't being dealt with properly. Get used to looking at maps of various distributions, until you start to have a feel for what kinds of people are doing what where. Then you'll have a head start when you are staring at a mysterious residual map.

One last thing: perhaps the most exciting avenue of progress is in the development of various models of local regression effects—that is, we allow the effect of some independent variable to change as we move across space. (This is often called "non-stationarity.") While you can always do a brute-force method where you use polynomial functions of variables tapping position (like longitude and latitude) as interactants with other variables, this isn't statistically defensible and so there are attempts to develop more rigorous approaches.

One way involves a moving window traversing the space, and is called "Geographically Weighted Regression" (Brunsdon, Fotheringham, and Charlton 1998). It turns out to be very bad at disentangling the local effects of different independent variables (Wheeler and Tiefelsdorf 2005). While there might be ways of using this judiciously (Páez, Farber, and Wheeler 2011), it may be a dead end. More complex ways of approaching this problem through spatial filtering (see note 17) might be worth pursuing, but for most of us, extensions of hierarchical linear modeling is probably more reliable, though also see Congdon's (2006) version of a mixture model for non-stationary effects. But what's essential for the practical user is not to jump too wholeheartedly on the most recent fix. Wait for things that have been proven in practice, because this is a hard problem to solve.

The First Law of Geography Repealed

I've emphasized switching to visualization as opposed to just charging ahead with modeling. It's important that you understand that there can be a major slippage between the two. Visualization shows you something about *outcomes*, while models are about the (supposed) *process*. Things that hold for the second don't necessarily appear in the first.

17. There is another approach to such data, known as "spatial filtering," which starts not so much by modeling processes as by purging the data of autocorrelation. This has been advocated, sometimes with great fervor, in place of the modeling approach of spatial econometrics. Because this approach has not yet made inroads into sociology sufficient for us to have a "feel" for its behavior, I do not devote space to this, but the interested reader may well take this seriously (Griffith 2008). But if my tolerance of fitting data without deriving hypotheses first has made you uncomfortable, hold on to your seat!

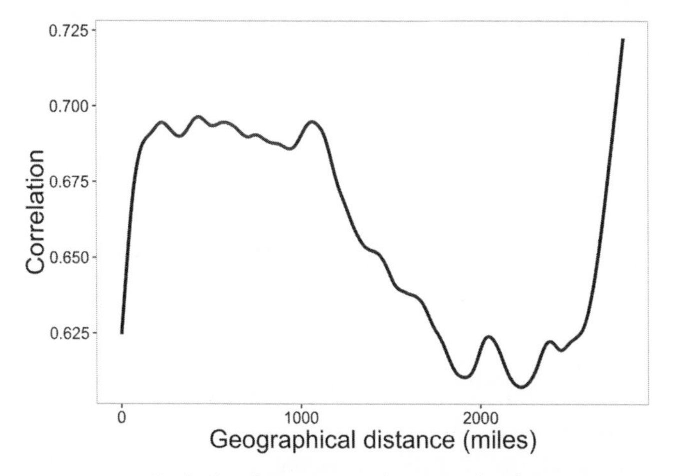

FIGURE 7.6. Similarity of Congressional Districts by Their Distance

Spatial statisticians are fond of citing Tobler's (1970) "first law of geography," namely "Everything is related to everything else, but near things are more related to each other." This isn't true, it turns out, or at least, not in the way we usually take it. It is certainly true that things *very* near are more closely related than practically anything else, but we often imagine that this implies that there should be a monotonic decrease in interrelation with distance; in fact, we often assume a pleasant Gaussian structure to this, a bell-curve falloff in all directions. This is wonderful, because it allows us to use some elegant statistics to solve problems in spatial statistics. Or at least, it would be wonderful, if it were true, but it's not . . . at least, not when it comes to *outcomes*.

Because it just isn't true that close places are more similar than far places. Let me steal a figure from Martin, Slez, and Borkenhagen (2016) demonstrating the overall similarity in a number of regression predictors across pairs of congressional districts (this is reproduced as fig. 7.6).[18] (Every pair of districts is an observation, which has an overall degree of similarity [the *y* axis], and a distance [the *x* axis]. We plotted all of these and then smoothed the line.) We might imagine that this will start high and trail off as districts get further apart. But in fact, as we see in the figure, although that's somewhat true for pairs of districts of medium range, districts that are very close are quite *dissimilar*, while those that are very far are *most* similar. Why? The only places where there are districts

18. Chad Borkenhagen made up this beauty and it's used by his kind permission. Thanks, Chad!

whose centroids are really close to one another are in cities, which tend to have very different people all crammed together, but also to be segregated.

Okay, but why the reversal on the right side too? This is because as a result of the doctrine of manifest destiny, the United States has the enviable position of occupying a nice horizontal swath of a continent with two large vertical coasts facing different oceans. Although the denizens of these areas are not wholly interchangeable, there is (as any Midwesterner can tell you) something decidedly similar about "coasties," whether they hail from the "east" or the "left" coast. Thus although the 48 contiguous states of the US look vaguely like a rectangle, in social terms it is perhaps more like a cylinder that we would form were we to detach the country from the Earth, and join Portland, Maine, to Portland, Oregon, on top, while overlaying Disneyland and Disney World on the bottom.

We can expect that most of this pattern is driven not by autocorrelation processes but by unobserved predictors—there's a lot that coasties have in common, and a lot that divides urban dwellers from one another. Where there *are* real autocorrelation processes, there's a good chance that there are many different ones all operating . . . and that they operate at different scales! To return to our example of predicting crime rates on some block, we'll see that even the same pairs of cases and variables are likely to be connected by very different processes, some of which even have opposite signs. If crime goes up anywhere in the city, that may divert police from our focal block, and lead to an increase in crime (or a decrease, if the police aren't around to take crime reports!). If there are shootings in adjacent blocks, that might lead people to avoid the streets at night, leading to increased car thefts. But if a professional car thief nabs a Honda from the block over, that might lead him not to steal one from our focal block.

In sum, when it comes to determining the range of order in some data, we can't simply go from visualization to process, nor can we make many clear a priori claims about the likely relation between different places (the **W** matrix). For some processes, contiguity makes a great deal of sense, as it correctly expresses conditional independences. If we are examining the spread of the ash borer beetle, we know that it can't get from Indiana to Iowa without passing through Illinois (most probably), or any of the other states that are contiguous to Iowa. But that doesn't work for, say, the diffusion of viruses. Because viruses can get from Los Angeles to Chicago without passing through the mountain states—they get on a plane. Here, we might need to treat all places as connected, only to a variable degree. We might even want to experiment by considering different ways of thinking through the logic of the weights matrix (for example, using transportation time, or volume of telephone calls, as the basis for a function). (For a nice example here, see the work of Ron Johnston, e.g., Johnston et al. 2004.) The closer we can get to knowing the concrete processes involved, the better.

How Do We Think of Space?

Now very often, we take our spatial data for granted: we have data "on" states, say, and so we go right ahead and analyze states. But the most exciting possibilities in spatial data analysis often involve combining data of very different types. Here I'd like to progress from the simpler to the more theoretically complex problems that arise in merging different types of data. To avoid confusion, let's distinguish between two ways of thinking about space. One is to treat it as a continuous surface of an infinite number of points, which, when used as a modeling strategy, is called a "field" approach (Anselin 2002; whom I follow here).

The other approach is to represent data as closed curves (which is called an "object" approach). For computational ease, these are generally turned into polygons—it's easier for a computer to keep track of a border that is a 94-sided figure composed of straight lines than one that involves complex curvature. This has a practical advantage over the field approach: if we don't have a clear functional relation between position in this field and the value on some variable (say, a function that relates population density to latitude and longitude), we need would need an infinite amount of storage to describe the infinity of points we have. Furthermore, much of our data comes in sets of points with boundaries—for example, census tracts, counties, or states.

But sometimes we have more than one way of dividing up a field into sets of polygons. We'll call each such set a *partition*, and it can be seen as a set of line segments ("boundaries") that divide the total area into smaller areas ("regions"). Your problem is that you can have data that come from different partitions. One partition *A* is *nested* within another *B* if none of *A*'s boundaries cross any of *B*'s regions. Thus counties are nested within states, and, therefore, variables that are measured at the county level (e.g., number of health clinics per county) become attached to a "finer" partition than those measured at the state level (e.g., does the state have an income tax?), state being a "coarser" partition. To include both in the same analysis, we need to reach a common partition, and there are two ways of doing this: lump or split. We can lump counties together into states, or split the state up into counties. Which should we choose? The answer is easy. We should only lump when our dependent variable comes from the coarser partition, and we aren't in a position to do any fancy adjustments for the non-independence across units in the finer partition. Otherwise, we split.

Non-nested Partitions

But more generally, we may face non-nested partitions. For example, you might get data on townships, school districts, and water districts, all of which are different and overlapping. What do you do? We still have the same basic choice: lump or split. But now the answer is simple: make like a banana and split!

So in fig. 7.7, top, we have two partitions, *A* and *B*, with *A*'s lines solid, and *B*'s dashed. To show where one partition's regions are cut by the other's boundary, I put the labels on the lines in such cases. So, what can we do to make these data analyzable?

For one, we could lump. (Here see Slez, O'Connell, and Curtis 2017 for a general method.) We make maximal units so that in our unit, no lines cross. In that case, it means producing the partition in fig. 7.7, middle. Clearly, we've lost a lot of information. Or, we can split. Then we produce fig. 7.7, bottom, where $C2 = A2 \cap B2$, $C3 = A3 \cap B3$; $C4 = A4 \cap B2$; and $C5 = A4 \cap B3$ (where $\cap$ means "intersection").

All of this works very well if our data is data that intrinsically applies to the level of the object. For example, the political party of the governor is the same in every little bit of California. But sometimes that isn't true. For example, one of our partitions might be the total number of hospitals in a county. How then would we split an area into two? *Adding* is no problem: we know how to add the number of hospitals in one to the number of hospitals in the second. But how do we know how to divide them up when we split?

The first answer (not necessarily the last) is that we simply divide things up proportional to area. If we split *B*2 into *B*2a and *B*2b, and *B*2a is twice as big as *B*2b, then we split up *B*2's "stuff" by giving two thirds to *B*2a and one third to *B*2b. Does this introduce error? Very little, for most of our purposes.[19] It's worth adding that kind of error if it allows us to avoid throwing away information by lumping. But we can do better by smooth-

19. Why? First, in most cases, we're going to weight our cases by the population in each. (If you're *not* going to, then you have to figure out other ways to keep the weighting stable, but you'll probably need to go back to rethink your units anyway. . . .) And that means that for many non-spatial analyses (for example, just running correlations across two variables), the results you'll get from data that come just from one partition will be exactly the same as those that will result from the smaller, split, version of your partition. (This enjoyable exercise is left to the reader for the case of the correlation coefficient.) It's true that if you are doing analyses involving weights by distance, your results will change, if only because you're now introducing a bit of distance inside a previous single unit, but they rarely change much.

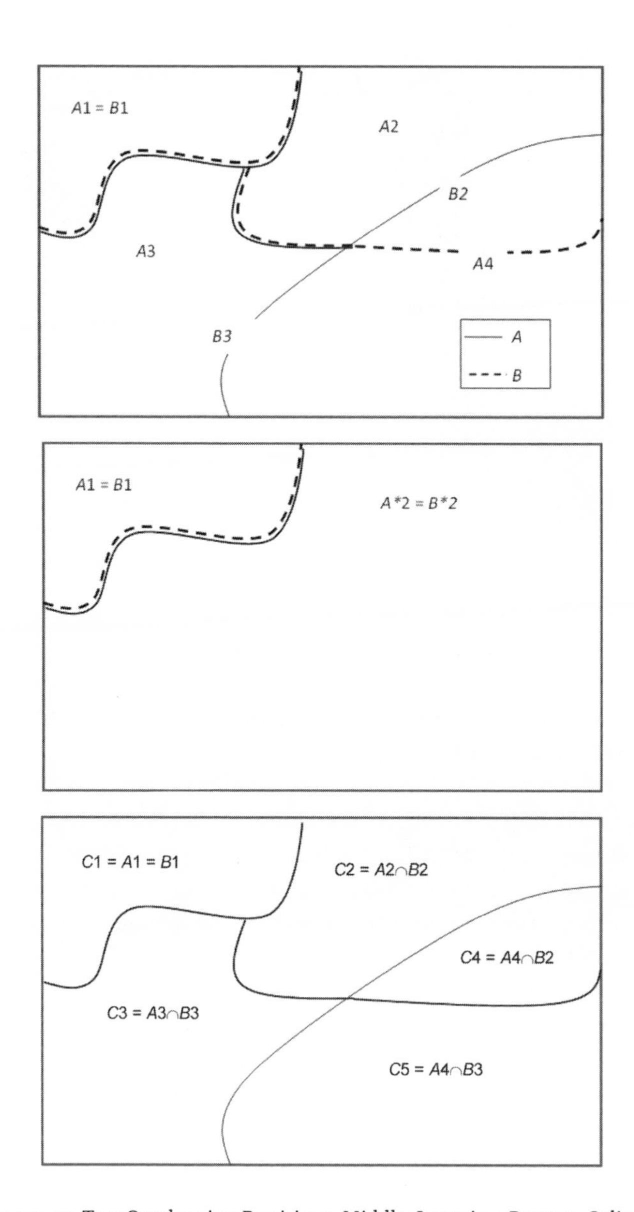

FIGURE 7.7. Top: Overlapping Partitions. Middle: Lumping. Bottom: Splitting

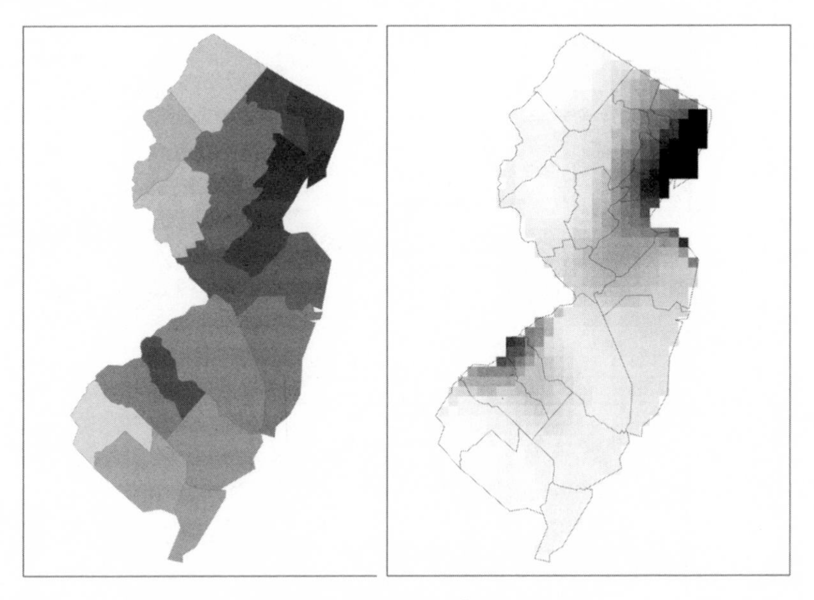

FIGURE 7.8. Two Maps of New Jersey

ing. That is, even though our data come in polygons, we approximate a continuous field underneath it.

The basic idea of smoothing is simple: if our polygons contain *aggregate* data, we're seeing jumps at the boundaries between objects that probably don't correspond to reality. Fig. 7.8, left, charts the population density of the state of New Jersey by county. Fig. 7.8, right, gives a map that seems to have a much finer dergree of resolution. But actually, the data is the same in both!

The data are smoothed according to a reasonable assumption, which is that (for example), the population density in Camden county (the darkest one in the southern half), is relatively high on its northwest side, where it abuts Philadelphia, which is an area of *higher* density than itself. Contrarily, its population is going to be relatively low on the southeast side, where it abuts Atlantic county, which has a lower population density. So if we assume that the population distribution doesn't care about county lines—an assumption we know isn't quite true—we can do a better job determining the probable density at any point.

How do we do this in practice? There are different techniques, depending on whether one has complete data or gaps, whether one has point data or polygons, and it's worth investigating a number of different techniques: not only simple smoothing (like a moving average), but also pycnophylactic interpolation (especially useful if you are merging different partitions), and Kriging (which has some ties to the random field per-

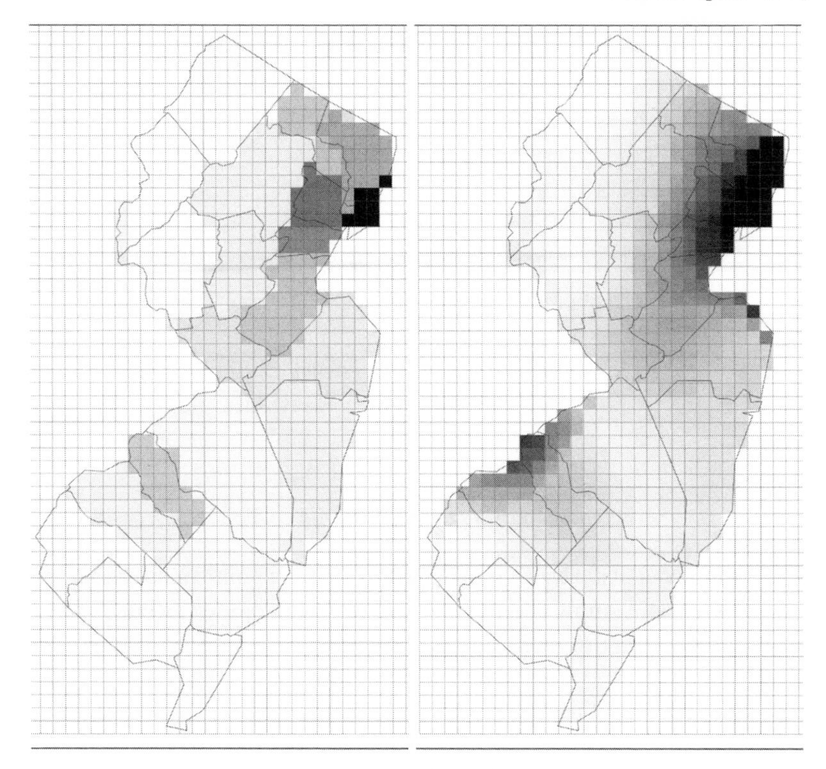

FIGURE 7.9. Rasterization and Smoothing

spective). If all you want to do is to smooth a map, you can use a packaged routine. But what we see in fig. 7.8 is an approach that can create a finer data set than we had going in.

We first deliberately introduce a finer partition than what we have in our data (or than we have as a result of "splitting" as above). This is called "rasterization"—we impose a grid of squares (sometimes dots), and attach our data to them. Fig. 7.9, left, shows a rasterization of the county level data: each raster is assigned the value of the population density of the county that contributes the majority of its area. We then replace the value in each raster with an average of the values around it (for example, the 8 "queen contiguous" members). Note that we have to include the counties in *other* states that border New Jersey! Then we re-weight so that the sum within any county is correct. Fig. 7.9, right, then shows the results.

This process of rasterization and smoothing can be used to re-allocate observations—most importantly, individuals—in a way that is more likely to be correct than assuming that they are uniformly distributed across our polygons. Compare our smoother versions to a finer map of the population density of New Jersey. It isn't perfect, but it's better than the

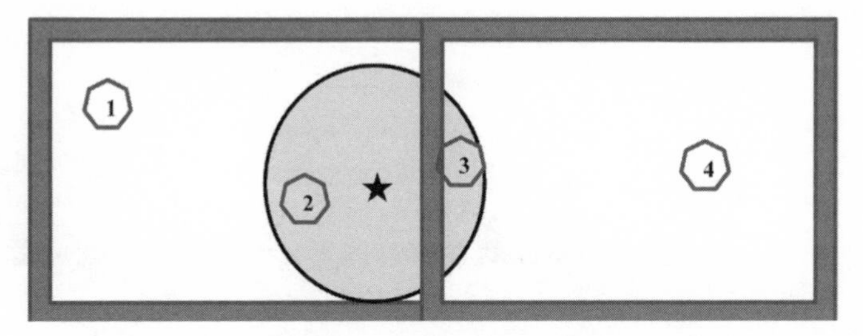

FIGURE 7.10. Catchment Areas and Neighboring Units

unsmoothed data. And that increased accuracy can be important when we are trying to merge two very different types of data: polygon data and point data.

Points and Polygons

Sometimes we have data that is linked to a single point in space. For example, we might know the location (in latitude and longitude) of grocery stores, or environmental contaminants, or where shootings took place. We are often interested in seeing how these are related to individual outcomes (for example, are people who are far from grocery stores, or close to contaminants or murders, of lower health than similar people elsewhere?).[20] For the purpose of brevity, I'll call these "amenities," although as the examples show, they aren't necessarily pleasant. In any case, we are often tempted to link point data on amenities to other data, like that on individuals, by counting up the number of the amenities in any polygon (for example, the number of churches in a city), and treat this as a covariate of all the people who are "in" the polygon.

In many cases, this can lead to unanticipated complications. Geographers consider each of these amenities to have a "catchment area," a surrounding region in which they are effective. For example, people might really only attend a church less than 20 minutes away, or go to a grocery store within five blocks. So now consider an amenity (fig. 7.10) with a catchment area—say, a grocery store. We're looking at two polygons (or "areal units"), and the store is represented by the star. Note that although

20. We also sometimes try to use them to predict other point-like variables, such as events. An often ignored problem is that social scientists are disproportionately interested in *portable* events, like demonstrations. A demonstration doesn't break out where the demonstrators happen to be. They get to choose where to have it—and the characteristics of the people who live in this place are often not the same as those of the demonstrators. Because this is such a problematic type of analysis, I won't even give it attention here.

the store is in the left unit, its catchment area spills over into the right one.

Let's take the first problem—if we only know which unit a person is in, and not her position within it, we are likely to have patterns of correlated errors if we use "number of stores in your area" to predict something (say, obesity). In this case, because we can't tell that person 1 isn't in the catchment area, while person 3 is, we're going to under-predict person 1's obesity, and over-predict the value for person 3. In other words, we'll tend to have negatively correlated residuals for neighboring areas.

And that might lead to certain soluble statistical problems, but that's not the sort of major headache that we're going to have.[21] A bigger problem is that in many cases, the location of our amenities is non-random across boundaries, and is related to variables you are interested in. You might wonder why, given how little of *anything* there is right around the town of Wendover, Utah, there would spring up *another* town, just a mile west of it (West Wendover). The answer is that *West* Wendover is over the Nevada border. While Utah was controlled by Mormons, who disapproved of drinking, gambling, and whoring, Nevada was controlled by Yosemite Sam, basically. All of Wendover was in the catchment area of West Wendover.

Thus the problem with simply assigning people and amenities to the same units is that this really makes sense only when our units have very "hard" borders—like countries (and not those in the EU!). Then, catchment areas may stop at borders. But more generally, the people who are affected by an amenity aren't necessarily in the same areal unit as the amenity.

Most important, a large number of Americans eat breakfast in one city or town, and lunch in a different one. Their jobs are in the second town, their homes in the first. Yet they are one person. Their work (and their value added) gets assigned to the first place; their personal income to the second. And these asymmetries are systematic, and related to the placement of amenities. Further, these places have a lot of people.

Where we have a very large number of amenities to our areal units, we are less concerned about this. But where the number is relatively low, we might want to take seriously the plausible catchment area of these amenities. And if we've used some form of smoothing to move our persons around across our units, we can probably get a better estimate of the

21. It seems possible that rasterizing at a scale that is so coarse that catchment areas are not much greater than the raster size could lead to a checkerboard pattern of over- and under-predicting, and hence resonances between different variables. But that seems an unlikely scenario.

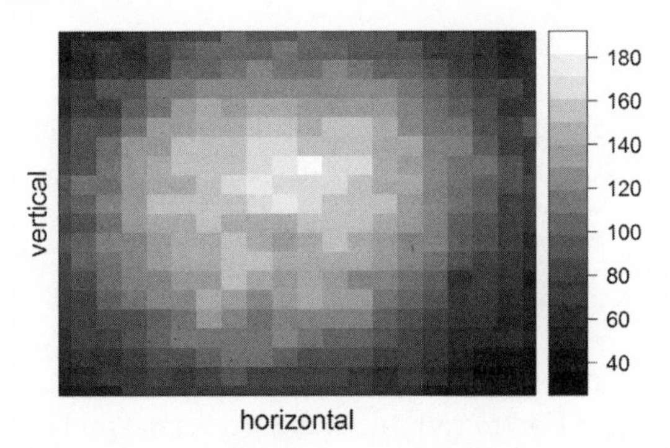

FIGURE 7.11. Edge Effects in Catchment Areas

actual degree to which the people in any unit are actually exposed to the effect of the amenity (able to get to the grocery store, say).

A final problem with the location of such amenities pertains to a class of problems known as "edge effects." However, some edge effects are substantive—they really happen out there in the world—and others are artifacts. Imagine people living in areal units arranged in a checkerboard, with amenities scattered across. (And you can see some simulated data [R 7.8] displayed as fig. 7.11 here.) The catchment areas of amenities are at least as big on average as the squares. Someone living in a central square can be in the catchment area of an amenity in her own square, or in one to the north, or one to the south, or to the west, or to the east. But someone living near the western edge necessarily loses the possibility of being in the catchment area of amenities to the west—because there's nothing there. And someone in the southwest corner loses both south *and* west.

The shading in fig. 7.11 is for the intensity of some dependent variable that is a function of two amenities. The intensity increases the closer one is to the center. It's worth emphasizing that this is a "real" effect, not an error. But now imagine that we are selecting some regions out of a larger set—for example, looking only at the counties in Louisiana. If we were to try to take the catchment areas of our amenities into account, we'd be introducing a potentially systematic bias, by giving all the counties (actually, in Louisiana, "parishes") that abut Texas, Arkansas, or Mississippi a deflated count.

And the same goes for, say, attempting to see the effects of shootings on school achievements in the city of Chicago. The neighborhood of West Pullman doesn't usually have the highest rate of violent crime. But it is very near the suburbs of Harvey and Riverdale (not *in* Chicago), which have extremely high crime rates. Ignoring that can lead to serious bias.

In sum, using spatial data is hugely cool, especially because it makes it possible to merge data of very different types. But it's worth thinking through the logic of what you are doing, and using interpolation and smoothing to come up with the best guess as to where people are before you go about making conclusions.

Conclusion

Temporal and spatial data are a field day for statisticians. They find all sorts of problems and come up with all sorts of adjustments. And good for them! But that's not your problem. Your problem is that these types of data are giving you hints as to likely unobserved predictors, and you need to listen to these hints, as opposed to ignoring them. Because the cool thing about being a social analyst is that you're studying something that has its own principles of organization, and which is constantly constructing and delivering data to you on the basis of these principles. That's an opportunity, although also a complication. In the next chapter, we pursue this to principles of organization that are contained in social structures themselves.

* 8 *

When the World Knows More about the Processes than You Do

Map: In the previous chapter, we dealt with the fact that there is information about the similarity between our cases that is held in their spatiotemporal location. In this chapter, we'll see something similar—how social structure holds information about our cases. That social structure could be *family*, or *friends*, or other interactions our cases have with the world. When we wrongly assume that our measures are perfect, and that we know all there is to know about the cases as *individuals*, we'll end up making very exciting—and totally untrue—claims about social structure.

After demonstrating this, and getting used to the idea that social structure contains information about individual characteristics, we'll consider what happens when we construct a sample using such structures, which is what we do under network sampling. And here I'll consider one set of claims about such sampling ("Respondent-Driven Sampling"). The nonindependence between our cases can mean that what seems to be knowledge about relations between variables has more to do with the relations between our cases . . . and this in turn has something to do with unobserved heterogeneity on the individual level.

We'll then turn to ways of dealing with this sort of interdependence in network analysis, quickly replicating the logic that led the subfield to be extremely enthusiastic about one approach to network analysis ("Exponential Random Graph Models"). We'll then see why this is a really poor choice for a ritualized, general solution. And then we'll close by reminding ourselves that the "simpler" network "measures" also don't work well when ritualized.

Making Knowledge out of Ignorance

We've seen that a number of our statistical procedures have a corruption, whereby we transmute ignorance into knowledge—the *less* we know, the more we *think* we know. (A nice example from the previous chapter is turning omitted predictors into evidence of diffusion via a spatial lag model.)

Now we're going to go on to consider an especially devilish set of related problems where this ignorance rises to the fore. These are ones that come not simply when we don't know something about the world, but when the actors who *do* know this are constructing the very data structure we use to make claims about them. I know that this is an awkward way of putting it. But I think it helps us see the similarities across different types of problems. We already saw one version of this in chapter 4, when we realized that using a principal's labeling of which kids are rotters might predict later arrests better than would more "objective" measures not because it is truer, but whatever makes the principal hate this kid might make cops hate him too. The complex structure of *relations* which we can't see gets misrecognized as something about *persons*.

Now we're going to turn to cases in which it isn't just that there are omitted variables that lead to bias. Rather, it's that the imperfections of our measures lead us to underestimate certain types of connections, and misattribute association. The problem is that often, the people going about doing things have better measures of what is important to us than we do. And if we don't realize that, we can come to truly wrong conclusions.

Let me give an example. Imagine that we are trying to figure out how "homogamous" by class marriage is. Somehow, we divide subjects into social classes, perhaps by asking them for their own subjective class identification, or perhaps by asking them their current occupation, and then coding this up into 8 categories, say. In either case, we then make each marriage a data point in a matrix with rows being husband's class and columns wife's class.

Let's briefly remind ourselves of why these measures are likely to be poor, at least, for contemporary US samples. First, lots of Americans don't like to talk about class. Second, even when they do, they have different ways of thinking about class—someone may identify as a "worker" when thinking about issues of production, but as "middle class" when thinking about consumption (Halle 1984). Third, if you force them to pick a class using *your* terms, some really don't know what you are talking about. Fourth, if instead of asking them for their own judgment, you try to pigeonhole them on the basis of "objective indicators," you're at the mercy of the quality of the information we have. Occupational codings, for one, are necessarily extremely inexact: in many cases, the head librarian for a research library is in the same category as the school librarian in a public elementary school; a day-trader and a hot dog vendor both end up as "self-employed," and so on. Further, there is a lot of volatility in these variables. Many people's income goes up and down. Some people cycle through periods of (usually unsuccessful) self-employment. And in most cases, we lack the information we need to really understand class (namely wealth, family wealth, relation with family, and so on). Fifth, if you are really trying to use these to get at class, static measures aren't where you want to be. We've tended to under-estimate the persistence of socioeconomic status by looking at men who are too young.[1]

And finally, for many purposes, we have to choose a number of classes. Who's to say there are eight classes anyway? Suppose there are only four! By splitting each one randomly (imagine in equal halves), we've just taken one half the homogamy in each and declared it to be non-homogamy!

So there are plenty of reasons to think that our measures of class are poor. How poor? We really don't know. Assessing this would require access to the "ground truth" that we just don't have . . . or haven't been willing to spend the time to collect. But we go ahead and make a matrix of the cross-classification of husband's "class" and wife's "class," and use *this* to determine the degree of homogamy. Chances are that we're not going to find perfect association (say, everyone on the diagonals of the matrix). But if we conclude that marriage in the United States isn't very homogamous, we're being illogical. Our results only set a *lower bound* on the degree of homogamy.

And notice that our misinterpretations come from an arrogant assumption that we have a better notion of our subjects' class than they themselves do. In no way do I doubt that Jack truly loves Jill, as a person, and Jill reciprocally. But if there *were* classes, I bet Jack can tell whether Jill is in his class, and she reciprocally, a lot better than we can from looking at survey data. That might be true even if Jack doesn't even believe that there *are* classes in American society! Actors can tell the difference between the nouveau riche and the declining gentry; between black sheep and climbers; and, to take Plato's metaphor, can identify the occasional gold child born by freak of nature to a bronze family. In fact, rather than assume that our data are right, and throw out the hypothesis of strong homogamy, we might be on somewhat better ground assuming homogamy and throwing out the hypothesis that our data are right. If Jack and Jill married, well then, they must be in the same class.

Of course, I'm not arguing that we should, on a priori grounds, assume perfect class homogamy. This is a thought experiment that reminds us that we have to bear in mind the imperfection of our measures. If there

1. Every Haverford graduate waiting tables and working on slam poetry, canvassing for a fundraiser, or working for a struggling NGO, seems to indicate that inheritance of socioeconomic status is low. But it's rare that we come back six years later and make sure that this guy hasn't thrown in the towel (and the mike) and joined Dad's investment firm. Getting them at 40 shows a higher estimate for intergenerational transmission.

are social processes that are driven by the *true* values, results based on the imperfect measures will imply biased conclusions. Let me now go on to show how this can really bite you in the ass.

Families

Race

Just as with class, people know more about their race/ethnicity than we do, or that researchers even *can*, with their crude measures. Especially given the one-drop rule in the United States (anyone with any recognizable sub-Saharan African ancestry *is* "black"), we have lots of data that is cruder than people's own way of thinking about race, which is, even in the US, attentive to many other things, including skin color (Monk 2015). This means that a sociologist can have data on a relationship between two people with the exact same biodescent, yet call it either "intraracial" or "interracial," depending on how the subjects choose to label themselves. And for the reverse, two people with the same racial label may have very different profiles of biodescent—profiles that might be correlated with other individual characteristics.

For a recent example, Davenport (2016) had the interesting idea of seeing how biracial youth identified themselves, and what characteristics of these predicted their choice. So she had data on incoming cohorts of undergraduates, who report not only their own race, but that of their parents. And she found that religious identification seems to predict which way these will go. Those with one black and one white parent lean black if they are Baptist, and those with one Latino and one non-Latino parent lean Latino if they are Catholic.

That might well be true. But going from correlation to causality would require that we really fix and reify race in a way that we saw in chapter 2 is a bit dangerous. We don't, in other words, have data on *grandparents*. Imagine that whenever there is an interracial marriage, there is also an inter-religious (or inter-denominational) marriage, and the religion of the children is a random choice between these, and imagine (for sake of simplicity) that, at the grandparents' generation, all blacks were Baptist, all Latinos Catholic, and all "whites" something else. That means that a youth with three black grandparents and one white (who might identify one parent as black and one white[2]) has a 75% chance of being Baptist,

2. You might think that because of the "one-drop rule" (see Washington 2011) this is implausible—the subject would identify both parents as black. But precisely because of the one-drop rule, the *parent* with one black parent and one white parent might have

while a different youth with one black grandparent and three white (who also might identify one parent as black and one white) has a 25% chance of being Baptist. The first might be more likely to identify as black than multiracial or white, but not necessarily *because* of the religion.

Indeed, sorting out different types of response error when it comes to racial identification is a total nightmare, and, after a great deal of interesting work, it seems that while there is evidence for context effects in terms of racial self-identification, we have a major David and Goliath problem. Random noise can swamp the signal here; all we can say is that people probably know more about their ancestry—and, in many cases, that of those they encounter—than we researchers do, even if we force them to supply us with check marks in our boxes.

For example, as one part of a fascinating line of work, Saperstein and Penner (2012) looked at the differences between how interviewers classified respondents in terms of race at two interviews a year apart. (The respondents were interviewed once a year; the models are looking at the labeling in one year, controlling for the previous year.) They argued that some of this difference was due to the stereotypes that the interviewers had about blacks and whites, and this (as well as ascriptive processes *outside* the interview) would guide how the interviewers "saw" race. That's because (for example) an interviewer who learned that in the past year, a respondent had been incarcerated was more likely to label the interviewee as black, controlling for the last year's labeling, than if the respondent hadn't reported incarceration. It's a fascinating notion, and it's supported by social psychological experiments. We shouldn't doubt that it *can* happen. The question, though, is how often it happens, and whether it explains the bulk of the change data they saw.[3]

As Kramer, DeFina, and Hannon (2016: 238) pointed out, we'd see similar results if (1) there is random error present and (2) the stereotypes point to statistical correlations.[4] That is, if blacks are more likely to be incarcerated than whites (and they are), subjects who were incarcerated might be

had three white grandparents. The thing about the one drop rule is that it's *both* rigid *and* impossible, as we all go back to Africa.

3. It's also the case that most of the racial fluidity happens in categories that actually are simply problematic for our vision of race, which is based on the white/black divide, then assimilating other categories (Asian? Or maybe, East Asian and South Asian? And, uh, what about linguistically defined Hispanic, which draws from two totally different biodescent groups? Is that a race . . . ?). Latinos sometimes are white, sometimes other, and those who switch from white to black and back are likely to have Afro-Caribbean ancestry (Alba, Insolera, and Lindemann 2016: 253).

4. One difficulty in adjudicating between different explanations is that all the critiques assume, if only for mathematical tractability, that each person has a "true" race—but this is what Saperstein and Penner forcefully deny, and on good theoretical grounds.

more likely to go from "white" to "black" than those who were never incarcerated not because the interviewer's "view" is "colored" by this knowledge, but because the initial report of "white" is more likely to be in error than is a report of "white" from someone who wasn't incarcerated.

It's a tough problem—how can we ever know? Kramer, DeFina, and Hannon (2016: 241f) have an idea—let's look for some variable that we would expect interviewers would *think* is associated with race, but isn't. But without doing specific research on interviewers' theories of race, it's a long shot to try to use a guess to peel these apart.[5] The key thing is to remember that the world contains information that sometimes isn't accessible to us. Just like Jack knows Jill's class, so the incarceral state might "know" race better than we do. And the same exact formal ambiguity arises when we look not across years, but across generations.

Non-Markov Time Processes

What we're learning is going to be a very general lesson: *any* time our data or our models are imperfect, and we think that the *world*—most obviously, the same actors we measure, but possibly other actors or institutions—might know more about things than we do, we can expect to find real-world associations popping up in our data as a *different* form of association.

First, let me give an example having to do with measurement error. Let's imagine that we have a sample of 1000 people from a world in which half the people are upper class, and half are lower class. The kids of upper class women tend to become upper class themselves, and ditto the lower class, though it isn't a perfect relation. In fact, 15% of the kids of upper class parents end up as lower class, and 15% of the kids of lower class parents end up as upper class. All sociologists have assumed that this is a "Markov process," meaning that children's status depends only on that of their parents, and not of their grandparents (let's assume perfect homogamy to make things easy on ourselves). You, however, suspect that social scientists have been wrong. And you've got this cool three-generation data that no one ever had before with which you can examine the inheritance of class, and see if there's a direct grandparent effect. So you do a logistic regression with both terms in, and get the results in table

5. They used marijuana consumption and sales, and show that whether the interviewer heard about this doesn't predict race the way it would if it activated a stereotype. But not only is there no reason to think that interviewers connect marijuana use to blacks (as opposed to white hippies), it turns out that the interviewers didn't have this information when they made the ascription (personal communication, Aliya Saperstein).

Table 8.1. Grandparent Effects in Class Inheritance

	Estimate	*Std. Error*	*z value*	*p-value*
(Intercept)	−.538	.105	−5.148	<0.001***
Parents	.649	.131	4.967	<0.001***
Grandparents	.351	.131	2.693	0.007**

8.1 (R 8.1). Holy smokes! That's a pretty strong result! This is going to really change our interpretation of class!

Except it isn't true at all. These data come from a process in which each generation replicates its state with 85% probability, just like I said. How does this come to be? Simple—there is reporting error. The data are only 75% accurate; 25% just get put down as in the wrong class. Whether they make a mistake in reporting, whether the recorder puts something down wrong, whether it happens in the data entry stage, doesn't matter: all I'm saying is that there's random error.

You might imagine that the inclusion of random error only biases our results to zero, not upwards! But as we see, that assumption isn't always true. That's because at time 2, some of the parents who say they're lower class are making a mistake. They're really upper class. And there's two other things about them that are important. First, they are more likely to have had upper class parents than the true lower class—and more likely to have parents who *identify* themselves as upper class. Second, they are more likely to have upper class children (and children who identify as upper class) than the true lower class. What seems to be an effect of grandparents on grandkids is nothing of the kind—it isn't tapping a *process* so much as *information* that is being stored in the world better than it is being organized in our data. I am convinced that most of our supposed "grandparent" effects are simply driven by the presence of error in our data . . . or misspecification. And that's true even though (as George W. Bush could probably tell you) there *are* certainly real grandparent effects out there in the world![6]

And these spurious effects can emerge when there is no *measurement* error . . . so long as there is error of another kind in our model. So let me give an example having to do with specification error. Usually we assume linear relationships in pretty much whatever arbitrary scale we happen to have on hand. There are exceptions—in many cases, we'll test quadratic terms for *age* variables, and often take the logarithm of *income* or *wealth*.

6. Indeed, as this goes to press, I find a wonderful study by Hällsten and Pfeffer (2017) which explicitly deals with these concerns (see pp. 340f, 344, 346) and develops an "upright" research design.

Table 8.2. Grandparents Effects Again

	Estimate	*Std. Error*	*t value*	*p-value*
(Intercept)	.164	.015	10.639	<0.001
Parents' Income	.586	.030	19.542	<0.001
Grandparents' Income	.084	.030	2.802	0.005

N = 1000; for all variables, mean = .50; sd = .30; R^2 = .408.

But usually we assume linear relations. If it's not so, we are, we figure, only biasing our slopes to zero, so it's a conservative error.

Imagine that as before, we have three generations of a Markov process—now we are measuring average income, which is partially hereditary (for whatever reason). The correlation between the first generation and the second is pretty high, r = .58 (R 8.2). And the correlation between the second generation and the third around the same, r = .64. If this were in fact a Markov process, we'd expect an r between the first and third times of .58 × .64 = .37. But you notice that the r between the first and third periods is a bit higher, .43. So you decide to include time 1 (grandparents' income) in a regression of time 3 (children) on time 2 (parents); see table 8.2. Yes, the coefficient for grandparents' income is a lot smaller than it was in our first exercise, but it's definitely significant, good for a *Social Forces* at any rate.

But hey, didn't I say this was a Markov process—that, by definition, time 1 only affects time 3 via time 2? So what's up? What's up is that the relation between these variables happens to be non-linear. Now in this case, if you had looked at scatterplots first, you would have seen this right away (take a look at fig. 8.1).

And, of course, if you didn't look at the scatterplots, you deserve anything you get. Further, bear in mind that the assumptions that are critical for your models don't usually have to do with the simple bivariate relations like this. They really have to do with the relations *conditional* on the other included predictors. You can have a relation that looks pretty linear, and a big fat cloud, at the bivariate level, but once you control for something highly associated with one of the variables, that relationship, insofar as the Mr. Computer sees it, can be different. That's why you might need to look at residuals, and not just the raw values. But the raw values are at least a place to start (see chapter 2).

We've looked at this in the form of relations between generations. But they also hold for relations between the same person over time. As Jeremy Freese pointed out to me, kids' first grade reading scores predict their fifth grade scores, even taking the third grade scores into account. That's because the A+ kid who was sick when he took the test in third grade, and

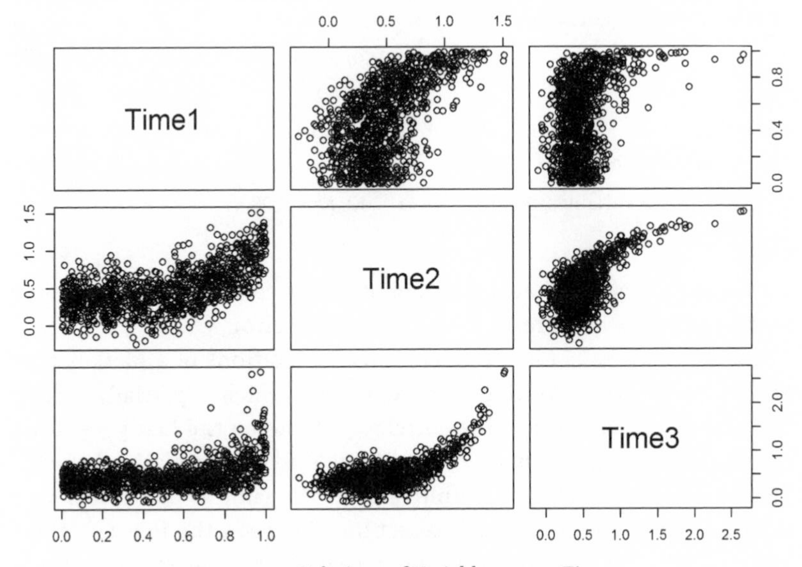

FIGURE 8.1. Relations of Variables across Times

messed up, probably won't be sick again in fifth grade. Get used to this idea, and look for it.

Social Networks: What They Know

Networks and Error

Thinking in terms of the information stored in social structure gives us a better way of approaching the complexities of network data than one will get from simply learning this as a problem of statistical autocorrelation. After introducing some of the problems that arise with network data, I'm going to argue that the way—in mainstream sociology—we've decided to think about this is really screwy. I want to walk through some of the mathematical issues, and then point to a serious slippage between what we should be aiming for and the current ritualized approaches we use. In a nutshell, just as with space, we've thrown together some different problems and slapped a label on them, and then paid a lot of attention to issues that are largely tangential for most of us. There will, however, be a happy ending here, which is that the people at the forefront of these methods have also pushed us to check our results carefully and attempted to incorporate into their approach just the sorts of critiques I'll be raising.

In any case, when it comes to worrying about network data, we often talk loosely about some sort of "non-independence" or "correlated errors" as if there were a single problem, for which there could be a single solution. That is, so far as I can tell, totally untrue. And I think the best way of thinking this through is to imagine that instead of social network data, we have geographical data. It's the same basic thing—a case-by-case distance matrix, such as we worked through in the previous chapter—but we have better intuitions for geography. Here we know there are three very different problems.

The first problem, we recall, had to do with the non-independence of our *sampling* of our cases. Here too. There are three very different ways of collecting data when it comes to network analysis. In the first, we sample a set of individuals, and then query each about his or her local network, which we call an "ego network." If we're just asking them about unidentified others, we often collapse this to individual-level variables, such as "number of friends" or "percent friends who are from work." Since we don't try to embed this information in a larger network, we don't think about it in network terms, and everything is simple.[7] However, in some cases, we ask people to construct a bit of a social network, by asking them to report on relations among the contacts (for example, "which of these people know each other?"). While there are some ways in which this induces network dependencies among the observations, most people still collapse this to an individual-level variable, because there's so far no evidence that it is worth doing anything special to deal with the statistical issues.[8]

In the second way of gathering data, we sample some sorts of *groups*, and then trace out the networks among the members. For example, we might choose classrooms in schools. One form of non-independence comes because two students in the same class either both do or both do not get into the sample. If we are trying to infer to the general population (e.g., all classrooms in the US), our estimates of *anything*—including network structural parameters—should be expected to have larger standard errors than we'd otherwise get. It's not intrinsically different from the case in which we pick one hundred towns at random from the entire United States, and then sample from these towns.

Now, if we're also using these groupings to collect network data, there's another way of forcing non-independence among the observations. If kids have friends, only some of whom are in their class, but we sample on

7. An exception is found in the "scale up" approach pioneered by Bernard and colleagues (1991), and still pursued today.

8. The fact that the ties between one's friends aren't independent might mean that the uncertainty of these measures is greater than you would compute for individual-level variables. But worrying about this . . . see the discussion of Diderot in chapter 1.

classes and only ask them about their friends, we're doing a sample of the full friendship network of the sampled egos.[9]

The third way of gathering data is when we trace out a network, doing what is called "snowball" sampling. Here the problems are much more severe, as we know that our sample isn't just less efficient than a random one (increasing standard errors), it is also biased. Start with a Catholic who likes Heavy Metal music and is a vegetarian, and chances are, you'll get more Catholics, Heavy Metal fans, and vegetarians than you would in a random sample. It's this kind of sampling (snowballing) that should really scare the bejesus out of us, because we're basically letting people choose the sample for us, and they haven't really told us their principles in doing so.

This sort of non-randomness of network sampling can lead to overly deflated estimates of standard errors even where we have the right model. But a bigger problem has to do with the likely correlation of unobserved, relevant predictors. Just as there are lots of attributes that people from places close to one another tend to be similar in, so too those who are tied are likely to be similar in ways that we haven't measured. And that can lead us to convert ignorance into (seeming) knowledge. In particular (and this is the second type of problem), attributes of individuals—for example, differences in thresholds at which they consider a tie to "exist"—lead to commonalities that can seem to be about relations or structural positions.

The third problem is when the existence or non-existence of one tie (or some subset of ties) influences the existence or non-existence of other ties. To separate this from *statistical* non-independence, I'm going to call it "existential non-independence." It can be a bit confusing that this problem can shade into the other problems, and so we have to tread carefully to make sure that we are focusing our attention to handle the substantive problems, and not assuming that there's an obvious fix. I think it helps to make a distinction between two kinds of existential non-independence, one having to do with error, especially due to nodal characteristics, and the other not. Let's start with the first: to take an example of a generic error process that leads to non-independence, think about having 40 people rating 10 films. We model their ratings as an effect of the prestige of the film directors and the budget. Well, statistically, we have fewer true independent observations than we would if 400 people each rated a

9. Finally, sometimes people see the fact that the same alters are linked by our set of egos as a form of non-independence of sample but I think that's confusing to most, as we can imagine having the entire population in our sample, without that fixing the problems these folks are worried about. So I'll explore this in terms of the existential interdependence of ties, not the interdependent sampling of nodes.

single film. So we often do some relatively minor fiddling to inflate our standard errors appropriately. There's a way in which this is true for network data as well.

When the errors due to nodal characteristics are interpretable, they're something we need to think through on the level of real-world processes. Imagine that people have different thresholds at which they call other people "friends." Those with lower thresholds make more nominations: those with higher ones, fewer (all other things being equal, which we'll assume). If we look at unreciprocated nominations, we're more likely to find someone with a high threshold receiving these than someone with a low threshold. So a lot of measures will tell us that those with a high threshold are more "popular" than those with a low one. Now it could be that this is correct—in fact, that might be why some people have a high threshold and others low ("beggars can't be choosers"). But it's really hard to figure out how we can ever tell the difference.[10] Any statistical technique that isn't based on a theoretical model of this process can't "correct" for the response differences without eliminating the popularity differences.[11]

Other forms of non-independence across ties come neither from errors nor from individual-level characteristics, but from actual interdependencies across dyad formation and retention. For example, if Archie asks Betty out to prom, he can't also ask Veronica, so these two dyads are interdependent. Even more, as we'll see below, if Betty says yes, Reggie therefore can't ask her out, and so if he asks Veronica instead, Archie's asking has flowed through Betty into affecting Veronica's prom choice, even though Archie had nothing to do with her! This is a statistical issue that means that we can't consider the factors that lead Reggie and Veronica to go out by just looking at *their* covariates.

The same thing arises if Reggie just doesn't answer the survey.[12] There's something we don't know about the Reggie-Betty tie, and it's related to the thing we don't understand about the Reggie-Veronica tie. And it

10. You might think that where there are limited numbers of choices one can make, we can get some traction on this by assuming that everyone has the same virtual threshold. And that maybe people are prioritizing their better friends. Good thinking! Unfortunately, most of the network data I know of doesn't have most people hitting the maximum of nominations, and it isn't clear how they proceed to order their nominations. But it's a good problem to work on. . . .

11. And by "modeling the process," I would include making a priori assumptions about the distribution of the reporting and/or popularity effects.

12. I'd like to note that there's recently been some excellent work on the effects of missing nodes (Smith and Moody 2013) and edges (Wang, Butts, Hipp, Jose, and Lakon 2016); these are the places to start if you are interested in pursuing how structural aspects of networks may be mis-characterized in the presence of error (also see Wang, Shi, McFarland, and Leskovec 2012).

arises if Reggie goofs around on the survey, or if he simply happens to be mean, and tends to downgrade his reports as to how smart everyone else is. It's these sorts of statistical problems that most of the mathematical sociologists have focused on. And we'll get to them. It's the cooler problem, I admit. But the first two forms of error—sampling and omitted predictors—can lead to big biases in our conclusions. So we don't want to ignore them.

But guess what? There's a way in which network studies easily reparameterize fully *random* error as it if were a network effect! It will make sense to you, given what we've seen already about the grandparent effects, but it's not something most network researchers have wanted to confront.

Influence and Mere Error

Let me give you an example from my own work, where I was a second author with King-To Yeung (Yeung and Martin 2003). Embarrassingly, I was the methodologist for one part and while he got his part right, I got mine wrong . . . or somewhat wrong. It looks like our conclusions are right—but now I see a problem so clearly that I didn't see (or maybe I didn't *let* myself see) before.

Here's the setup—we had 60 groups of people who lived together. In each group, every person was asked to say which other people in the group were any of a long list of adjectives: strong, loving, passive, charismatic, and so on. Then they were also to say which of these applied to *themselves*. Here was King's idea: let's compare self-image to how others see one, and see if there's a relation—and also see whether this relation might vary in the ways that the great social psychologist Charles Horton Cooley had predicted. What a cool way to see if there's an internalization process.

Of course, you might say, just because there's an agreement between self and others doesn't means it goes from the others to the self. Maybe the person they see as dominant says he is dominant because he really is, and he knows it, and they know it. Well, this is where King's idea was so cool—we had multiple waves of data, three in fact (though only two with a lot of information). Do we see people changing to follow the ideas of the group? That is, imagine that you don't think that you are a "loving" person, but everyone else in your group thinks that you *are* loving. Over the course of a year, between when we do these measures and we re-survey your group, you come to accept their view of yourself and you report it.

That's what we found. What I didn't really accept at the time was that this sort of pattern could also be due to random error. That is, suppose someone really is loving, but she accidentally answers wrong the first time. She's more likely *both* to be seen as loving by everyone else in the group, *and* to transition to seeing herself as loving next time she is asked, than would be someone who had reported that she wasn't loving but the group *agreed* with her. In this case, the group "holds" the knowledge about the person's *actual* state, in the same way that your grandmother's religion holds the information about your *real* ethnicity, or your wife holds the information about your *real* class.

This is a very general feature of peer influence models—they tend to take any measurement or specification error and turn it into influence. You are a Deadhead, say, and your friends are. First time you get the survey, you actually say you *don't* like the Grateful Dead (maybe you are stoned at the time and not thinking at your sharpest). Yet all your friends like the Grateful Dead! The next time you're asked, you do too! Aha! Influence!

For another example, Seth Abrutyn and Anna Mueller (2014) were interested in the contagion of suicide attempts and thought ("ideation"). Now there's really very good reason to think that there *is* a diffusion here (in fact, it's basically undeniable that there's *some*), but they understood that there might be selectivity at work. So they tried to look only at those who didn't have any suicide ideation or activity at time 1, to see whether those who had friends who experienced this ideation between time 1 and time 2 would be more likely to be suicidal at time 3. Sounds good, huh?

Well, it was good, as good as what others have done, but still not good enough for us to be sure that there is contagion in these data. The problem we have is that the "false negative"—the person who really did think about suicide, but forgot she did, or punched the wrong answer, or lied—is more likely to (seemingly) transition to the suicidal category. And she's almost certainly more likely to have friends who are suicidal. Or just depressed (which we haven't selected on, note well). Her friends, then, store some information about her that we don't have. Now in this case—because rather than compare those with and without time 1 ideation, they simply selected only those who reported absence of ideation—we might imagine that there will be countervailing errors from people who, at time 2, report ideation though they don't actually have any. False positives and false negatives may cancel out, but there's no reason to assume that they will, or that the probability of making a false positive is the same as that of making a false negative. For a number of measures (especially sensitive and stigmatizing information, such as events indicating mental distress), we can be pretty sure that these will be different.

Is it hopeless? Actually, there's one cute simple test that can help you in these circumstances. If the seeming influence is really just a pattern of random error, those who disagreed with the group or their friends about their tastes or traits or ideas at time 1 are more likely to switch to be in line with their friends at time 2, because they're not likely to make the same

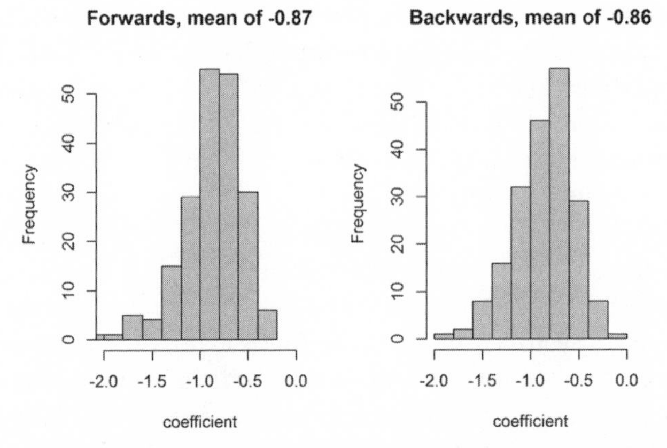

FIGURE 8.2. Forwards and Backwards Time with Change Due to Error

error twice. So if you compare their tendency to switch to the difference between their report at time 1, and the opinions of their group, you'll get a positive coefficient.[13]

However, in this case, you'll get the exact same results if you run time backward, the trick we worked out in the previous chapter. Fig. 8.2 shows results from 200 trials in which we generate 200 people, all of whom have a true degree of some trait or taste (say, "enthusiasm for jug band music") that their friends know about, but their initial answer is only probabilistically related to the degree of the trait, at both time 1 and time 2 (R 8.3). What we do is predict their time 2 response from their time 1 response, and the residual of the regression of the time 1 response on their friends' reports. That's the distribution on the left. On the right is what we get when we run time backward. It's the same. Ouch. So if you have an influence story, and it works the same way when you run time backward, you'd better retract it.

In contrast, if you have a process whereby only those who disagree with their friends change, then the results going backwards can be quite different. Fig. 8.3 presents roughly analogous results for a case manufactured to have real influence (R 8.4). In this case, anyone whose friends are over 85% united simply switches to join them at time 2, if she weren't already with them at time 1. Because here I'm modeling this process as a deterministic

13. For example, you can regress their time 1 self-report on the group's report on them, as we did in Yeung and Martin (2003), and use this residual to predict time 2 response, taking time 1 self-report into account as well. While this isn't guaranteed to give you a great causal estimate, if that's what you want (because in the residual, we also have ego's report), rejection of the null hypothesis indicates that there is a tendency of disproportionate change towards (if the coefficient is negative) or away from (positive) the group's view.

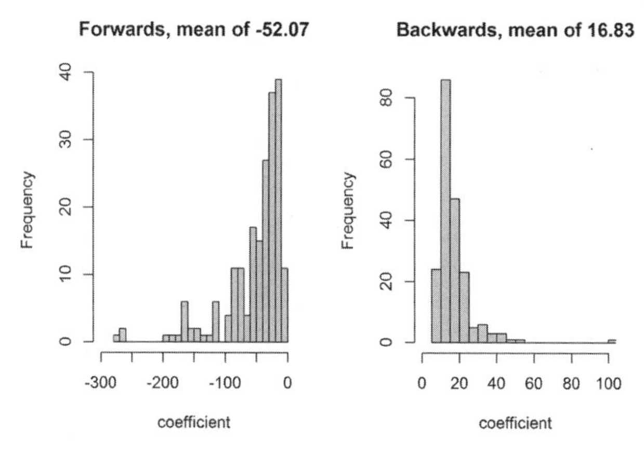

FIGURE 8.3. Forwards and Backwards Time with Change Due to Influence

one, the coefficients portrayed in fig. 8.3 are incredibly large, but more important, they're clearly different when we run time forwards from when it goes backward; indeed, they're in different directions.

Now I'm not saying that real influence processes can't produce significant coefficients or ratios when you run time backward. But for the simple ones I've investigated, they always have a *greater* coefficient when time runs forwards than when it runs backward. Temporal relations can be asymmetric for reasons other than influence, too: think of subject maturation, changing interview conditions, and, in many cases, consistency effects. But still, something that fails this simple test has two strikes against it. And all models for influence in non-experimental data start out with two strikes, one for unobserved heterogeneity and the other for selectivity.

Now there are three reasons King's and my work still looks like it was right all along. The first is that we found that time flowed one way—the group became more unified in their judgments about one another over time. The second is that when we realized this problem (after publication), we *did* run time backward, just as I outlined here, and found no evidence of a tendency toward convergence with the group. Yay! But the third reason is that we found patterns that had to do with the *content* of the relation—such as that while for most relations, those who held a discrepant view about themselves compared to their peer image tended to win their peers over to some extent, this wasn't true for the quality of passivity. If you can't actively convince people you're passive, that fits the idea that there is real change going on here, not simply measurement error.

And I'd like to take a moment to emphasize the importance of this point. Often, we assume that the only solutions to our *statistical* problems are *statistical* solutions—the sorts of things that are totally indifferent to the content of what we are studying. But that's not quite true. Sometimes we can adjudicate reasonably well between possible explanations for our data by being attentive to the meaning of the terms we have, most simply, the difference between positive and negative valence.

Unobserved Predictors and Influence

We've seen how various forms of error can lead to spurious conclusions about structural effects. Can I go a bit further? I wager that most network findings are simply unobserved individual-level heterogeneity, repackaged as if it were dyadic. I'm not denying the importance of social structure! Quite the contrary, I think it flows into everything we do . . . and everything we are, including our individual-level covariates. So when we go around acting, we do so in ways that are part of a larger social pattern. That goes for both what music we choose to listen to, and what people we choose to be friends with, for example. Absent information on these individual characteristics, statistical analysts will make all sorts of misattributions, thinking that, say, friendships *cause* certain outcomes, when really, the friendships are just giving us *clues* about the nature of the individuals involved.

We can think about this in terms of selectivity bias. But when we do, we often think we know the next step—model the selectivity! Take it into account, or get around it with pre-treatment measures. If we're trying to see whether liking Dub Step music diffuses along friendship networks, we might *first* ask people if they like Dub Step. And then we can model how their friendships change their tastes.

That's the wrong response. The proper response is—throw up your hands and go home. The reason is, if certain types of people are more likely to enjoy Dub Step the first time they hear it, they're also more likely to become friends—whether or not they've heard Dub Step before. You're *not* going to measure these subtle, unnamed, untheorized personality traits that affect both friendship and taste, and so you're not going to take them into account. Think of it (and treat it) like the plague.

So the story of the plague goes something like this. Folks could see that the plague would sweep into an area, coming closer and closer, and that people would die together. They knew that many diseases were contagious, so they assumed that plague spread from one person (or body) to another. So they'd shun the sick and anyone who had anything to do with them. But that didn't help. And the reason was, it turned out that the plague *wasn't* spread by interpersonal contagion. Rather, fleas that rode around on rats would bite people, transmitting the disease to them.

Now here's the thing: If we imagined having longitudinal data on people, their interpersonal connections, and their plague status, we'd see people converting to the plague in a way correlated with network ties. Before and after data just isn't enough, if you don't know about the rats and the fleas. The same goes for the spread of cultural elements. The role of the rats in this case are played by the conventional mass media, by institutions like schools and employers, by informal interaction, and by similarities in social interests. The fleas are advertisements, radio shows, announcements, informal discussions, and so on. Unless you really know about *all* of these, you don't have much chance of identifying the effect of influence using non-experimental data.

Sure, there are a few studies in which there is a randomization involved. But you're not going to like them. Why? Because they won't tell you what you want. The most rigorous ones (for example, Kevin Lewis's great work [see Lewis et al. 2012]) won't show you any significant social network effects. (Lewis's work takes advantage of the fact that roommates are randomly assigned at some colleges.) But even if you were willing to risk things on experimental trials, most social network effects can't be studied this way. A friend experimentally allocated to you by a researcher isn't the same as the friends you make yourself. And if the social world contains more information about who likes whom than you have in your survey, you probably can't transform non-experimental data into a simulation of an experiment, even by adding controls for selectivity.

Getting at Selectivity

Let's briefly walk through a justly famous debate over the attempt to get at influence via network analysis, namely Christakis and Fowler's (2007) claim to have demonstrated that obesity spread along a social network, which I brought up in chapter 3. They had some pretty cool data—over 12,000 people studied up to seven times over *thirty years* as part of a health study. Because this was tracing out families from original subjects first sampled in 1948, they had information on families. But more important, the people keeping the data had tried to make sure they could always refind their subjects. And to do this, they had asked about friends. That way, if a subject had moved, they might be able to find the new address by asking a friend.

With this data, we can see friendships being born, coming to an end, and being revived. We can also see people growing up . . . and gaining weight, shedding the pounds, and putting them back on again. Christakis and Fowler's notion was that, with this temporal ordering, we could see whether having fat friends leads people to gain weight, or whether gaining weight leads them to add fat friends, or both, or neither. Indeed, we could see whether if your existing *friends* get fatter, you will too. And what they found was that yes, yes you will.

Their work attracted attention not simply because they sought it,[14] but because it didn't pass the smell test, as they were finding effects of one's neighbors—even "neighbors' neighbors"—that no other social network researcher had ever found. In any case, they may have been hasty, but they had tried to rigorously deal with the most obvious problem for such an influence account—unmeasured heterogeneity in the form of selectivity.[15] Obese people may have obese friends, but that doesn't mean that the friendship causes them to become obese. It could be that obese people prefer having obese friends, or that non-obese people prefer having non-obese friends (and therefore aren't available to be friends with the obese). It's clearly a hard problem to solve.

Christakis and Fowler tried to use the temporal nature of the data to solve the problem. If we look in any person's *change* in obesity, that won't be explained by individual-level factors, right? Now we can see whether previously non-obese people who were friends with obese people *became* obese. Sounds cool, right? I hope you are thinking about measurement error—but we are going to trust that in this sort of medical study, there's a lot less of this than in a conventional survey. I want to pay attention to other difficulties—but not because I want to pile on. Rather, because there was so much attention, and because some very good critics did a gang beat from all sides, we get to realize the basically impossible nature of the sort of investigations that Christakis and Fowler were trying to make. And we get to learn about pretty much everything that can go wrong with this sort of endeavor. I'm not here to tell how to answer their question—I'm here to tell you not to ask it in the first place. Because the better the work you do, the harder it will be to come up with a plausible answer.

Okay, let's start with the issue of selectivity. Recall from chapter 3, Noel and Nyhan (2011) pointed out that Christakis and Fowler had forgotten about differential selection *out* of friendships! It's another of those cases of our tendency to focus on the cognitively marked side of things. Whenever you measure relationships, remember from chapter 2, you're examining not just the results of *creation* processes, but also survivals.[16]

14. Sorry! I call 'em like I see 'em.

15. They also took into account possible transmission via smoking cessation, which makes a great deal of sense, as this could also spread on (or near) social networks, and affect weight change.

16. Now Christakis and Fowler (2013) replied that the turnover in friendship isn't big enough for this to have been a practical problem—but then again, that suggests that there is a lack of longitudinal variation in one half of the independent variable (obese friends).

But social network researchers will also be concerned about unmeasured similarity across dyads. Christakis and Fowler (2007) tried to use the asymmetry in their data to support a claim of influence (which would be directional, and hence asymmetric) over some sort of homophilic selectivity (which would be symmetric). Archie and Reggie might indeed be closer to one another in some "social space," and that might imply *both* similar weight status, *and* their propensity to make a friendship. That's the selectivity explanation that would be raised as an alternative explanation to the influence that Christakis and Fowler hypothesized. But if Archie chose Reggie as a friend, even though Reggie didn't choose Archie, and Reggie's weight tended to affect Archie's, and not vice-versa, it can't be due to their closeness, since this is symmetric! And indeed, they found that the seeming influence on ego occurred more when ego picked alter as a friend than vice-versa.

That's not bad reasoning. But it's not, as Lyons (2011: 9) pointed out, airtight. Imagine, for a moment, that everyone can choose only one friend, and that Archie, Reggie, Jughead, and Moose are distributed on a single circular dimension with locations in degrees of 0, 100, 180, and 230; with distances accordingly. Archie would choose Reggie because he is the closest neighbor, while Reggie would choose Jughead, and Jughead, Moose (who would reciprocate). If these asymmetries produced influence, Archie would become like Reggie, and Reggie would become like Jughead, while Jughead and Moose would converge. Now let's imagine that each's position corresponds to some quantitative attribute such that the closer people are in position, the more similar they are in this attribute. Then Archie is most like Reggie, Reggie is most like Jughead, and Jughead and Moose are both most like one another. Thus the correlation of asymmetry and similarities in the dependent variable don't necessarily imply any influence at all![17]

Finally, there are indeed problems with estimating the sorts of autoregressive models that Christakis and Fowler first used—because they had one person's weight on the left side when they were a focal subject, and then on the right side when they were a friend (after this was pointed out, they re-estimated with only lagged versions on the right side). Some critics have made it seem as if it is intrinsically impossible to have the same variable on the left and the right side of an equation, which is not correct (we've seen these models in the previous chapter for, say, the spatial lag

17. It's for this reason that data that arises from people distributed in a space of likeness, with more people concentrated in the center, and a process whereby people pick those most like them as friends, actually leads to data that appears to be generated by a vertical status-oriented procedure!

approach). However, it is true that you don't want to use conventional *dyadic* models to examine this sort of influence process, for the simple reason that ego's value is tied to that of *several* alters, but ego has only *one* value. But there isn't inherently anything wrong with an autoregressive model. (And some models like this have a pseudolikelihood ERGM [Exponential Random Graph Model] interpretation.) On the other hand, it isn't necessarily easy to get defensible estimates from them, and you can't estimate the parameters through an OLS regression.

When you add all this up, the chance of successfully identifying an influence relation isn't really worth it for you to go down that road. There are going to be more holes in this dyke than you have fingers to plug them up. That's because while you don't know all that much about the subjects—what they are really like, and what they really like—*they* can find this out about one another, and use it to form and break relationships. And that means that anything you *don't* know about is likely to reappear as influence. So, if you're getting positive results, it's probably because there's a hole you forgot to plug up.

That's sad, because there really *is* interpersonal influence out there. It's one of the coolest, and most sociological, dynamics out there. No wonder we all want to study it! But our methods just aren't good at reaching it, and, since it has "downwrong" incentives—the worse our measures and specifications of control variables, the stronger our effects—it probably isn't where we should be working.

Context Effects

We've been examining one way that the world can know more than we do. Here, each focal person A's friends or fellow group members contained information about A's true state on some variable y that was not contained in A's report y_A. What seemed to be influence was simply correction. We see the exact same problem coming in many studies of what are believed to be "context effects." Here, we attempt to see whether the values of the people around you influence your own outcomes. That is, even when we take into account your value on some predictor x, we find that the average value of y among those around you also predicts your outcome. And the chances are good that it will. But *predict* doesn't mean *cause*. Because . . . what if there is error in our measure of y? Then probably the residual between our prediction and the true value is correlated with the average y around you.

That is, imagine that we have two people, *A* and *B*, who, on the basis of our measures, we declare to be equally poor ($y_A = y_B$). But *A* is surrounded by *other* poor people (whether on a block, or as fellow students in school, or whatever), while *B* is not. *A* is, I'll wager, more likely to be really poor than *B*. Why the error? Perhaps we asked about last month's earnings, and *B* doesn't work summers; perhaps *B* has a college degree and is staying with a friend since he just moved to town and is looking for a job; perhaps *B* is a little paranoid and didn't want to say his true income; or perhaps the data entry operator was listening to music and tapping the keyboard and pressed 2 instead of 8. *B*'s value ("poor") can, of course, be absolutely correct, but I'm guessing that it's more likely to be wrong than *A*'s.

The take-away is that there certainly can be consistency effects, network effects, and context effects. In fact, if there weren't, social life would break down. But any approach to estimating these that assumes perfect measurement is likely to over-estimate these effects, and the more error, the greater the over-estimation.

A Coda on Social Capital and Asymmetries

Okay, you say, perhaps we should stop thinking of influence flowing through social networks. But with all the groovy theory you've read, you feel confident taking a more "agentic" view, and seeing networks as a resource that actors will use, not just pipes that bring stuff into their heads. You will look at these relations as a form of "social capital." Is this a solution? I'm skeptical. Treating this or that as a form of capital isn't the worst thing in the world, though it's rarely a major breakthrough. And reifying social networks also isn't dreadful. But putting reification and capitalization together has led to some of the weakest thinking in sociology.

Most important, the way we think about this "social capital" is intrinsically contradictory, especially when taken to the context of the urban poor. It's that typical form of cognitive asymmetry that we've discussed under the umbrella of "marking." We think that it's *good* to have ties. All those people you're tied to can do you favors! You can, say, borrow money from them. But hey, . . . wait a second. . . . For every tie you have, someone has a tie to you! It's all well and good to borrow money from someone else, but when they borrow money from *you*, that's not social capital! That just plain stinks!

Not all aspects of social relations are zero-sum. But lots are. If *you* can pull yourself *up* with them, *others* can pull you *down*, and that's what happens when they're below you and yank on that tie. There's no action without an equal and opposite reaction. Ethnographers and interviewers have been pointing to this (e.g., Smith 2005, Desmond 2012), but formal data analysts haven't caught up with them. Why? Largely because having few ties *does* predict all sorts of negative outcomes.

Yet that doesn't mean that the ties are doing anything but holding the information that regular folks have about one another and that we don't. According to William S. Burroughs, Ernest Hemingway had the capacity to detect the stench of death on soldiers. As in, precognition. That's a bit hard for me to believe, but I'm completely convinced that people can smell the stench of, for example, *career* death on others. Or *popularity* death. Or *economic* death. And, in some cases, actual upcoming physical mortality, as in, soon that person will be dead. And people often decrease their interactions with those who have this stench. Sometimes this is mediated by the glum, hang-dog look that future losers often take on. Other times it's deliberate calculation. Other times it's an awkwardness (what do you talk about with the person about to be fired?). And sometimes, it's just that feeling that if Jove is about to chuck down a thunderbolt at that poor bastard, you don't necessarily want to be standing next to him.

And even when there isn't some future bad event that we're predicting—and that other people are also predicting—we can find that ties predict certain outcomes without their indicating anything about influence. Instead, they're giving us hints about unmeasured individual heterogeneity of at least one type. Refresh yourself on fig. 6.5: Your Brain Is Here, but Your Data Are There. Many social network results are driven not by the presence of ties, but by their absence. As in, their *complete* absence. But, as we've seen, what we *call* our index can orient us to the wrong side.

Because there's a group of people who define the [0,0] cell in many of our relationships between "social capital" and various positive outcomes—they may be mentally ill, down on their luck, or just plain crabby. Once again, *we* don't know from our survey who's incredibly crabby. But other people do, and they tend not to be friends with them, or to give them jobs, and so on. Our tendency to be cognitively distracted by the marked end of the spectrum means that we have a story about how lots of ties help you. But in these cases, the actual relation is about something more important—loneliness. And that's because, for most humans, social isolation is distressing. This isn't a happy story about ties . . . it's a heartbreakingly sad one about isolation.

Isolates can drive a lot of our findings. Yet the popular can drive our sampling procedure if we snowball . . . and can drive us crazy. Let's go on to look at the problems that arise here.

Letting the Network Determine Your Sample

Here I'm going to discuss what is called respondent-driven sampling (RDS) by Heckathorn (1997) and co. RDS is traditional snowball sampling, but with two important differences. The first is that one generation of respondents *themselves* is given an incentive to get new people into the sampling frame, as opposed to giving their names to the researchers, who then try to get these named persons to cooperate. This is very good. The second is that it was initially wedded to false assurances as to the properties of the resulting sample. That is very bad.

The idea of RDS is that sometimes there are populations that are too small to be accurately sampled from a general frame, and we have no clear definition of the population from which to sample. For example, there is no "Guide to Heroin Users" phonebook we can sample from. So, we

1. Start with a few heroin users in an area.
2. We give each seven (say) tickets with each's identification number on it and our address. We say, "give these out to heroin users you know, and for each one that shows up and takes the survey, assuming they haven't already done it, we'll give you $20."
3. We repeat (2) with the all those who are recruited, and so on and so forth.

Referral samples like this are often criticized for being "non-representative," as if that were a plausible outcome or even important. If you've read *TTM*, you know that I don't think this is a big deal. The more fundamental problem is that chain referrals are *systematically* non-representative in very, very worrisome ways. To appreciate the strength of this phenomenon, we must bear in mind that social referral processes will invariably seek out the popular, for, as Feld (1991) wonderfully put it, your friends have more friends than you do. The average person's friends will have more friends than the average person, since the person disproportionately picked as a friend is one with many friends. (We'll follow conventional usage and call the number of ties a node in a network has the "degree" of that node; ties going *out*, like *choosing*, lead to an "out-degree" and those going in, like *being chosen*, an "in-degree.") Tracing a friendship network randomly from any point leads one to progress toward more and more "popular" people. These people are different from others in all sorts of ways, and that's almost certainly *especially* true when it comes to the "hidden populations" that RDS was advocated for.

I've constructed (R 8.5) a super simple example of a network with only a mildly skewed degree distribution: 100 times we drop into the middle of the network and trace a chain randomly choosing the next place to go.[18]

18. The network is a Barabasi type random graph, meaning that it is constructed via the successive addition of new nodes who decide who to pair with, with a moderate (less-than-linear) degree of preferential attachment. Note that some chains end earlier, because we get to a dead end, in which there is no one to add who hasn't already been added.

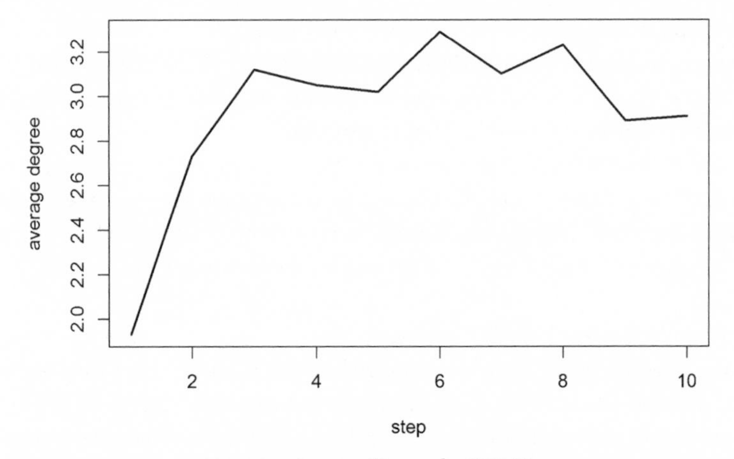

FIGURE 8.4. Average Degree by RDS Step

Fig. 8.4 graphs the average degree of the people we bring in. In the first five steps, we see a steadily increasing degree of the nodes we are reaching. Up to a point. As Matthew Salganik pointed out to me, you can run out of high degree people rather quickly. In this network of 500 people, you can see that by 6 links in, the degree is no longer increasing. If we keep going, we're going to need to start finding less and less popular people. Of course, they're harder to find, and, in many cases, harder to recruit. (There's a reason why no one likes them!)

So we are really worried that the further we go in a snowball sample, the more unusual our people are—up to a point. Heckathorn's (1997) big claim was that this was *not* true for RDS. Here he relied on some information that turned out to be largely irrelevant,[19] but even more, on somewhat misleading mathematics. We're going to look at that math carefully.

First, Heckathorn focused on *observable* states (like race, or town), which usually aren't what we worry about anyway. We should be more concerned with harder-to-measure personality traits that will affect network formation and other actions. Second, he assumed that the process was a Markov one, which implied that it would reach an equilibrium, as opposed to getting worse and worse and worse. You'll note that this is a simple mathematical assumption: almost *any* Markov process *eventually* does this (subject to a few assumptions about the connectivity of the underlying network—some unusual types can lead to endless cycles). The

19. Thus (1997: 179) he said that there was no statistical association between the number of heroin users a nominator knew, and the number of referrals s/he made. But the Feld point is about the degree of the *referred*, not the referring. The only way to have a sense of the magnitude of the bias is to have the actual population, which of course, we don't.

fact of Markovity *can't* be a solution to the problem of snowball sampling. There *was* reason to think that in a number of cases, this equilibrium was reached reasonably quickly, though of course that doesn't mean that it converges with a distribution that approximates the population.

But so far, what we have is a good snowball procedure, a nice way of trying to examine the relation between each wave, and some incorrect assurances that the process did not leave important pockets of the hidden population behind; in the words of Heckathorn et al. (1998: 177), "Network interventions work best when they are most needed, that is, when network structures facilitate the spread of HIV" (for why this is unlikely to be true, see Martin, Wiley, and Osborn 2003).

At this point, enter Matthew Salganik, a bright young mathematical sociologist. (I'm going to be very critical of the work here, but I am a huge fan of Salganik's, and he has successively moved this research to more and more rigorous approximations of reality; in fact, if you are seriously interested in these issues, his work is the most important stuff I think we have going.) Salganik and Heckathorn (2004) teamed up and made an award-winning article. For mathematical reasons, they assumed, first of all, that they had sampling *with replacement*. That means if a person gets in the sample, he can get in the sample again. If a second person is chosen, the first person is—in terms of the cognitive processes in the head of the respondent—thrown back into the pool for this respondent, and for any other respondent.[20] But of course, since the researchers are paying subjects to refer others, they're absolutely intolerant of sampling with replacement, or folks could continually bleed money out of them! Indeed, rather than allowing multiple referrals, RDS always involves some clever methods to prevent people from getting in more than once under different names.

So this is a nice opportunity to think about assumptions, and when they go from being convenient and non-remarkable to predetermining a perverse conclusion. It's not that simplification is a crime, but sometimes—and I think, in this case—the strong claims turn out to come from the assumptions being made, assumptions that we would not justify on substantive grounds (though there's still work on the relative importance

20. They did note that this wasn't true, but suggested in a note that preliminary investigations trying sampling *without* replacement didn't make much of a difference. That never seemed plausible to me. You can see what that does in our simulation by taking out the part that checks if a respondent has already been "visited." It seems to me to make a *gigantic* difference! Every person tied to someone of degree 1 is sampled twice, once coming in and once leaving; more generally, one's probability of being double-included goes up with the number of low-degree ties one has. So the sample moves to hubs.

of different assumptions). The next consequential assumption is that respondents choose *randomly* from among their contacts. Even Salganik and Heckathorn (2004: 210) said that "some readers may doubt this random recruitment assumption, but there is some empirical evidence that is consistent with this assumption." (The piece cited here actually offers no evidence about the randomness of the recall and name-volunteering process at all, though it's *consistent* with this assumption.) Certainly, we can imagine that if our concern is whether or not the snowballing procedure successively introduces more and more bias, as most researchers have assumed, the *assumption* that the nomination procedure is random is certainly a consequential one pushing the findings in a convenient direction.

Unfortunately, there is no reason to think that the process of recruitment is random, and very good reason to believe that it is not. If you were paid $200 to name a psychologist who would prove willing to be surveyed, would you randomly pick from your mental list of psychologists? You would try to get someone you thought most likely to comply.

But still, we generally start with convenient assumptions, and then see if we can successively relax them. And that's just what Salganik and Goel started to do: to see what happens if you throw away some of the most pernicious assumptions: First, Salganik (2006) noted that it was important to have a decent sense of the research design in order to produce plausible estimates of standard errors of parameters, and proposed a bootstrap approach. But even more important was his focus on the types of network structures that would introduce serious problems to the RDS scheme, and a sense of when these methods are likely to be inapplicable.

Then, Goel and Salganik (2009) acknowledged that actual survey sampling doesn't involve replacement, and more generally, an understanding that any true estimate requires a model of the weight structure of the network. For this reason, this work came closer to a plausible model that could produce reasonable estimates. But they still needed assumptions that, while perhaps necessary for the model, were implausible and were not sufficiently discussed. In particular, they preserved not only the assumption of randomness of recall, but also that ties are symmetric. (This might sound reasonable, but it isn't when it comes to recall processes: do you know the name of everyone who knows your name?)

The next year, Goel and Salganik (2010) showed the implausible nature of the standard errors that Salganik had previously derived, although they still retained the problematic assumptions as to the network processes (symmetry, randomness, stable and known degree counts). But what is most noteworthy is the following conclusion: "RDS as currently practiced may be poorly suited for important aspects of public health surveillance where it is now extensively applied." Finally, in a 2012 *Epidemiology* comment, Salganik emphasized the degree of imprecision that the estimates produced were likely to have.[21]

And that's why Salganik comes off as a hero for this book. It took a lot of work to basically force the retraction of the grand claims that had been made for RDS in the early days, and he himself did a lot of it. I don't think that work should have even been necessary—it was too clear from the beginning that the assumptions being made, and the conclusions being drawn, were too strong. That's the kind of thing we accept when it *doesn't matter*, but for public health, it does (the Jim Wiley dictum: "when we're wrong, people die"). Once it became clear that things were not going well for RDS as originally spiffed up and sold, Heckathorn didn't really respond. He left the early paper claims up on a RDS website, and didn't update it. It was like he wanted to freeze things back when they all felt more optimistic. But Salganik didn't. He pushed further, brought the limitations to the front, and kept on trying to figure out more about how people actually named others, and how we could get as much from this as possible. Because we're going to need to use this technique, we want realistic appraisals conducted *sine ira et studio*.

Interestingly enough, Heckathorn maybe gave up too soon (and he's back at work now with some recent collaborations on trying to deal with methodological limitations). In a recent paper, Forrest Crawford (2016) realized that everyone had been overlooking one important bit of information in the data—the relative *timing* of the referrals. By making assumptions about the distribution of waiting times, one could make some probabilistic statements about the set of possible networks compatible with the observed sequence of referrals.[22] The lesson is, with network structures, you can't safely assume *anything* except that people will *not* make your life easy on you. But instead of assuming, you can study. And every now and then, thinking a bit harder about what's really happening, plus some good math, might solve a problem, and not just gloss over it.

21. Recently, Baraffa, McCormick, and Raftery (2016) came up with a technique for estimating the true imprecision. We may be wrong, but at least we're not going to be confident that we're right!

22. Further, Mouw and Verdery (2012) show that we'll probably do a *lot* better if we can first get respondents to give us as much of their social network as we can get, and then we strategically sample, given our reproduction of the full network. Of course, this requires our imagining that there is "a" network, which isn't really true, but it still pushes us to interact with respondents in a way that is going to reduce the bias. We'll have to see if either of these really do well in the field with hard-to-reach populations. They aren't always as tractable as the simulated people in most articles!

Interdependencies

Against Ritualization

The last inherent statistical puzzle posed by network data comes from the structural interdependence of ties. We have a disciplinary problem right now in that a rumor has gone out that there is one way of dealing with this (Exponential Random Graph Models, or ERGMs) that everyone should do. That's hurt our collective ability to learn from data. Let's walk through some of the key issues, and see what we get from this approach, and what not.

When we have a non-independence between ties, we might want to do two things. One, we might want to *neutralize* it, so it doesn't affect the estimates of the coefficients (and that of their standard errors) that we are interested in, coefficients that may be at the individual (and, in some cases, dyadic) level. Second, there are times when we want to *describe* the type of network structure we see. Third, there are times when we want to *account for* this structure—either by reducing it to certain fundamental structural principles, or by linking it to a process that we believe generated this structure. One problem comes in the fact that people who are really oriented to the goal of *neutralization* believe that they must use techniques that were developed for the other goals.

When it comes to neutralization, there are a number of other (imperfect) solutions. Which one is appropriate actually can have to do with the precise nature of the process of tie formation—something we rarely know about. A very popular one for some time was a permutation test, called, perhaps for indefensible historical reasons, a QAP [Quadratic Assignment Procedure] (Baker and Hubert 1981; Hubert 1985; Hubert and Schultz 1976).[23] The test was introduced in regression form by David Krackhardt (see Krackhardt 1987, 1988, 1992). This took advantage of the fact that for OLS regressions, the mere fact of the network interdependence didn't affect the estimates of a correctly specified model, only the standard errors. (This isn't true for non-linear models, where the "properly specified misspecification" never works. Still, we often go ahead and use a permutation test for non-linear models, in the hopes that it still provides reasonable rules of thumb for model selection.) Unfortunately, it turns out that this isn't very good for distinguishing the effects of correlated independent variables. Dekker, Krackhardt, and Snijders (2007), however, recently introduced an approach that seems to work quite well for OLS models.

23. I used to call it QAD—Quick and Dirty—because it's so simple. But truth is, it was never all that quick to run.

If all you care about is neutralizing the dyadic nature of the data, this is a fine way to go. Another approach is to use the multiway non-nested clustering to correct standard errors, as proposed by Cameron, Gelbach, and Miller (2011). And by the time this comes out, I would be confident that people have started using existing routines for mixed models, now that they can fit cross-nested random effects, to neutralize such dependencies in network data (see Hoff 2003). It's worth noting, however, that these approaches are assuming that you have a properly specified model. What they're not doing—and what you of course need to be doing—is thinking about how the fact of a network structure can lead relevant omitted predictors to pull your estimates in certain ways. In particular, they aren't necessarily good for correctly estimating the effect of dyadic variables that have dependencies other than those coming from individual-level differences (e.g., different reporting thresholds in making a friendship nomination).

But what if you are trying to "explain" the network structure itself—for example, you want to explain not just, in general, why one person is friends with another, but why the network has the particular sort of configuration it does? It is this that takes us to ERGMs, and I'm going to move moderately slowly here, because it's worth understanding both why there was (and, to some degree, still is) such enthusiasm for this approach, and why it can be absolutely wrong for you.

From Description to Accounting

What we now call ERGMs arise from a cool solution to a problem in nonsocial statistics, which is that our units receiving treatments often aren't really independent from one another. For example, corn plants might be arranged in a checkerboard type lattice, such that each one has eight neighbors. We might find that the outcome for any plant was similar to those of its neighbors (on average), even after we take treatment into account. Fertilizer put on one might leach over to influence the growth rate of a supposedly untreated plant. One solution to this rather uninteresting statistical issue was, long after it was developed, seized on as the key to network analysis. Why?

Here we need to briefly review the groundbreaking work done by Paul Holland, Samuel Leinhardt, and Jim Davis (e.g., Holland and Leinhardt 1970, 1971, 1976; Davis and Leinhardt 1972) . They were interested in characterizing certain social networks, which they had in the form of directed graphs, to examine the applicability of some strong theories of social structure. Since I've reviewed those (in *Social Structures*, chapter 2), I won't do that here, except to say that the problem is that if you assume that all the measures are perfect, almost every nice structural model is going to be rejected by some stray piece of data. So we'd like to be able to talk about *tendencies* toward certain structures. But because dyadic reports aren't independent, it's not an easy matter to quantify such a tendency. One needs a clear null model of randomness.

What Holland and company did was propose that we could examine a set of random graphs that all shared a certain distribution on some very simple statistics—the number of null dyads, the number of asymmetric dyads, and the number of mutual dyads. Within this set, we would examine that distribution of all the triadic configurations (sets of three nodes). With this information, they could see which sorts of triadic configurations seemed to be favored (present above chance) or disfavored (present below chance). These configurations could then, in Levi-Straussian fashion, be used to generate an ideal structure. This work is astoundingly important and has not yet been surpassed.

But there were drawbacks—it was ungainly, and rigid, and there were reasonable questions about the extent to which it was really accurate for observed networks. Well, Holland and Leinhardt (1981a, b) didn't stop there, but worked to parameterize the same basic distribution. They ended up with a model that decomposed the probability of a tie into one portion having to do with the *sender* of ties ("expansiveness") and another having to do with the *receiver* of ties ("attractiveness"), as well as a general tendency for a tie to be present, and another for it to be reciprocated, in the network. This approach was simpler conceptually, but it was restricted to thinking in terms of *dyads*. That was better than treating a network as a bag of individuals, but network researchers didn't believe that this could capture the logic of the non-independence in networks.

The reason that the results about the non-independence of corn plants became so exciting was that some researchers realized that these could be used to derive an approach that was much more general than the Holland-Leinhardt approach. That's why Wasserman and Pattison (1996; Pattison and Wasserman 1999; Wasserman, Anderson, and Crouch 1999) originally called it the p^* model; Holland and Leinhardt had called their parameterized model p_1, meaning the first of many probability models, and p^* meant a wide supervening class of probability distributions—in fact, in a way, all of them. The key was, as it often is, making a problem (in this case, the non-independence) a solution, by looking for solutions to dependency graphs.

In the next section, I want to work through, in a non-technical fashion, the logic that led to the development of the ERGM approach. And that's because I want you to see where the ERGM approach starts, and why it was so promising, and then how further assumptions and simplifications have entered, and where it is now. The point isn't going to be to bust on them, but to allow us to realize that they don't have any particular magic.

Dependency Graphs and the Hammersley-Clifford Theorem

Let's remind ourselves about the spatial statistics we looked at in the last chapter. The problem that we encountered there, which was formulated in terms of "error," can also be understood as one of dependence. Let's say that our model is $\hat{y} = bx + c$. The regression model can be read as that any y_1 and y_2 should be conditionally independent . . . conditional on the values of x_1 and x_2 (given b and c, that is). If these two observations are similar to each other, it's either because *all* the dependent variables are relatively similar, or because they have similar x values. The problem we found in geographic data was that this often wasn't true—the two observations were still dependent in a way that had to do with spatial position. Thus, one way of thinking about a good model is that it explains the dependence in our observations.

Interestingly, in chapter 7 we saw that we might, if we knew everything, be able to improve some of our estimates by *modeling* the dependencies. When we entered the weights matrix in the spatial error model, we were basically saying, "we know that these two observations still aren't conditionally independent given the covariates, but we know to what degree they're still dependent." It's this same basic idea that the statisticians were puzzling over when it came to the corn yields.[24]

What we need, then, is to be able to figure out what the network of dependencies is between all of our units. And this brings us to dependency graphs. The trick is that we're going to be applying this notion not to *units*, but to *relations between the units*. That's what a network is. In our case, the dependencies are themselves edges or ties between *other* edges or ties (our observed network). That can be a little puzzling, so we'll go slowly.

The key notion is that we can get our answers right *if* we know the dependency structure between our observations. When it comes to space, we often have a very theoretically compelling guess—things that touch are dependent, and others are conditionally independent. But what about when it comes to networks? Which *ties* are dependent? Maybe those that "touch" each other. How do ties "touch" one another? By sharing the same node.

To help visualize this, fig. 8.5, left, gives an example of a network. On

24. It's for this reason that you should expect a wholesale importation of techniques developed in spatial science to network research over the next ten years. Just make sure that your weights matrix doesn't imply complex solutions!

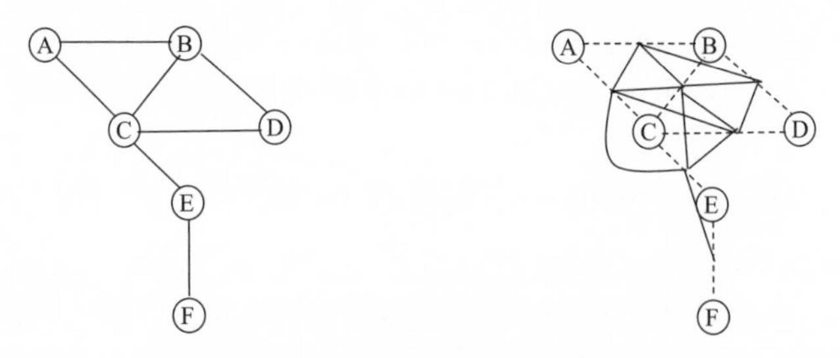

FIGURE 8.5. Network and Dependencies

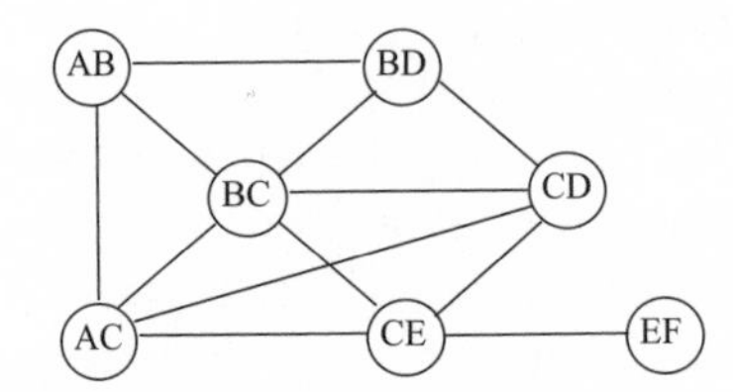

FIGURE 8.6. Dependency Graph

the right, it is re-drawn with lines between the ties that share a node like this. Or better, we can make a whole new graph (fig. 8.6) where the nodes are our edges (that is, the "ties") from the original network.

Now here's the cool thing. There's a fundamental piece of math, which is called the Hammersley-Clifford theorem (Besag 1974), that links a form of expansion of the expression for conditional probabilities in the dependence graph to the expression for the probability distribution of the original graph. What is key in making this linkage is what is called the "Markov" assumption, which is equivalent to the corn plants being affected only by their neighbors. In the network case, it's the assumption that we made that edges are, given the rest of the graph, interdependent only if they share a node.

What the Hammersley-Clifford theorem shows is that, given this assumption, we can factor the probability (or, better, the odds of any state as opposed to a null state) into a long chain of all possible subsets where all the terms that don't refer to *cliques* in this dependency graph will drop out! A clique is a set of nodes all of which are tied to one another. The largest clique in the graph in fig. 8.6 is the one with four members: {(AC),(BC),(CD),(CE)}. You can see that this translates to node C and all its neighbors in fig. 8.5. For a symmetric graph like we have here, the cliques translate into "stars" of different orders—a central node and its neighbors, like the "four-star" around C—and triangles (like the clique {(AB),(AC),(BC)} in fig. 8.6). With this, we can write out an expression for the probability of any observed graph in terms of these cliques, a lot like a regression equation with a bunch of dummy variables. Instead of a massive complex expression for the probability of this graph as a whole, we can actually decompose it into the probability of each part of it (each tie). This is a mathematical *truism*. It's not an assumption. That is . . . *if* the Markov property holds for this graph.

Let me stop for a quick second to emphasize while the Markov assumption might be a safe assumption for some networks, it isn't for all. Imagine that it is a few weeks before the time of a survey of high school dating, and Archie (i) asks Betty (j) to start dating him, and she happily agrees (thus x_{ij} is now 1 instead of 0, where x measures a dating relation). A few weeks later, Archie asks Veronica (k) if she will also begin dating him, and she too accepts (thus x_{ik} is now 1 instead of 0). Betty finds out about this, and confronts Archie. "Archie, when you asked me out, I just assumed that we were going to be exclusive. I guess that's not what you were thinking. Well, if you're going to also date Veronica, I'm going to start going out with Reggie (h) as well" (thus x_{jh} would now become 1 instead of 0).

"You can't do that!" Archie protests. "If so, you'll be making the ties x_{ik} and x_{jh} non-independent! Don't you realize that will violate the Markov assumption? If you do this, no social scientist will be able to use Markov exponential random graphs on the resulting data!" Do you think Betty was convinced by this logic? If you don't, you should be worried. (And let me hasten to add that in recent years analysts have worked on ways to avoid making the assumption of a Markov relation.)

But let's stick with the Markov assumption. If that's true, the Hammersley-Clifford theorem tells us that we can factor the probability into terms having to do with the cliques. Except that we're going to find that we have more terms on the right side than we do observations. For the case in fig. 8.5, we have 25 different cliques (one 4-clique, six 3-cliques, eleven 2-cliques, and seven 1-cliques) and only 15 ($= 6 \times 5/2$) observations. So what we need to do is to make the further assumption that all the structurally identical terms contribute the same probability—that is, all of the three-cliques that represent three-stars are the same, and so on. Then we have only one four-star, one three-star, one triangle, one two-star, and one edge parameter to estimate. And since the sum of the number of observed structures of each type is a sufficient statistic for the estimation of this effect, we just have to count them up. That sort of assumption seems pretty safe, if all we have is the graph (that is, we don't have any information about nodal or . . . edgal?. . heterogeneity).

So now we have a second assumption, though it doesn't seem particularly worrisome. This sort of model, however, is a bit hard to get a maximum likelihood estimate for. So researchers instead got *pseudo*-likelihood estimates—basically, by maximizing the value not of quite the likelihood, but the likelihood of all the observations conditional on the other ob-

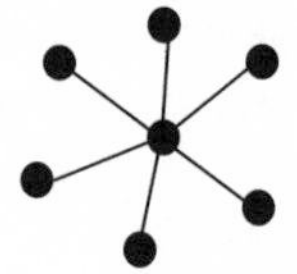

FIGURE 8.7. A Simple Star Graph

servations with which they are interdependent. In practical terms, this meant that simply counting up the number of observed local structures of any kind containing each dyad, and putting these into a logistic regression for the presence as opposed to absence of a dyad. This produced estimates of the importance of these effects in making *this* the graph that we observed. But this left statisticians unsatisfied—because we didn't have the best estimates. That's their job, remember? They got to work.

The Bad News

And so they used Bayesian methods, which can squeeze maximum likelihood estimates out of all sorts of complex models. And the Bayesian models started doing very unpleasant things, like coming up with parameters that suggested that the network really should have *no* edges at all, or be completely saturated. (In technical terms, the models were degenerate [see Handcock 2003].) And even when the models did, by some miracle, converge on non-extreme values, they often didn't make any sense.

Let's begin our considerations with the question of how to interpret the three-star effect (that which corresponds to the number of observed structures with a focal node that has three spokes). Remember that just as a three-clique includes three two-cliques, so a three-star involves three two-stars. And each two-star involves two one-stars. So if the one-star is the probability of a tie being present, a two-star, *given that we also have a term for one-stars in the model*, a two-star indicates the tendency for these ties to glomp on to the same node—that is, the tendency for some people to be more popular than others.

That means that if the one-star is giving us the *average* of tie probability, the two-star is giving us something like the standard deviation does in conventional statistics, as this is the second moment of a distribution. Now we can figure out what the three-star term is. It isn't really the probability of having three stars. For example, consider fig. 8.7. Imagine we fit a model with 1-stars, 2-stars, 3-stars, 4-stars, 5-stars, and 6-stars. We'd find a coefficient for the one-stars, telling us that 6 out of $7 \times 6/2 = 21$ possible ties were present. And we'd find a coefficient for the two-stars, telling us there were *lots* of two-stars. In fact, there's $6 \times 5/2 = 15$ of them! (Of course, we wouldn't find all these higher effects for this pretend example because our distribution is too simple, but the logic is easier to illustrate, and it holds in more complex graphs.) The three-stars would really tell us about the *third* moment of our distribution. And so on.

But unhappily, in practice, we tend to find for almost all plausible distributions that these parameters alternate in sign—they oscillate and damp down. And that was one of the factors contributing to the weird fits. So a number of statisticians thought, if you can't beat 'em, join 'em. Why not work *with* this tendency, instead of fighting against it? Rather than have fifteen different parameters to describe the degree distribution, we'll have one that *expects* this sort of oscillation, and has, say, a free parameter that describes the overall shape (on the Bayesian methods see Handcock 2003, and Snijders 2002). This worked well in practice, at least in a large number of cases. Problem solved? It depends on what you are trying to do.

Up until now, we haven't said anything about external covariates, remember? Well, how does all that enter here? The answer is a bit unsettling—they're usually just jammed right on in. That is, if you take seriously the starting points, and the Hammersley-Clifford theorem, the only obvious way of maintaining the logic is to have external covariates change the assumptions of equality of structural effects. That is, rather than thinking that the number of triangles observed is a sufficient statistic for the graph in a classroom of boys and girls, you might decide you needed *four* parameters: one for three boys, one for two boys and one girl, one for one boy and two girls, and one for three girls.

Could you think that perhaps a covariate leads to a *net* increase in the odds of a tie independent of all the structural effects? And so we'd like to partial this out in a model by adding it to the structural parameters. You certainly *can* think that. And that's been the approach that people have settled on. Then the notion of the structural effects is that they are getting at all the conditional-dependence net of the covariates. This isn't a crazy idea—but it's certainly not necessarily the case. It might be true, or true enough, in some cases, but it's not going to be true in all. It means that we're back in the same world as everyone else—making assumptions about models and if the assumptions are wrong, the conclusions are wrong. The "special" status that we had with the Hammersley-Clifford theorem has long been left behind.[25]

25. Almquist and Butts (2014: 278) point out that the ERGM approach can be considered a framework for "representing distributions on graph sets" that is *complete* for distributions with countable support. More simply, we can imagine it as a way of generating classes of graphs to compare our observed one to. That's a very worthy activity, but, as I'll emphasize in the next chapter, the fact that a model generates data like yours isn't necessarily strong evidence that it's the true model.

But the biggest problem is that even with the combined super-parameters, we often find that we just can't actually parameterize the model we're interested in. And so we have to throw out some parameters, in order to get the model to converge (or even to start up). In some cases, we have more structural parameters than we can independently estimate. Of course, if some subset is good enough for soaking up the conditional-non-independence, and we are mainly interested in the coefficients for external covariates, we might not worry about this inability to identify all of our parameters. But if all we're caring about are the *non*-structural parameters, and getting their coefficients right, there are, as we saw, simpler ways of going about doing this.

Yet those who use other methods may be rapped on the knuckles, and told by a reviewer that they should do an "ERGM." In most cases, the reviewer doesn't even actually mean this—he is confusing the method of *estimating* the model with the model itself, which is a statement linking parameters to observations. As a result, many models that obviously don't come close to parameterizing the dependency structure, such as a regression for the log-odds of ties that happens to be fit by Markov chain Monte Carlo methods, are not only referred to as "ERGMs" (which is technically true if pretty meaningless—most distributions we use are members of the exponential family) but assumed to be "right." Someone who uses pseudo-likelihood methods[26] to fit a well-defined Markov model will be assumed to be doing something wrong, while someone who uses a Bayesian MCMC method to fit a model for ties with just a few odd structural parameters thrown in, is assumed to be doing it right. Unless the ERGM parameterization happens to coincide with the processes of the world, what the reviewers are basically saying is, please get the best estimates you can of the *wrong* model. That doesn't seem to make a lot of sense to me.

Interestingly, I think a lot of the people who have been working on these models will agree with me. They know that it takes a lot of work to fit a single model, and it is a bit of an art form. And they listen when the data tells them, "I can't answer that question." Finally, some (see, especially, Pattison and Snijders 2013: 293, 295) have tried to bring order to the chaos of different models, and have introduced the notion of a general null "clique" model; further, they (299f) also point to the problems of the homogeneity assumptions and the unknown relation of biases in external covariates introduced by misspecification of the dependence structure.

26. These still make sense for certain complex models, like temporal models (Leifeld, Cranmer, and Desmarais 2017).

Judging Fit

Another thing that experts know is that we need to use different criteria for judging fit depending on whether we do or don't have dyadic covariates. If we don't know anything about the relations or the actors, it's a bit much to demand that our model guess *which* relations will exist. In this case, we're going to want to see whether the model fits overall structural characteristics of the graph in question—how *many* popular people there are, say—but not care whether it fits particular dyads. In other words, if the ERGM parameters generate networks that *look like* ours, that's good enough. However, if we have dyadic- or individual-level covariates, then the capacity of the model to reproduce the overall graph statistics isn't necessarily very comforting. In this case, we want to hold the ERGM to the same standards we hold our other models to—does it fit *this* particular data set? (Again I note that many leading researchers here make exactly the points I am making here.)

Unfortunately, with the ritualization of ERGMs, there's been an assumption that it isn't really the responsibility of the model to fit the data; rather, the data had better fit the model. But often when one actually tries to encompass the dependency structure and fit the data, ERGMs often just won't converge. So then what do you do?

Sometimes there are only three choices. The first is that you can throw away the data that wrecks the estimation routine. For example, in a wonderful recent piece of work that I'd consider an example of current best practices, S. Smith et al. (2016) tried to investigate ethnic composition and friendship patterns, a difficult problem, as we remember from chapter 6. They used ERGMs to look at these friendship patterns in a large number of school classrooms from four countries. But when it came time to run the models, they ground to a halt for half of their classrooms. So the authors shelved the data that the model couldn't fit (1237, 1259f), concluding that it was the lesser of two evils (since they figured that alternative methods would have led to more bias than the sample selection resulting from omitting those cases).

The second would be to change the model—to ignore theoretically relevant predictors. We resign ourselves to getting the best estimates of parameters from the wrong model. Of course, it isn't only in this case of ERGMs that we find we can't fit the model that we want. But many network analysts have struggled with data sets in which the ERGM can fit only a model that throws out important predictors, and not because those predictors don't seem to matter. But other methods are able to keep those predictors in the model, and so a third possibility would be to use a different method. But most journals today will simply refuse to publish any network analysis that doesn't use (what the editor believes to be) a state of the art method. A simpler method will be considered "wrong," but not the first two approaches. Really, this isn't about one right and two wrong, but the plusses and minuses of different choices. This fetishization of the approach is what sociologists call goal displacement—when what was once a means to an end becomes an end in itself. It's ritualization, not science.

A Simple Example

Let me close by demonstrating something, a nice finding that fits with the core arguments of this book. And it's this: simple methods can help you build theory even if they don't give you bestimates, and complex methods can leave you with nothing. I've simulated data on friendships (in a long, rambling program [R 8.6]). What's important here is that I've written the simulation *not* in terms of true parameters in a linear model, but instead, as a set of plausible processes. Let me tell you about this world.

There are 100 students, who take classes, four per day. In each period, they're split up into eight different school rooms. The students vary in "status" (some vertical position) and also along a single personality dimension. They join activities depending on their personalities—those on one side tend to be jocks and join sports teams. Those on the other side tend to join the math team. They each start with two friends from grade school, but they want around eight real friends. Every period, they look around to see who is like them in personality, and who is of decent status, and make friendship overtures (never making more in any one class than they need to fill up their docket).

Then, at lunch time, the kids begin accepting mutual nominations. Each kid adds only one per day. The kids choose in random order, and if your choice is taken, too bad. Then, after school, they hang out. Each kid who has more than one friend decides to invite two of his friends over. If both of his friends come over after school, they might then become friends with each other. This goes on for 100 days. Then they get the survey. They "forget" about some of their friends, and they disproportionately forget low status friends, and they name some of the people they've recently *wanted* to add as friends, but haven't yet really. That's where we come in. We don't know the true model, but we know what kids are like.

Let's say that we're going to start with regular old logistic regression, to model the presence or absence of a tie in each dyad. Let's start with a model that looks for popularity effects. How can we do that? Why not make a sum of the number of people *other than* ego who chose the alter in question? We can look for reciprocity by looking at whether alter chose

Table 8.3. Naïve Logistic Models

	Model 1	*Model 2*	*Model 3*	*Model 4*	*Model 5*
Reciprocity	5.453***	5.926***	—	5.865***	5.933***
	(.144)	(.200)		(.196)	(.197)
Alter Popularity	.225***	.188***	.040***	.188***	.190***
	(.013)	(.014)	(.010)	(.014)	(.014)
Ego Popularity		−.212***	−.012	−.213***	−.214***
		(.016)	(.010)	(.017)	(.017)
Transitivity		.606***	.910***	.567***	.550***
		(.039)	(.030)	(.040)	(.041)
Both Math Team				.421***	−.449
				(.152)	(.340)
Both Sports Team				.496***	−.367
				(.166)	(.345)
Both No Teams					−.398
					(.702)
Different Teams					−.562***
					(.199)
Intercept	−5.899	−5.018	−3.835	−5.224	−4.296

Standard error from naïve model in parentheses

ego. Yes, this means that we have the same thing on the left side and the right side, but in the pseudolikelihood interpretation this isn't a problem, since it doesn't entangle one dyad with another. More important, we shouldn't leave it out for the simple reason that getting the specification correct is far more important than having the best estimate, and I can tell you from plenty of experience that you can't ignore reciprocity when it's there without all your other results going to kaput.

So when we run this model, we find (model 1 in table 8.3) that both coefficients are positive and significant. Is that significance correct? We've just used a logistic regression, and haven't even done the sort of QAP test that I've relied on in the past, and there's a problem of consistency. So the answer is no, absolutely not. They're probably more optimistic than other tests would give.

Anyway, the reciprocity coefficient is huge, and tells us that at least the kids agree on who their friends are. The coefficient for alter's status is positive, and highly statistically significant, so it looks like kids do tend to choose the more popular. But we ask, "and do more popular kids *make* more nominations? Or just *get* them?" We enter ego's popularity, along with a parameter for the tendency to be friends with your friends' friends (transitive closure) in model 2.

First, this term for whether a dyad is transitively implied is quite large, and it takes away some of the effects of alter's popularity. That makes a lot of sense. In a linear hierarchy, those at the top complete lots of transitive triads. But of course, there's also a tendency for transitivity because students are in classes together, and hang out together. Second, we see a strong tendency for the more popular kids to be *less* likely to make nominations. In fact, it's so strong, we're probably starting to worry that the data can't be believed. Why would this effect be so significant?

Fortunately, a friend reminds us that with RECIPROCITY in the model, maybe we're really only finding that more popular kids are less likely to make *unreciprocated* nominations. So to test this out, in model 3, we drop reciprocity. And indeed, not only does the coefficient for ego's popularity become much smaller, it isn't, even according to our over-optimistic standards, statistically significant. So those asterisks might be an okay guide for us.

Now of course, we don't have a lot of data here, but we know who is on the chess team and who the math team, and so we examine whether being in these same teams predicts friendship in model 4.[27] As we see, it sure does. But then we wonder whether this comes because of opportunities for socialization, or whether it is because it is tapping similarity in personality. To push further on this, in model 5, we also enter terms for both being on *no* teams, and for them being on *different* teams. We add up every time that one member of the dyad is in a team and the other isn't, so this can be 0, 1, or 2, and enter that.

The results are surprisingly clear—what really matters isn't being on the *same* team, it's being on *different* teams. Of course, being on the same team and being on different teams are closely related to one another; we're parameterizing a set of options, leaving out only two cases, one being on a team and the other not, or vice-versa. But the difference between the parameters is substantively significant—it suggests that this is more

27. If I were really doing this, I'd always try including terms for EGO_CHESS, ALTER_CHESS, EGO_MATH, and ALTER_MATH to make sure that what I am considering to be a homophily parameter isn't actually due to nodal heterogeneity. But I want to avoid cluttering the model for reasons you'll see later.

about tapping difference in personality than opportunities for meeting. And in fact, that's exactly how the simulation worked, if you remember.

In other words, using the wrong model got us what we wanted—a sense of what was going on. We aren't trying to get *the* right parameter of, say, the extent to which being more popular affects choice, or being on different teams affects choice, *because those don't exist*. They don't correspond to the actual processes generating the data. So we shouldn't worry about beating ourselves up to get the exact right estimates. There aren't any.

Now let's replicate this model with results from ERGMs.[28] In model 1 in table 8.4, the constant is replaced by the "edges" parameter (tapping the baseline probability of an edge), and alter's popularity with the geometrically weighted in-degree. This is that cleverly designed parameter that deals with the tendency of the degree parameters to pogo up and down. This model didn't quite converge; while we could fine-tune it to get the standard error of the in-degree parameter, given that it's so hard to interpret, we can just move on to model 2. Again, I didn't get convergence, and again, because there was no variation in the degree parameters across the chains, we don't have significance tests for them. I re-started the algorithm at the values shown, but it crashed even worse, unable to get variation on any of the parameters. But don't worry, things look up after this. We don't always need to get extremely good fits for our provisional models; it's the final ones we need to have "well done." Even with the rough findings, we see a similarity to the previous table.

With model 3, we were attempting to see whether ego's degree was related only to *unreciprocated* ties. Unfortunately, here the routine also didn't converge properly and in fact later values had worse Akaike Information Criterion (AIC) than earlier, so I present those reached after the first 20 iterations. (That five-digit AIC is not a misprint.) That also happened for model 4, despite trying to tweak the fitting routine. Further, the results for model 4 are quite shocking: the transitivity parameter is found to be insignificant (though it's close, $p = .059$). (I haven't been showing the standard errors of these parameters to keep the table brief.)

But now let's ask that same question that we did in table 8.3, apparently successfully, trying to see whether the "same team" effect is about opportunity versus homophily (model 5). Unfortunately, the model didn't converge; it kept falling into singularities, as if there were a linear relation

28. Here I am using the wonderful package ergm written for R. I've used baseline defaults, and stopped after 20 iterations. However, I would re-start after this, but sometimes give the earlier results, as the longer the routine went on, the more problematic the results.

Table 8.4. Exponential Random Graph Models

	Model 1	*Model 2*	*Model 3*	*Model 4*	*Model 5*
Reciprocity	8.450***	5.013***	—	4.637***	4.835[a]
	(1.026)	(.133)		(.230)	(—)
Alter Popularity	-8.919[a]	-4.000[a]	-1.868[a]	-1.272***	-3.113[a]
(GWID)	(—)	(—)	(—)	(.386)	(—)
Decay value	-1.010	.389	-.697	1.59	.860
Ego Popularity		2.882[a]	-.697[a]	4.630***	3.736[a]
(GWOD)		(—)	(—)	(1.191)	(—)
Decay value		-.242	1.865	.769	-.007
Transitivity		.426***	.922***	.413	.336[a]
(GWESP)		(.086)	(.174)	(.219)	(—)
Alpha		1.652***	2.128	1.400	1.524
Both Math Team				.464***	.195[a]
				(.141)	(—)
Both Sports Team				.566***	.339[a]
				(.1333)	(—)
Both No Teams					-.111[a]
					(—)
Different Teams					-.274[a]
					(—)
Intercept (Edges)	-.911	-4.464	-.3.745	-5.022	-4.160[a]
AIC	54005	3894	31008	3684	3600

[a] Problem with the model; unestimatable standard error because there was no divergence in the estimates across steps of the algorithm.

or near-linear between the predictors. Because we are parameterizing the team possibilities nearly completely, I thought I'd drop one of those, and re-run, but no dice. (And it's because ERGMs tend to have trouble with lots of terms like this that I hadn't added in the EGO_MATH, etc., terms, as described in note 27.)

Let me emphasize that I didn't set out trying to find data that would crash the ERGM. It took me so long to finally get my stupid simulation up and working properly that I sure wasn't going to change it for *anything*.

The naïve logistic worked, and the ERGM crashed, for the most straightforward friendship data I could simulate.

I'm sure that with sufficient ingenuity and duct tape, someone could coax something closer to a plausible model 5 using an ERGM. More commonly, one would just throw out our parameters—and ask a different question, one that this estimating routine can answer. But it's not clear to me why even if we could get MCMC estimates of this model, that would be such a good idea. The ERGM *won't* be "the" right model. It'll take a lot more time to do . . . and that can be good, if it forces us to get to know our data, as we're debugging one failed model after another. But some of that time could be better used exploring alternate specifications. I worry that by the time most users get an ERGM to cross the finish line, they've become very cautious about trying different ways of accounting for the interdependence!

I'm not saying that, given a choice of two equivalent models, one that can be fit with a procedure that will deliver maximum likelihood estimates, and one that won't, you shouldn't prefer the former. You should—it means you have a better sense of what the data are telling you in answer to your question. But what's most important is that you are able to get robust answers to the right questions, not the best estimates of the parameters for the wrong question. Because, when you think about it, *none* of the parameters in *either* of our tables could be "right," in the sense that they didn't map onto real-world (that is, fake real world) properties. Alter's popularity *doesn't* increase choices; being on different sports teams *doesn't* decrease choices. These are indirectly tapping something that *is* happening, but only indirectly. What we need is to use statistics to reject false interpretations, not refine our estimates.

In sum, the ERGMs are exciting, but they're brittle and they don't have obvious epistemic privilege. For this reason, for a long time, I was urging people to drop this approach entirely. The models no longer have the rigorous connection to the Hammersley-Clifford theorem, nor do they have any behavioral interpretation, which is almost always what sociologists want. That is, we don't want a story about abstract probability distributions, which is what we get. Given that Tom Snijders had been working on an alternate way of approaching the same problem that *did* have a behavioral interpretation ("stochastic actor-oriented models"), I've been telling everyone that this is the way forwards.

You'll find out in the next chapter why I was wrong.

Throw a Centrality on That Baby

There's nothing inherently wrong with ERGMs (or the stochastic actor-oriented models [SAOMs] we'll look at in the next chapter, for that matter). But there is with ritualism . . . here, the complacent conviction of the non-technically expert that ERGMs or SAOMs are "the" right way of approaching network data. Still, I think this might be better than what we had before, which was a ritualism on the part of users who relied habitually on canned network "measures." Let me close this chapter with a brief discussion of them, since we're on the subject of network rituals.

There are two problems with the use of such so-called network measures. The first is that the meaning of these statistics depends heavily on the type of tie that is being examined. The second is that the measures don't really behave as you might think.

Regarding the first problem, we need to bear in mind that these numbers aren't really *measures*. They are the results of computations that may produce explicable quantities in some circumstance. But we need to keep distinct the nature of actual computations (on the one hand) and what they are generally *called* (on the other). In particular, there is a great deal of confusion over the term "centrality," in large part due to an extremely irritating tendency for people to speak of centrality as if it were some master concept, but that it came in different flavors. Centrality makes sense only in a network that *has* a tendency toward centralization. When people talk about "degree centrality," what they actually mean is . . . degree. Period. The number of ties a node has.

In some cases, a measure of degree might tap popularity. But popular people aren't always "central" in any way that has anything to do with the meaning of the word. There can be central people who are unpopular, and popular people who are peripheral. Where a network *has* a center-periphery structure, you can classify nodes only once you get a handle on this structure, which you might do through a qualitative model like a blockmodel, or a quantitative one, like a multidimensional scaling model. Centrality is an important idea, and we don't want to lose it by confusing it with other good ideas.

One of these other good ideas is the notion that it matters less how *many* people you know; what is important is *who* you know. So let's give you a status based on your popularity with the *cool kids*. How do we determine who is cool? Let's assume that you're cool if you're friends with the cool kids! But wait, how do we know who is cool? Because *they're* tied to *other* cool kids! It's a circle, but just the sort of circle that can be solved in mathematics. It's just like the spatial lag model where each place is affected by its neighbors. It leads to an eigenvector representation. Hence

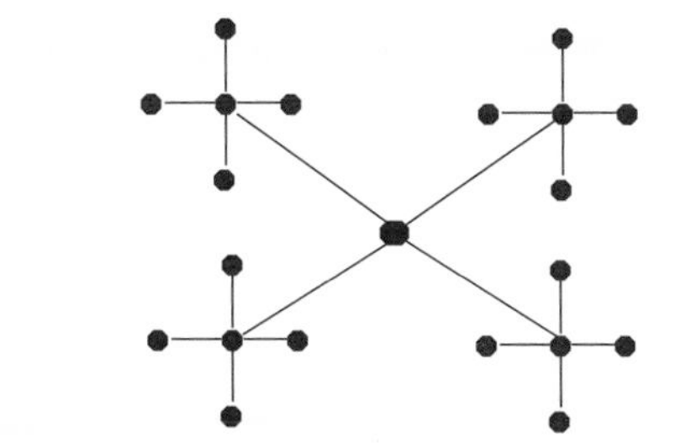

FIGURE 8.8. A Broker?

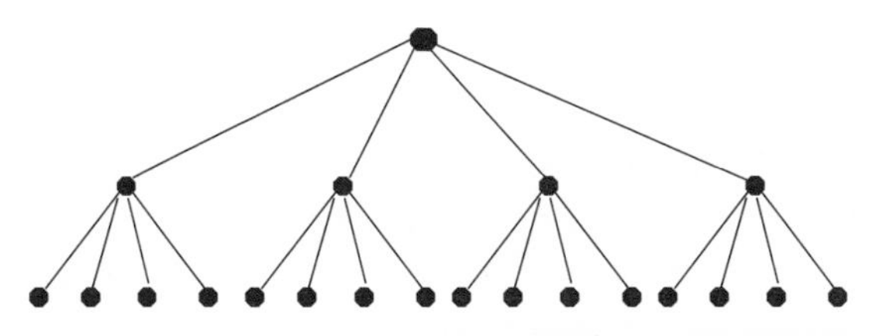

FIGURE 8.9. A Leader?

"eigenvector" centrality is where we weight each person's contribution to anyone else's status by the status of those who are tied to them (Bonacich 1987). The problem is that, for most social networks, it almost never makes any difference—this measure usually correlates quite highly with degree, and when it doesn't, it usually means that the assumptions that justify the measure aren't met.[29]

A second good idea is that we can look at those who tend to be "in-between" others.[30] That's closer to what we mean by centrality, but it's not necessarily the same thing as a *global* centrality. And it can also be totally misleading in certain networks. For example, let's look at the network graphed in fig. 8.8. Here it seems reasonable to speak of betweenness, and the central node certainly seems most between others—because for most of the dyads, any connection has to pass through this node. If we were thinking in terms of brokerage, we might think that this is a good example of a broker. But that only makes sense if this is a symmetric set of relations. If the relations are actually antisymmetric, this sort of structure might be better understood in the form displayed in fig. 8.9.

29. It is Scott Feld who pointed this out in a talk, but so far, he hasn't written this up. You rock, Scott!

30. Mathematically, what we do is, for each node, look at all *other* pairs of nodes, and see what proportion of the shortest paths connecting them go through our focal node. It turns out to be a bit less ad-hoc than it sounds, but there are information theoretic versions, ones that tie to approaches taken in operations research and fluid dynamics, that are even cooler.

In this case, what appeared as "betweenness" in the first graph was actually hierarchical position; and if we were looking for brokers, they wouldn't be the person on the top, they'd be the four people who sit in between that node and the bottom nodes. The same computation has different meanings given different types of relationships.

The second problem with the use of these canned statistics is that they're often included in traditional linear models (as in, "I'll take each person's centrality, status, and betweenness into account!"). But if you are looking at networks, you have just opened up a 50-gallon drum of nonlinearity. This leads to problems of comparison both of one network to another (*across*-network comparisons), and of one node to another (*within*-network comparisons).

Let's start with the first issue. Most of these so-called measures are going to be linked to the *size* of the network in ways that are hard to disentangle. I'll forebear from working through any examples since the general principle is so simple: if you have an idea, see if you can construct a model, and then run it on simulated networks that are similar in structure, but of different sizes. For example, if you are interested in a changing tendency for published papers to cite the same predecessors (such that some early papers get elevated to a position of "canonical"), write that down as a choice process. And then apply it to networks of different sizes (since if you are looking at a citation network, that necessarily grows over time). If you see that your results vary non-linearly with N, as they almost always do, you realize that you probably can't make strong claims about change just by using a conventional linear model.

Some great recent work (A. Smith, Calder, and Browning 2016) has started grappling with these issues. They're showing that rather than control by N, or even try to standardize the results in terms of an empirical distribution, we can be better off using a full parameterization of the *type* of graph involved. That's definitely true . . . except that in most cases, it isn't at all obvious what type of graph we have, if any. It's for this reason that we are tempted to try to use ERGMs, and to interpret differences in parameters across graphs (for example, this one seems to have a stronger tendency toward transitive closure than that).

But we can also use the ERGMs not to quantify effects, but to make different, nested, families of graphs, and compare the distribution of different, more interpretable, graph measures in each. That can be an advantage when it comes to complex models, because the "partialed out" coefficients from an ERGM—one that has evolved as a compromise between what you are interested in and what you can fit—can be harder to interpret than you might want. And you can also use a non-parametric approach to generating theoretically interpretable, nested families of random graphs using the *dk* series (Orsini et al. 2015); explaining that would take us too far aside, but you can see a use in Martin (forthcoming).

Okay, so we've got to be cautious in making cross-network comparisons. But it's also very easy for us to screw up cross-individual comparisons within the same network. We often try to do this by making individual-level network measures and tossing them into an individual-level regression. The statistician may focus on some of the ways in which this violates this or that OLS assumption, but that's almost certainly not your problem. First of all, most of the node-level graph statistics we use are usually all correlated highly with the degree (the number of ties). That means that we can't just throw in "betweenness" and interpret the effect as one of brokerage, because it could just as well be due to popularity.

So you might think, I'll control for popularity. But the two measures are tangled up in ways that make this a risky bet. I understand that the processes that lead to one's education and one's income are also tangled, yet we use linear models to disentangle them all the time, and to apparent moderate success. And where the correlation between betweenness and degree is low enough (say, $r < .75$), that might work. But you'd want to look at the scatterplots to see where your variation is. And don't trust an assumption as to what sort of graph you have.

Fig. 8.10 gives scatterplots (and correlations) between three common node-level statistics for random graphs of 500 nodes (R 8.6). On the upper left, we have a totally random graph (an Erdös graph of a constant tie probability), and on the upper right, a preferential-attachment graph (a Barabasi/Albert graph). On the lower left we have a small world graph (a Watts graph). And on the lower right, we have that funny high school that I simulated! All are treated as undirected graphs. What you should notice is that there are different patterns of correlations, and also that the one that was simulated on the basis of a plausible guess at actual dynamics, as opposed to mathematical elegance, has higher correlations between degree and the other measures than the preferential or small world graphs. And in real world data on interpersonal networks, I tend to see those sorts of high correlations.

So you need to think through the question you want to ask of the data, and make sure it isn't "double barreled" (that is, you combine two questions that get one answer). Don't trust what everyone *calls* the measure. Study your data.

Conclusion

Sociology is all about trying to study social structure, and statistics wouldn't be of much use to us if they couldn't help in that project. They

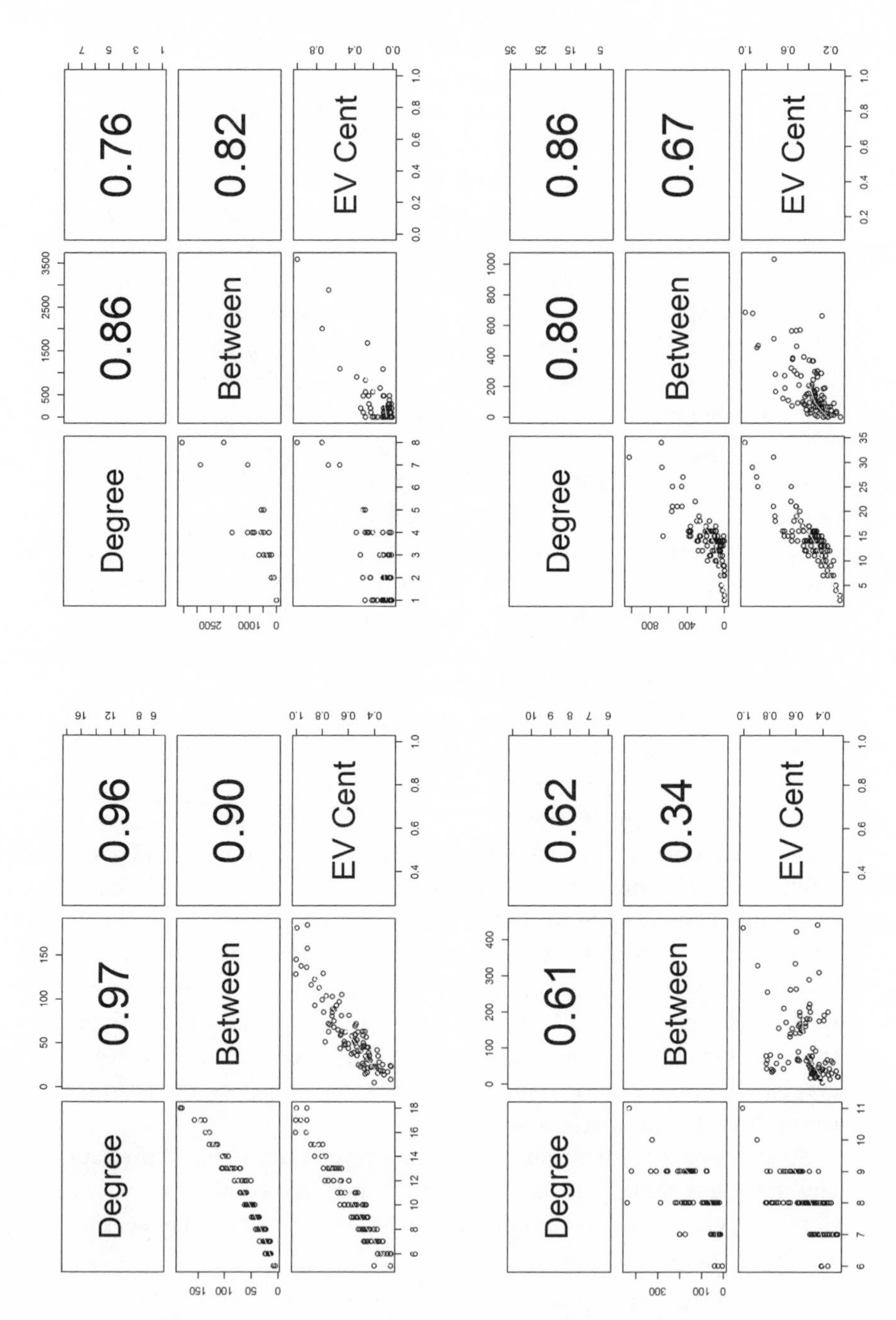

FIGURE 8.10. Interrelations of Some Nodal Level Statistics

can. But most of the time we think our model is giving us evidence of some interesting structural effect, that's because we have, in effect, assumed that people are complete idiots. They don't know what sort of music they like, which other people are jerks, what their race is, what is a smart paper, and so on. That's why they need to be told by their friends or their parents or whatever. If we try to take into account the fact that two friends may be friends precisely *because* they have similar tastes, we make equally insulting assumptions (they both like "books," and since, after all, all books are the same, that can be trusted to control for taste heterogeneity). Of course there's tons of social influence all over the place. But that doesn't mean our statistics are good at finding it.

Don't give up! But we must avoid ritualization at all costs, and focus on the *substantive* implications of our own questions, and less on what people tell us is the flavor of the month in terms of models, or, equally bad, go with what strikes us as "the traditional" approach.

* 9 *
Too Good to Be True

Overview

In the previous two chapters, we explored cases in which we think we're getting good results, but they all involve a similar form of misspecification. There is information about our cases stored in the world—*where* the cases are, *when* they are (or were), or *how they relate* to one another—that we don't know about. Yet they drive our findings. Now I want to turn to a different way in which we can get results that are deceptively good. This is when what seems to be information about the world is actually built into our methods—that is, our assumptions are too strong.

Map: I'm going to focus on three classes of methodological approaches that are, in a nutshell, usually too good to be true (TGTBT). They give you more information than is plausible given the data you have. That's because they have too many assumptions, and whenever you test a model, remember, you are testing not just your *substantive* interests, but the assumptions of the model. If those assumptions aren't safe ones, you'll never know you're wrong.

The first kind—latent class analysis and its kin—involves a very *formally* exciting and cool method. And it feels like one is doing that sort of inductive data-exploration which I've advocated. But we'll see that to interpret the results, we almost always undo all the formally interesting parts. Since the results are so unstable anyway, it invariably makes more sense to do a simpler analysis. The second kind of TGTBT method—QCA (Qualitative Comparative Analysis) in its various incarnations—requires perhaps fewer assumptions, but these assumptions are implausibly strong, such that you'll always be handed a strong finding on a platter. And the third variant—the conduction of microsimulations based on a priori reasoning—leads us to multiply parameters *ad astra*, which is going in precisely the wrong direction.

And one small word here: I've decided not to discuss the well-known failure of structural equation models (SEMs) to identify reciprocal causation from cross-sectional data, for three reasons. The first is that this approach is rarely used in sociology (though implausible SEMs may still be seen in psychological social-psychology). The second reason is that these problems are well understood and taught in every SEM course, along with ways of checking the robustness of estimates. And the third reason is that SEM was too quickly dismissed along with these problems . . . and, with it, a serious investigation of error in our measurements. But as we've seen, this sidelining of measurement error was a disaster for social statistics. Even for cases where you aren't going to rely on the findings from a complex SEM, without actually running one, you can still get a great deal of insight from considering one as a thought experiment—*could* my data be generated in such-and-such a way?

Also, to be honest, SEMs got the reputation of being sort of fuddy-duddy. (That should be a signal to take the method seriously—in social statistics, "old fashioned" often really means "robust.") In contrast, the methods here all seem new and exciting (even if they're actually no newer than other ones). They can get you to invest a lot of time trying to solve their problems. And chances are, you aren't going to get much back in return.

'Twould Be Pretty to Think So

Latent Mixture Models

Consider the data in table 9.1, on attitudes about abortion and the death penalty (real data, from the 2000 General Social Survey). There are a few different ways we that we could think about this table. One is to imagine a generic person, and introject the interpersonal comparisons into his/her head, turning them into intramental conceptions (Martin 2000; Borsboom, Mellenbergh, and Van Heerden 2003). Thus we imagine that this is about a *dependence* between these two ideas—holding one makes you more or less likely to hold the other. If you thought the death penalty was basically okay, and I convinced you that it wasn't, you'd probably change your mind about abortion also, and this degree of association is nicely given in the odds-ratio of this table. Of course, as we see here, there's little relation between the two (the odds ratio is .95, very close to 1), which would imply that these beliefs aren't related in people's heads at all. But that seems a bit strange, given that they are both about highly politicized life-and-death issues.

If it sounds odd to you that there is "no relation" between the two sets

Table 9.1. Cross Classification, Attitudes toward Abortion and the Death Penalty

		Death Penalty	
		Favor	*Oppose*
Abortion Legal For Any Reason	*Favor*	447	210
	Oppose	670	298

of beliefs, I feel for you. You might think that it's because there are two different associations that just happen to cancel out. On the one hand, there is a positive relation due to feelings about life; on the other hand, there is a negative association due to politics.

Or you could propose that rather than there being just *one* type of person, and us therefore using all four numbers to make a single odds ratio, there are *different types* of people, and we need to split them up. There might be two, three, or four types. Well, in one way, we know that there *are* four types. Those types are found in our four cells: some favor abortion but not the death penalty, some the death penalty but not abortion, some both, and some neither. But accepting that kinda kills our sociological investigation. Because within any of these categories, there is no variance at all on either item. For that reason, the items are *conditionally independent* (that is, they are independent *conditional* upon what cell a person is in). So there's nothing left to say.

But we don't have to choose between either one class or four classes. We might, for example, propose that actually, there are two kinds of people with real attitudes, pro-life and pro-death. The former are against both abortion and the death penalty, and the latter support both. How do we explain the presence of people who are for one, but against the other? Well, some of the folks with real attitudes just give a wrong answer (or are recorded as doing so). And then, maybe, there are some people who don't have any attitudes at all—they just answer randomly. That would actually be a third class.

Okay, maybe. But here's another argument. Sure, there are two kinds of people with real attitudes, but they're not pro-lifers and pro-deathers. Rather, they're liberals and conservatives. The former are for abortion and against the death penalty, and the latter are against abortion but for the death penalty. How do we explain the presence of people who are for both or for neither? Well, some of the folks with real attitudes just give a wrong answer (or are recorded as doing so). And then, maybe, there are some people who don't have any attitudes at all—they just answer randomly. But still, in either case, we propose that there are only three classes, not four. Two consistent classes, and one "trash" class of bad respondents.[1]

These last two theories can be formulated as the extremely lovely latent class model (Lazarsfeld and Henry 1968; Goodman 1974). (It's usually called LCA for short.) It works like this. We have J different classes, each with a proportion π_j of the total population. (For simplicity's sake, I'll assume dichotomous observed items, though we can generalize.) For each k of K items, the members of class j (C_j) have a probability of answering k in a positive direction denoted p_{jk}. So what is the overall probability that the ith person answers the kth item, x_{ik}, positively? It's just the sum of the conditional probability of a positive response *given* that i is in class j times the probability that i actually *is* in class j (π_j), or

$$\Pr[x_{ik} = 1] = \sum_j \Pr[x_{ik} = 1 \mid i \in C_j] \times \Pr[i \in C_j] = \sum_j p_{jk}\pi_j . \qquad (9.1)$$

And since the items are conditionally independent within any class, the probability of any observed response vector $\mathbf{x}_i$ is $[x_{i1}, x_{i2}, \ldots, x_{ik}]$ is

$$\Pr[\mathbf{x}_i] = \prod_k \sum_j p_{jk}\pi_j . \qquad (9.2)$$

Neat! Isn't that elegant? Well, the problem is that the three class model has 2 + 3 + 3 = 8 parameters, twice the number of cells (there are two π_j parameters, since the three π's are constrained to sum to 1, and then a p_{jk} parameter for each class-item combination). Even if we had a third variable, we'd still have more parameters (11) than cells (8). If we had four items, and 16 cells, we could identify our parameters, just barely. But chances are, we'd still be making nonsense.

Why? Basically, because (speaking somewhat informally) we take our observations—the *one* thing, namely respondents saying yes or no to an item—and try to get *two* things out of it. One is the conditional probability of people in the class agreeing to the item, and the second is how many are in the class in the first place. We can put more people in the class . . . but then we'd have to lower our estimate of the conditional probability; or we could go the other way.

Of course, the reason the model can be identified is that we're doing this for a number of observations we're fitting all at once. But the more general point is valid: we have *two* moving parts where we really have one piece of information. Borrowing my terminology for theoretical argu-

1. This is what we actually called that sort of miscellaneous class back in the heyday of LCA. . . . but I anticipate. . . .

ments from *Thinking Through Theory*, I'm going to call these approaches "Tweedles," because, like Tweedledee and Tweedledum, there are always pairs with their arms around each other. These techniques look for invisible groups that, *if* they existed, and *if* they had certain properties, would have produced these data.[2]

The result of this is that there are often multiple solutions that are nearly equally good. The way to think about it is to imagine that there are two parameters—say, two regression coefficients—that we are attempting to estimate. We want to pick the ones that maximize the likelihood of our observations, right? So envision a hilly terrain or a mountain range, with latitude being the value on one coefficient, and longitude the other. Our different techniques will all do well if there is one smooth hill, and you just need to march "up" from wherever you are. Where there are crags and different peaks, if you start out on the wrong peak, you can start walking uphill and end up atop Mount Pokalde, never dreaming that if you went downhill and back up, you could end up almost two miles higher at Mount Everest. So most routines need to drop you pretty close to the main mountain.

In other words, our algorithm can get stuck on a local maximum, as opposed to the global maximum. In this case, our solution is very sensitive to the start values we use. Now let me emphasize that sensitivity to start values doesn't necessarily indicate that we've asked the wrong question. But in some cases, it does. If there are multiple ways the world can work, and they have substantively very different interpretations, and they all explain the data equally well, it doesn't make much sense to see whether one solution's log-likelihood is 0.00004 lower than another's, does it? Why should you care? In many cases, this is just what LCA has in store for you. You get a local maximum likelihood estimate of the parameters, and it might even be a global one, but what you don't know is whether there's just as much support for the theory that is *antithetical* to your own.[3]

In other cases, the problem is that the likelihood surface is very flat (which is also seen in SEMs that incorporate too much endogeneity). You might have a class that is 24% of the population and answers item *A* positively with a 82% probability. But the model where this class is 58% of the population and has a 45% probability might fit basically the same. That's a

2. If you are mathematically sophisticated, you're thinking, "but most of the time parameter estimates are intertwined, to an extent given in the information matrix!" That's true, and my exposition here is inexact, but I want you to think in Tweedle terms for a moment, because it will help us understand the problem of local solutions.

3. I'm not going to demonstrate this, because every introductory piece on LCA will warn you about this, and suggest ways to deal with it . . . numerically, though. They won't suggest that this calls the whole enterprise into question.

Tweedle for you. You've asked it to walk uphill, and you're in the middle of Illinois. There ain't no hills here, far as I can see. Finally, in a few cases, the surface isn't quite flat, but there is a long ridge—there is one direction in this space in which it doesn't matter which way you go. You can generally find this by inspection of standard errors of the parameters.

Let me say, once again, these problems aren't restricted to LCAs, and there are ways of avoiding local maxima (multiple start values, simulated annealing, re-starting away from a first convergence, and so on). But with unconstrained LCA models, these problems should be enough to make you very cautious about making an existential claim—there "are" five classes—without actually checking out alternative configurations, including alternative systems *of five classes*, by adding arbitrary constraints. More important, in other words, than the difference in the likelihoods is the difference in *substantive implications* of your different maxima.

It's the open-endedness, the "tell me a story" nature, that leads fitting problems to become theoretical problems. You can, however, impose constraints (such as telling the routine, "I want only classes that are nearly uniform in accepting or rejecting some attitude") or even use LCA in a confirmatory fashion. (For example, Jim Wiley and I [Wiley and Martin 1999; Martin and Wiley 2000] used LCA to incorporate response error in algebraic models that we still stand behind.) But when it comes to an *exploratory* investigation, where we are allowing the model to choose both parts of the Tweedle, . . . no. We have too many moving parts.

Does LCA ever help you see things in data? Of course it does! Does it even help you see *true* things in the data? Almost certainly. Are these things classes? Usually not. Often, the classes turn out to be classes in the sense that we might speak of "high," "medium," and "low" (e.g., Bonikowski and DiMaggio 2016: 958). The classes vary, that is, simply in terms of their members' tendency to respond to a clump of items one way as opposed to another. They introduce qualitative simplifications to quantitative data where these just aren't needed (we usually have no trouble thinking in continuous terms of more-to-less), nor are they justified by the data. Or in other cases, the classes simply point to clusters of items (like "these people like religion" versus "these people like politics"—we can see the clumpiness in the substance of the items). In such cases, the LCA doesn't add anything to more straightforward approaches except confusion.

LCA might sound a lot like conventional factor analysis, about which I've spoken quite positively (see chapter 1). But factor analysis and related Principle Components techniques are, in practice, different from LCA, because they're intensely stable. Like OLS, they tend to fall into the same configuration no matter what you do (though how you interpret/rotate is up to you!). Factor analysis works well for exploratory analysis because all it's doing is twirling the data space and projecting it down into a space of reduced dimensionality. Tweedles are different. They are incredibly sensitive. They look beautiful. And they work fine if all you need to do is to reduce your data (and not make meaningful claims about the world). But watch out—you can spend a long time twiddling with a Tweedle and only end up with twaddle.

Finally, I'd like to point out that the original inspiration for this approach comes precisely from a line of thinking that proved to be famously unreliable. Lazarsfeld had been inspired by Freud's distinction between the manifest and the latent content of dreams (Lazarsfeld's mother was a practicing psychoanalyst, a student of Adler's!). But this so-called latent content of the dream was, to quote one insightful analyst, "simply Freud's own associations authoritatively introjected backwards in time into the patient's own dream" (Martin 2011: 103). It was actually like a Rorschach test that the *analyst carried out on himself*, turning his own projections into "scientific fact." And that's, sadly, usually what happens with LCA.

Sequence

The latent class analysis is only the simplest case of a wide variety of methods that have the same basic Tweedle nature. They look great on paper, and we can spend a lot of time making algorithms and pictures and all that, without really having any way of getting a sense of whether they're true or not.

One very interesting class of problems has to do with trajectories and sequences. In many forms of data, we have the same units (for example, persons), observed on the same variable (for example, weight, occupation) over time. We'd be interested to see whether there are a few different types of trajectories. Recently, Warren et al. (2015) compared a bunch of the most widely used methods—optimal matching, grade of membership models, and finite-mixture models—using both real data, and simulated data, to see which methods were best. It's that sort of horse race that methodologists love. Guess which horse won?

It's a trick question. They all lost. To stick with the metaphor, they basically all wandered in different directions. The methods not only rarely got the right answer, but, just as bad, they *never* agreed with one another. The most consistently picked solutions were wrong. They might be more likely to pick the true value than any other value (not always), but still, it was a minority decision for the routines. They just didn't do well enough that you'd rely on them if you cared about being right.

But most of us never know that we're wrong. One of the problems with these extremely cool methods is that they don't give off obvious squeaks and groans when they're failing. We ask them a question that they can't plausibly answer, and they chug along merrily. Because the math is so elegant, and the connection to the data so complex, it's very hard to know what is actually happening.

And it turns out that most sequence data is being crammed into algorithms that don't fit it. Sequences turn out to differ in substantively meaningful ways. Some have fixed times when each state is sampled (e.g., yearly reports on BMI from a health study), some don't (e.g., data on career taken from a resume). Some have states that one can only occupy once (e.g., being an undergraduate) and others have states that one can occupy over and over again (e.g., being unemployed). For some, states are ranked (e.g., position in the military) and in others they aren't (e.g., pregnancy, lactation, fertility, and infertility). And many are confusingly in between (e.g., conventional careers).

Very often, the work that it takes to squeeze the data into the form that one of these techniques requires is going to introduce so much new pseudo-information that it isn't worth doing the technique. If you think about the substantive nature of the data in question, the chances are you can ask a clearer question (such as "time to promotion to lieutenant general") that won't leave you in that position of choosing the "best" world (according to a largely arbitrary criterion) and ignoring the many opposite worlds, one of which might be the real one.

Get Your Ass Lost

Am I saying that you shouldn't use these exciting exploratory methods? Not exactly. But in almost all cases, your readers should never see them. They should be used to generate ideas that you then present more directly. Because you almost certainly will interpret your results as if you had used *manifest* variables. That's because no one cares that "there are eight kinds of people" in the world. Now that I know that, what do I do? Most probably, once you've assigned people to these eight categories, you're going to try to learn something about them. You might find that those in categories 1, 2, 3, and 4 are mostly men, while those in categories 5, 6, 7, and 8 are mostly women. You might find that those in categories 1, 3, 5, and 7 are mostly college graduates, while those in categories 2, 4, 6, and 8 are mostly not. You might find that those in categories 1, 2, 5, and 6 are mostly Republicans, while those in categories 3, 4, 7 and 8 are mostly Democrats. Great! *Now throw the latent classes away*. Present your results in terms of the eight *manifest* classes formed by the cross-classification of these three dichotomies. That's what your reader is going to be thinking about, so that's what you need to share.

Of course, that's not what anyone wants to do. After all that work, who wants to go back to crosstabs? For example, in a recent article using sequence analysis, Aisenbrey and Fasang (2017) clustered life sequences in terms of work and family in Germany and the US. For Germany, for example, they came up with 8 classes based on job prestige and number of children. One class is called "couple, childless; Upward mobility." They say (1470) this "accounts for 10% of the population and consists of respondents who live with a partner and remain childless by age 44." But if you look at the graphs, you should be able to tell that in fact a fair number of those in this cluster *do* have children. While the cluster is disproportionately childless, this isn't the set of those indicated by the labels. There is something else that the members of the cluster have in common . . . but we don't have any theoretical grasp on it. So if we want to talk and think about the childless, we should construct a group that *actually* represents the childless—otherwise, our statements and our methods don't match up.

Now there are times when we do have a stronger theoretical interest in the clusters as clusters. Most notably, Amir Goldberg worked damn hard on coming up with a set of interconnected algorithms that would, he hoped, find that there were latent groups in the population who "agreed to disagree"—who agreed as to what items implied which others, even if they didn't agree with one another about which package to accept. It's a hugely cool idea. And if there were separate fights going on at the same time, conventional techniques would all be giving us the wrong answers![4] He and Delia Baldassarri (Baldassarri and Goldberg 2014) worked to apply the algorithm to data on public opinion. And what they found was that it looked a lot like there were three classes—one that fit the usual conception of how political ideology works, and a second that matched economic and moral issues differently. The third class was just the "trash" class of folks they couldn't make sense of.

But because these classes are combinations of *enemies*, they are going to necessarily include people who are different. Describing each class therefore requires describing the polarities, not the modal states, which they went on to do. But this is in effect first to "back up"—to make classes and see what characterizes members—and then to "go forwards," trying to figure out within each class, what characteristics put a person on one side versus another. In other words, our normal sociological way of thinking

4. It turns out even if this is what we want, there's a better way to do it; see Boutyline (2017).

is, for better or worse, one that starts with explanations [*explanantia*] and ends with the results [*explananda*]. Most latent class techniques divide people up on the basis of the latter, which is, well, not really cheating, but still in a way short-circuits our normal ways of testing the strength of our assumptions. For that reason, it's crucial to compare the predictive power of a latent class formulation to a more straightforward approach using the manifest variables. In this case, as far as I can tell, the core of Baldassarri and Goldberg's results are related to a classic notion in American politics, namely that—to focus on the ideal typical representatives—private sector managers are socially liberal but economically conservative; public sector managers are socially and economically liberal; union workers are socially conservative but economically liberal; and the self-employed are socially and economically conservative. Rather than use the latent classes, we could use more refined information on *manifest* classes to try to go consistently forwards, and see whether we can explain who ends up with what opinions.

One last thing: a whole new set of related methods is making inroads into the social sciences due to the fad for textual analysis. The simplest of these, Latent Dirichlet Allocation (LDA), is basically latent class analysis for words instead of responses. Although the math is different, it's trying to do the same sort of thing as LCA, looking to make Tweedles. You might think that I'll say it's more garbage to stay totally clear from. Well, actually, it's probably better than most of the other techniques, for three reasons. The first is that it's usually applied to very large amounts of data, where it's harder to overfit, and, more important, we have fewer other options—we're going to need to reduce the data *somehow*. The second is that from the get-go, there's been a lot of critical attention and practitioners were sensitized quite soon to all the problems that they were likely to encounter. But most important, LDA and its kin are about *words*. It's sort of like doing LCA on people you know well. It's one thing to look at a category and say, "well, these people are 34% middle class, 8% obese, and 53% college educated . . . I'll call them the hipsters." It's another to say, "oh, look, it's Barry and Harrison and Peter and Ron! I know these guys!" That's what we can do with the words. That helps.

QCA—It's Just like Statistics . . . Only without the Statistics!

Case-Based Analysis

Wouldn't it be cool if we had a new method called "RWOST": Regression WithOut Statistical Tests? We'd all get results, and we'd never have to say "I don't know." That's basically what we have now in the technique called Qualitative Comparative Analysis (QCA). Indeed, it's the fact that QCA turned into RWOST which explains why this very simple and old-fashioned approach has become so widely loved. Now I have to say, not only am I a fan of the use of logic and Boolean thought in sociology, but I loved this work as it was unfolding by Charles Ragin, someone I think is a real treasure for the discipline. This is what we need more of—trying out new approaches and branching out, instead of putting all our eggs in one basket. But that strategy of diversification requires that we also throw things away when they disappoint us. QCA is one of those. So Ragin's done a lot of things; I doubt he'll be broken up if this one turns out not to work. Most things in science leave no descendants—if they did, we'd get nowhere. The system works by diversifying and culling, diversifying and culling, over and over.

I'm going to start with the basic logic, and then consider some more recent extensions. So do you remember that in chapter 4 we saw that focusing on interactions can be an easy way to make big mistakes? Well, QCA is basically an interaction-monger. It goes where our methods are the weakest, and sets up camp.

QCA is based on the simple logic that comparative historical researchers used, especially in the 1980s US. They tended to think in terms of *necessity*—to say that "the pattern of alliances was responsible for the first world war" was as much as to say that had there been no alliances, there would have been no war. The problem is that, as I said in *TTM*, necessity is spoiled by a single counter-example. If you're trying to explain only one event, sure, you might be able to claim that the pattern of alliances was necessary for *this* war. But if you had evidence on twenty or even ten wars, the chances are that you'd find a disconfirmation of *any* claim you made.

What QCA does is to attempt to salvage the basic logic in the face of more cases, as opposed to grappling with the underlying lesson (as we'll see). It allows that there can be more than one path to the outcome. Each path, considered on its own, is generally *sufficient*, and it is only necessary that *one* such path be taken.

And, at the same time as it multiplies explanations, QCA throws out the few constraints we have on making claims, namely conventional statistics. The essence of QCA is looking in a lattice of all possible states for certain patterns of variables that uniquely separate the cases that are high on a dependent variable from those that are low. In itself, nothing wrong with this, and it's something that we often do informally, as we look over cases. But by automating this, QCA stands as a shining example of how we can turn ignorance into knowledge, and rely on assumptions that are far too strong.

QCA is sometimes claimed to be different from conventional analysis, because it is "case-based" in some way, as opposed to variable-based. This is not true (Lucas and Szatrowski 2014: 63). The logic is the same as regression. Each "path" that QCA suggests is, you will notice, *not* given in terms of cases, but in terms of variables. And each variable, as Breiger (2000) emphasizes, is really a bundle of cases (and non-existence of cases). Usually what people mean when they say that QCA is case-based is simply that they have *fewer* cases, and that the cases loom large in the mind of the analyst. For this reason, QCA is often used with "large" cases, like countries. As I noted in *TTM*, this already involves a pretty strong assumption—that big effects have big causes. QCA is one of those approaches that assumes that the causes are what I called "elephant" causes, not "flea" causes. But we're going to leave that aside for the moment.

Even if we are focusing on elephant causes, because we have few cases, we can only include a few variables. But since each "variable" is a set of cases, when one has few cases, there are probably a lot of relevant variables that group the cases in the exact same way. Whatever you happen to measure and include in the data you feed into QCA, that's what it seizes upon. So say you made a variable "weak welfare state," which produces a vector across your 9 cases of [0,0,1,0,1,1,0,0,1]. Now if you had instead constructed a variable "weak Catholic church" and this led to the identical vector [0,0,1,0,1,1,0,0,1], you understand that anything you said about the welfare state in the first analysis, you'd now say about the Catholic church in the second. And the chances are, you'd be reassured in the first analysis that the welfare state was a causal factor for your outcome, but learn in the second case, that actually it was the Catholic church that had a crucial role.

If you have been following my logic, you should object, "but this is also true of conventional regression! If you have two perfectly correlated variables, they are functionally interchangeable!" *Exactly*. The logic of QCA is *exactly* the same as regression. Here are the only differences: QCA is characterized by

1) An automated greedy search for any patterns that might exist, that is . . .
2) nearly guaranteed to produce a perfect finding and as much explanatory power as you desire, in the . . .
3) absence of any statistical tests.

If you had data on 3400 people and threw 30 variables into an algorithm and let this choose not only main effects *but up to four-way interactions*, and then you reported this as a strong conclusion *without statisti-*

cal tests you would be *fired*. The idea that this is defensible so long as you have only *one hundredth* the number of cases is . . . insane. Statistics is not a cab that one can have stopped at one's pleasure.[5]

What QCA is good at is turning random patterns in data into unrejectable claims of deterministic causality. That is literally the last thing you want as a data analyst. So if you were wondering why, given the immense sophistication of current statistics, and the radical increases in computing power, there was such enthusiasm for a simplistic technique that could have been done in the eighteenth century, the answer is the same as if you wonder why a Nigerian prince needs *your* help smuggling a million dollars out of Russia: because it is too good to be true.

The Assumption of Determinism

It is the assumption of determinism that leads to the complex patterns produced by a QCA analysis. Very often, a QCA will return something like this: if you want to have a successful revolution, either be in an export-oriented country that has low state repression and a multilinguistic polity, or an export-oriented Catholic country, or an import-oriented country with high state repression and monolinguistic polity. We busily set our minds to work explaining why the same thing has one effect in one context, and a different one in a different context. But we're assuming what we should be testing, which is that all these attributes actually *are* causes in the first place. They almost never are. It's like trying to explain how hood ornaments make cars go. We know they go *with* the cars, but we don't learn how cars work by *assuming* this.[6]

I am not denying, as you will see below, that QCA *is* appropriate for situations in which we have complete, error-free data on deterministic processes. I'm just saying that I've not yet seen it applied to such a situation. And that's because the only such situations that I know of are trivial. Any layperson can figure out the answer informally; no need to turn on the computer. Our actual problems involve incomplete, error-prone data on processes that are epistemically, if not ontologically, stochastic. Adjusting our assumptions to justify applying a favored technique is reasoning backward: "we need to use QCA, so let's assume no error and determinism." That's like an engineer saying, "we need to launch this rocket, so the o-rings are safe."

5. The relations between QCA and regression type techniques have been most brilliantly and sympathetically analyzed by Breiger (2009) in another piece that should be required reading for all sociologists.

6. Mahoney (2008: 415f) simply declares that the actual probability of an event that occurs is 1 [!]. With probabilism like that, who needs determinism?

Table 9.2. Data from Hicks et al.

	Liberal	*Catholic*	*Patriarchal*	*Unitary*	*Workers*	*Consolidated*
Denmark	1	0	1	1	1	1
UK	1	0	0	1	1	1
Germany	0	0	1	0	1	1
Italy	0	0	1	0	1	1
Belgium	0	1	1	1	1	1
Netherlands	0	1	1	1	0	1
Australia	0	0	0	0	1	0
USA	1	0	0	0	0	0
France	1	0	1	1	0	0
Norway	1	0	0	1	0	0

Now many QCA'ers think that they can avoid this problem of assuming deterministic causality by changing their interpretation. We aren't, they say, assuming determinism at all! We're just looking at which *strategies* or recipes seem to work and which don't. For example, does a revolutionary movement triumph by becoming violent or by remaining nonviolent? By making coalitions or by undermining other organizations? And so on. But note that such strategies, unlike variables such as CATHOLIC COUNTRY, are something that actors can adopt or reject as it suits them in the moment. The problem then is that rather than *explaining* the success (or lack of it), the choice of strategies often *come from it*.

This is something that I talked about in *TTM*. Adolf Hitler did a simple sort of QCA when he was arguing with his generals. His comparison of his previous successes in Poland and other places, and the generals' failure in Russia, supported his contention that the explanation of victory was the combination SPEEDY ASSAULT and CRUSH OPPOSITION. His idiotic generals, he thought, were refusing to implement his wise plans. And frankly, if you are thinking about using QCA to understand political strategy and outcomes, it's rather like taking a set of chess games, choosing to code up 8 moves, and then seeing whether "Q-KB5" leads to checkmate so long as it isn't combined with "P-K3" and "K-KB3."

An Example

Things certainly can't be that bad, you think. We wouldn't institutionalize the use of a method that is so problematic, would we? Okay, well here's an example of QCA from Hicks, Misra and Ng (1995). The data are displayed in table 9.2. The question was, what factors are associated with a nation's state being "consolidated" (the rightmost column).

I'll present the formal results in QCA conventions, in which upper case means that a condition is true, and lower case that it is false. So applying QCA to this table gives us the following solutions:

CATHOLIC*patriarchal*UNITARY*WORKERS +
liberal*CATHOLIC*UNITARY*WORKERS +
liberal*catholic*PATRIARCHAL*UNITARY*workers +
liberal*CATHOLIC*PATRIARCHAL*unitary*workers

That means that consolidation could come from Catholic countries, so long as they are have unitary democracies and either strong working class movements in non-patriarchal systems, or at least non-liberal systems, or they can also have non-unitary democracies, weak working class systems, and patriarchal, non-liberal constitutions. Non-catholic countries can also consolidate, so long as they are patriarchal, unitary, conservative democracies with weak working class movements. Got it? It does make sense, when you think about it.

"Wait," some fan of QCA is saying. "That's not what I remember from the original article. Didn't they find that consolidation in 1920 was due to either (a) non-liberal, patriarchal, unitary democracies in Catholic countries, and then, in non-Catholic countries, either (b) liberal constitutions, unitary democracies and strong workers' movements, or (c) patriarchal governance in non-unitary democracies with strong workers' movements?" Oh, shoot, yes, you're right. I must have *randomly permuted the data in each row*. And yet I got an answer, and this answer seemed to be about *necessary causal relations*.

This is, in essence, the problem with QCA. With most of our techniques, if you put noise in, you get nothing out. With QCA, when you put noise in, you get out deterministic causality—the gold standard of science for many of us (Lucas and Szatrowski 2014: 20). In terms of reliable techniques for building a social science, it doesn't get much worse than that. Let's look at why.

Recall that above I emphasized that QCA is variable based, and that a variable *is* a set of cases. Thus note the column "Catholic"—this column, which is a variable, may also be more formally defined on "that which unites Belgium and the Netherlands and divides the two from all other cases." Here, this is labeled "Catholic." And I'm sure there's a sense in which that's true, but there's also actually a sense in which it's not obviously true (there are plenty of Protestants in the Netherlands, and plenty of Catholics in France). Further, *anything else* that united these two countries—such as, "stemming from an independence movement against the Hapsburg dynasty"—would be equally plausible as a label for these cases.

You will also note that there are 5 explanatory variables. That implies that there are $2^5 = 32$ possible arrangements of these variables. However, we have only 10 observations, two of which (Germany and Italy) have identical distributions on the variables. So one route to consolidation supposedly involved patriarchal governance in non-unitary, non-Catholic, democracies with strong workers' movements. We have two such rows in our data, one for Germany, and one for Italy. But they are non-liberal. If we assume that there could be liberal regimes with patriarchal governance in non-unitary, non-Catholic, democracies with strong workers' movements that could have experienced consolidation, it would imply that we cannot reduce the complexity and declare that it is simply (if it's actually okay to call this "simple") that one route to consolidation involved patriarchal governance in non-unitary, non-Catholic, democracies with strong workers' movements. Rather, we would say that one route to consolidation involved patriarchal governance in non-unitary, non-Catholic, democracies with strong workers' movements *in non-liberal governments*.[7] We can't figure out what we know about the cases we *do* observe unless we make a decision about those we *don't*.

How do we think about the 22 possible unobserved types of cases? In essence, there are three ways. The first, which was once advocated by Ragin, is to *assume* that for all of these, the value on the dependent variable *would have been zero*. This might seem an odd assumption, but it flows from the fact that, as a logic-based technique, QCA cares more about *not* having statements *disconfirmed* than in seeing evidence *for* them. Forcing unobserved patterns to be failures therefore may seem like a conservative move. For example, let us imagine that we have two explanatory variables, *A* and *B*, and they are linked to *C*, where we have observed three of the four possible states of $A \times B$, with values [1,0,1], [0,1,0], and [1,1,0]. The third pattern shows us that *A* being positive is not sufficient for *C*, but we have no evidence against the possibility that *B* being *zero* is sufficient for *C*. That could be used to make the strong statement that the lack of *B* is sufficient for *C*. However, if we force the outcome (*C*) for $A = B = 0$ to be 0, then we would disconfirm this simple solution, and have to stick with the less parsimonious theory that it is the presence of *A* jointly with the absence of *B* that is sufficient for *C*.[8]

7. Hicks et al. (1995) say that they followed Ragin's advice and set all unobserved patterns to a value of zero on the outcomes, but when I do that, I get different results. Hmmm. . . .

8. For example, you might notice that Hicks et al's (1995: 340) solution that covers Germany and Italy is PATRIARCHAL*WORKERS*catholic*unitary. However, there is no case that is positive on both PATRIARCHAL and WORKERS that isn't positive on the outcome. Why not go for the simpler solution? Because there's no observed case that is high on both of these but also high on Catholic but low on unitary democracy . . . and what if

The problem with this seemingly conservative solution is that there is no reason why our events have an obvious coding asymmetry. Are we predicting peace, or war? Depending on how we choose to code our outcomes, this rule leads to different results. The justifications that have been given for this asymmetry are rather lame and will be dealt with below in the section on "clinical pathologies." The acknowledgment of this problem led to a proposed second way of dealing with unobserved patterns, namely to treat the unobserved (or rarely observed) cases as possible counterfactuals (Ragin 2008: 173). "The resulting solution incorporates any counterfactual combination that yields a simpler solution." In a way, this *is* conservative; it pushes us to simpler and more parsimonious solutions as opposed to overfitting. The downside is that, given QCA's ease in turning noise into strong claims, we're more likely to *believe* false conclusions if they are parsimonious! So this in a way hides some of the evidence that QCA is spinning out imaginary stories on too little data.

The third way is to conveniently assume that the unobserved cases happen to take whatever form would give us the results we would prefer—either ones that fit our prejudices, or those that make for the simpler solution. It is one thing to say that one is absolutely sure about all of one's coding of *existent* cases. But to say that one is *also* sure about the outcomes of *nonexistent* cases seems to be going a bit too far! Yet this is the "intermediate" solution that Ragin prefers. I can think of no better rule of thumb for the identification of a pathological technique than that it *requires* that we make up data to fit our conceptions. But that's what Ragin advocates, noting only that it "requires utilization of the investigator's theoretical and substantive knowledge" (2008: 174). We'll return to this issue below. But first, let's stick with a simpler issue: the sensitivity of the results to the choice of cases, and to the presence of error.

Sensitivity

As a number of researchers have already demonstrated, the results of QCA are necessarily extremely sensitive to the choice of cases to be considered (Lucas and Szatrowski 2014: 55)—too sensitive to produce results that we can trust. Even in circumstances that satisfy QCA's ontological requirements of deterministic causality, QCA just does terribly (Hug 2013; Bowers 2014). If one uses the principle of setting unobserved cases to zero, then in many ways, we really have leverage only from the number of *posi-*

---

there was one and it wasn't consolidated. . . . ? However, no case was observed that had this combination but was LIBERAL, so in my mind, this combination really is liberal*PATRIARCHAL*WORKERS*catholic*unitary.

tive outcomes (six, in our example). That's not a lot to leverage claims involving five-way interactions! When you have so few cases, dropping any one will lead your "theoretical" conclusions to change, as will adding any. Which means that your results are very fragile.

But even worse, the results are incredibly sensitive to what most of us would call measurement error (Lucas and Szatrowski 2014: 9). Because this is such an obvious problem, practitioners have largely tried to argue away the very possibility of such error. And indeed, it might seem silly to imply that we are making "errors" determining whether or not Russia had a revolution (though when you come to think of it, did it have one revolution or two? Or more?). But when it comes to things like having a strong workers' movement, or being patriarchal, there is always room for disagreement.

All statistics involves assumptions, yet the assumptions of deterministic causality and error-free measurement are different from others, because they deliberately throw out most of the capacity of statistics to tell us that nothing in particular is happening. As we have seen, even with totally random data, QCA can hand you reassurances that you have very strong findings (Marx and Dusa 2011).[9]

Finally, it should be worrisome that the response of QCA'ers has been to emphasize their *ontology*—that their approach has different assumptions about the world. That has served as a justification for ignoring the evidence of fragility. Indeed, they dismiss the relevance of the simulations that show QCA behaving badly because their assumptions aren't met. Thus in a recent piece, Thiem and Baumgartner (2016: 348) insist that "As QCA searches for INUS conditions, it must be guaranteed in tests of its power of discrimination with respect to causally irrelevant factors that the data to which QCA is applied do not contain pairs of conjunctions of the aforementioned type if the factor in question is not to be identified as somehow causally relevant to the outcome." Do you understand how to read this? They are saying that *QCA should be given only tests that we know in advance it can pass.*

They recently went on (Baumgartner and Thiem 2017) to give QCA more stringent tests, but argued on principled grounds that *no* technique could plausibly be tested on data that did not fit its own assumptions. That *can't* possibly be right; if logistic regression, say, gave false positive results whenever the actual link function was a probit function or a complimentary log-log function, we'd all be told not to use it. Our concern is

9. QCA *can* lead to a total failure to come up with solutions. Sadly, the way for researchers to get out of this bind is to *throw away data* until they can "discover" the "true" causes. That's the definition of corrupt incentives (chapter 4).

that when applied to data where there *isn't* deterministic causality, QCA will tell us there *is*. But Baumgartner and Thiem say that we can test QCA only where there *is* deterministic causality—again, assuming precisely what is in question. They also, like earlier comparativists (e.g., Skocpol 1984), insist that the results not be held to have *out-of-sample implications*. But, following the pragmatist conception we started with, knowledge that does not support inference is not knowledge at all. They give us false reassurances that classic QCA works as intended, beginning with simulations in which *it is literally impossible* that QCA could fail (as in, if it didn't work, logic would be false). That was never in question. The question is, when applied to data from the real world, is QCA likely to mislead us? The answer is: very much so.[10]

Get Fuzzy

Many QCA'ers will say "Bosh! By focusing on the early incarnation of QCA, you've missed—or *pretended* to miss—how far the technique has been developed. It's been adapted to handle all these problems via the introduction of 'fuzzy set' logic. It's now more sophisticated and we can be more confident in using it." That isn't true. It's gotten *less* defensible.

Let's reproduce the basic problem that Ragin faced. He had started thinking in terms of variables that might be thought of as intrinsically dichotomous (e.g., MONARCHY, REVOLUTION), and worked out a logical technique for that. But soon he and others wanted to include variables that, while we might treat them as dichotomous in our simplified theoretical renditions, can't really be seen as intrinsically binary (e.g., POWERFUL_NOBILITY, RICH_COUNTRY). Rather than try to handle quantitative data as quantities, practitioners had just been dichotomizing, and basically putting the dividing line wherever it felt right. Ragin thought there could be a more rigorous version of this, and he drew on fuzzy set theory. This was a really good idea—and he's right. It makes more sense than just dichotomizing. At least, in principle.

The underlying idea is that there really are "sets" out there (like "had a revolution" and "didn't have a revolution," as well as things like "strong labor movement" and "weak labor movement"), but the boundaries of any set can be fuzzy (though they don't all have to be). So while we might be able to say that this case really *didn't* have a revolution, we're not so sure

10. Baumgartner and Thiem (2017) also have a notion of the consistency of different solutions that doesn't make sense to me. They claim that the solution "either [*A*-and-*B*] or *C* is sufficient for *Y*" isn't contradicted by the solution "either [*A*-and-*B*] or [*C*-and-*D*] is sufficient for *Y*." But the second of these means that *C isn't* sufficient for *Y*.

about whether we can just put it in the set of "strong labor movement." So we come up with a number (based on the data, not pulled from our heads) as to how much any case is in any set. Note that this isn't the *probability* that it really *is* in that set (a way of thinking more compatible with latent class analysis), but a quantification of the degree of participation in this set.

Now first of all, I have to say, this way of thinking about the data will— if we stop to think about it—puzzle most of us. That doesn't mean it's wrong. In fact, I'd say that I get it. We can imagine sets of ideal typical processes that happen and we can really have countervailing processes in the same case. That is, there might be times when you want to say that something is .50, and other times when it is half 0 and half 1—and these two ways of thinking can be different. So putting forward a fuzzy logic was an interesting, perhaps important, extension of our way of mental habits, and it's one that has been used to good effect for theoretical extensions (for example, see Montgomery 2000). But I haven't yet seen anyone use it for QCA in any way that wasn't actually probabilistic thinking dressed up as something else.

I'll spare you more simulation, because others have done them. Fuzzy set QCA (fsQCA) doesn't solve the main problems of QCA—the tendency to take noise and turn it into deterministic causality. It just allows these problematic techniques to be applied to a wider range of data— ones where the *justification* for taking the QCA approach in the first place makes less and less sense. Krogslund, Choi, and Poertner (2014) show that fsQCA has an unacceptably strong tendency to turn noise into false positives, and is extremely sensitive to minor changes in parameters, especially near the points at which a case is seen as being more likely to be in one category than the other. By generating simulations based on actually existing data, but adding a totally random variable, Krogslund et al. found that 75%–99% of the time, fsQCA returned a causal configuration swearing as to the importance of the random variable. There's no way to put a good spin on this.[11]

Tests for Fuzzy Thinking

Practitioners claimed that they developed numerical measures of the goodness of fit of QCA solutions, especially when it comes to fsQCA.

11. Given QCA's terrible performance with simulated data that should be right up its alley, Seawright (2014: 121) concludes that "given the reality of limited diversity, QCA is in practice not particularly powerful at detecting genuine causal complexity, yet it is fairly prone to false-positive results."

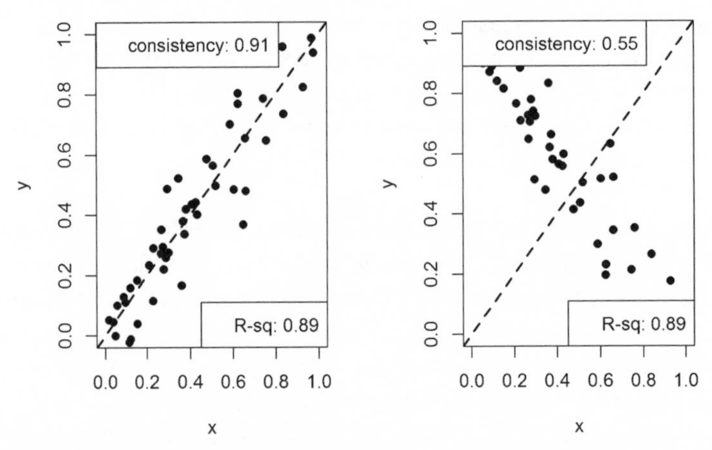

FIGURE 9.1. Two Different Fuzzy Set Distributions That Motivate Consistency Scores

We've been asked to treat these reasonable, but still ad hoc, measures as if they were statistics. One first notion was to claim that the cases are consistent with the causal claim that, say, *x* is sufficient for *y* if the fuzzy score of *y* is above that of *x* for every case; dually, the cases are consistent with the causal claim that *x* is necessary for *y* if the fuzzy score of *y* is below that of *x* for every case.[12] But this implies that the two distributions in fig. 9.1 (R 9.1) provide equally strong evidence of the hypothesis of a connection between *x* and *y*!

For this reason, there were attempts to come up with weighted indices of the strength of the association. First, there is the "consistency" for the claim that *x* is sufficient for *y* (and hence that the set membership in *y* should never be less than the membership in *x*), which is $= \sum_i [\min(x_i, y_i)] / \sum_i [x_i]$, the logic being that if *x* is always less than *y*, this statistic will be 1, but whenever *x* is greater, we'll see something tacked onto this sum that involves the degree to which it's greater than the corresponding *y* (Ragin 2008: 52). (I've put the consistency for the cases in fig. 9.1 in legends there, so you can see how it behaves.)

And then there is the "coverage," which is the exact same numerator, but instead, divided by the sum of the *y* values (Ragin 2008: 57). Low coverage (but high consistency) means that while *x* may indeed be a sufficient path to *y*, there aren't enough high values of *x* to produce enough *y*'s to really care too much about. Ragin (2008: 58, 62) nicely works through the logic with Venn diagrams. What he's doing is basically reproducing the

12. In the dichotomous world, for *x* to be sufficient for *y*, there can be no case where *y* is 0 and *x* is 1; translated to the fuzzy measures, that means that we expect all the cases to be above the $y = x$ line.

logic of statistics like phi that were popular in the 1940s and 1950s. Why don't we learn about these in statistics class anymore? Because they're the sort of rough rules-of-thumb that you use in the absence of theoretically defensible statistics.

So you can imagine the reason for coming up with a number like this. It's not a bad idea. But neither is it a *statistic* in any sense related to the science as it's developed over the past 150 years. And it's not like that's because no one knows what a proper statistical test would be for a QCA type hypothesis. The way to test small-sample claims through permutation approximations to exact tests is well known (see Braumoeller 2015 for statistics for QCA). Why isn't this done by adherents? Because almost no QCA procedure—which conducts a large number of simultaneous tests, usually far more than the number of observations—would *ever* come *remotely* close to statistical significance. So instead, there have been attempts at easier tests, a sort of Monopoly Jr, for QCA.[13]

For one, Eliason and Stryker (2009) took the charge seriously, and attempted to provide some statistical tests for fuzzy set relations. They started from the assumption that if the theoretical argument is, for example, A is necessary and sufficient for B, which implies you would find no cases $A \cap {\sim}B$, and none ${\sim}A \cap B$, then the cases shouldn't be far from a line in which the degree of membership of A (as opposed to ${\sim}A$) equals the degree of membership of B (as opposed to ${\sim}B$). That is, deviation from this diagonal should count as evidence away from at least a sufficient-and-necessary relationship. Well, that *sorta* makes sense, but if we actually believed the logic that was used to derive fsQCA, does it follow?[14] If I am "a little bit of a jerk," and being a jerk causes you to get a promotion, should I get only a "little bit" of a promotion? Maybe, but if so, that's a

13. Indeed, Mahoney (2003: 75) has claimed that QCA tests, because they test one conjunction as opposed to all others, one at a time, don't face the same need for degrees of freedom as standard multivariate analysis. This is not true. Anyone who a priori declares some group of cases to be different from the others is free to use up a degree of freedom fitting them; anyone who runs multiple tests burns up degrees of freedom. I think Mahoney's notion of degrees of freedom is a bit like the US banking industry's idea of debt—because there's no way that we have this much money, we might as well spend twice as much.

14. This isn't compatible with Ragin's (2008: 129) more consistent algebraic interpretation of various measures. Thus he argues that the fuzzy membership (call it *fm*) of being in set AB is min $[fm(A), fm(B)]$. Thus if one case has $[fm(A), fm(B)] = [.6,.6]$, its membership in AB is .6, as is the membership of a different case with values [.6,1.0]. As he (2008: 137) courageously emphasizes, this means that the same combination can be seen as a predictor of both an outcome, and the *inverse* of an outcome, or, at least, that there is no reason to think that consistency scores for an outcomes are negatively correlated with those for its opposite.

great reason not to use fuzzy set logic, but traditional statistics. It is interesting that to lay out their logic, Eliason and Stryker (2009: 105) have to use probabilistic imagery that they note is "not fully consistent with strict adherence to fuzzy-set logic." Can you see where we're going?

Back to Square One

So Eliason and Stryker proposed that solutions from fsQCA could be tested statistically if we basically treat the degree of membership more like a probability. This seems to me a lot like the playwright Woody Allen speaks of, who bought the rights to *My Fair Lady* and was working on taking the music *out* and turning it back into *Pygmalion*. (This is even more painfully obvious when people attempt to extend QCA to variables that take on multiple values. I am not making this up.)

But fuzzy set logic is *not* the same as probability. For one, according to fuzzy logic, there is *some* overall membership in a particular combination of variables even if there is no single case that is considered to have its maximum membership there. That doesn't really fit the way we think, and for this reason, Ragin (2008: 131ff) thinks maybe it just makes sense to throw that combination out. But no one seems to be able to actually consistently think in fuzzy terms.

If we're going to treat the number .83 in the row/column combination of SPAIN and RICH COUNTRY as an 83% chance of Spain being a rich country, why not go all the way and turn it back into what it *was*? Because at some point, you had a real number, like the actual per capita GNP of Spain. Of course, the real world numbers might not fall into a bimodal distribution that would support our thinking in dichotomies like RICH/POOR. Fuzzy set scores are especially prized when they spread close numbers further apart, thereby disguising the unimodal nature of the data. Remember that QCA "measures" that are used to judge fit are very simple ways of describing the cloud of points (look at fig. 9.1, right). If you're allowed to stretch out the *x* axis, maybe move the points up and down on the *y* axis, sure, you have an easier time proving what you "know" to be true.

If we actually *have* numbers, we should *use* numbers. If so, maybe what we want to do is . . . uh . . . OLS? QCA'ers may *wish* that there were really these Platonic Ideas behind the data of "the ideal rich country" and the "ideal poor" but that's being a bit silly, don't you think? How can we be scientists, and be unwilling to call things what they really are?

In other words, sometimes QCA'ers use fuzzy logic, sometimes crisp logic, and sometimes no logic at all, just their theoretical and substantive ideas to determine how to massage the data. Shockingly, advocates have considered this a *selling point* of the technique.

Clinical Pathologies

Faced with these rather serious indications of a method that is out of control, adherents coolly reply that they have *never* advocated "mindless" application of these methods; rather, these are tools to be used by skilled practitioners, by those who already know a great deal about their subject matter, and can be trusted to calibrate their variables, cherry-pick their cases, and so on.[15] As Lieberson (2001: 332) put it, they seem to have an "assumption that smart people do smarter things when working on a scale, say, using fuzzy sets than when they work on the same scale using conventional procedures." Charmingly, QCA'ers refer to their techniques, based on simple procedures applied to yes/no thinking, as "complex," and call those who think in more conventional, probabilistic terms, "simplistic" (see, e.g., Berg-Schlosser et al. 2008: 9; also see Ragin 2008: 178). At the bottom of these arguments lies an unpalatable conviction that more run-of-the-mill statistical types are dumber than they; while the big *N* researchers *need* the intellectual training wheels of *p*-values, the *übermenschenliche* QCA'ers do not (cf. Lee and Martin 2015). Stanley Lieberson was known for his bluntness, but let me be blunter. I say: nuts!

"Uncalibrated measures . . . are clearly inferior to calibrated measures" (Ragin 2008: 72). That's not true.[16] The notion that a calibrated (cooked-up) quantity is *superior* because it includes "theory and substantive knowledge" opens up the door to all sorts of pseudo-science. As Braumoeller (2015) says, it is a bit like the clinicians who say, "Trust me, as a clinician, I've seen this before and I know what I'm doing. Don't trust those randomized trials. Trust me." I'm sure there are trustworthy clinicians who have contributed to knowledge. But—as I think I showed in *ESA* (Martin 2011)—the greatest affronts to logic, truth, and human decency in social science have been supported by an epistemology that allows some people,

15. Vaisey (2014: 111) manfully says, "I concede, however, that the *naïve* use of QCA poses a greater risk for false-positive results than the *naïve* use of regression, because QCA's complex solutions can be driven entirely by cell missingness. But experienced and knowledgeable users will not have this problem."

16. Ragin's explanation for this statement is that without calibrating a thermometer, we might know that *A* is hotter than *B*, but not whether *A* is hot or cold. This notion that our measures must meet medieval standards ("from here on down is the true presence of that quality, coldness"), which have been rejected since the Renaissance, nicely demonstrates the regressive approach of QCA!

on the basis of their self-awarded superiority, to declare that they are above all normal criteria required of truth statements.

Further, we know that pathological science develops when researchers get rewarded for their capacity to produce subtle findings that are not required to be reproducible. If you don't believe me, think about Prosper-René Blondlot's Nobel Prize for his discovery of N-rays. Oh wait, he didn't get a Nobel Prize, did he? That's because he was chasing a ghost. It wasn't real, but he dismissed the failure of anyone else to replicate, because he saw his capacity to get results where others couldn't as evidence of his greater acumen and sensitivity.

We do the same thing when we declare that, because we have theoretical guidance, we don't need to worry about crude, atheoretical numbers. Actually, we *don't want* theory in our measures. That's called *cheating*. When you want to prove that trade-based early modern polities are free from revolutions, but find that England's trade-basis is too high, and you ding it a few points based on "theory" so that it now occupies the right cell, you aren't being more theoretically sophisticated, just less ethically defensible. I can think of no stronger proof of the scientific corruption of QCA than that adherents fall back on claims that they, on the basis of their substantive knowledge, do not need to meet the standards that the dumbass big-*N* people do. These are *exactly* the types who need to formally register their hypotheses before touching the data. They cannot be trusted to hand themselves a "get out of statistics free" card.

Or, at any rate, I would say that whether or not you believe someone who assures you that because they know their cases, they can judge this, that, and the other, that's up to you. But what this boils down to in essence is to say that QCA should be used only by those who don't need it. I just don't understand how someone can know *so much* about complex cases—for example, which countries really have "strong" feminist movements, which have "contested" regimes, and so on, *and* that the causes of regime breakdown are deterministic and to be found among the elephant causes they have measured—yet still not know just one thing . . . precisely *which* combination of causes led to the outcome in question.

In other words, the logic might be fine, that a really well-advanced theory should lead us to be able to do better than simple procedures. But there's all the world between having a *true* theory and having a *strong* theory. Those booming a theory are not to be allowed to judge how true their theory is. That's just not how science works.

Conclusions

I wasn't kidding above when I said that I basically liked the QCA approach. I was very enthusiastic when it started out, and I maintained interest for a while. I've worked, and continue to work, on algebraic methods, both those that have rather defensible error structures attached (Wiley and Martin 1999; Martin and Wiley 2000) and those that, just like QCA, work only for small data sets assuming no error (Martin 2002; 2006; 2014; 2016a). So why spend so much time trying to show the problems with this one? Simple.

"With fuzzy sets," promises Ragin (2008: 82), "it is possible to have the best of both worlds, namely, the precision that is prized by quantitative researchers and the use of substantive knowledge to calibrate measures that is central to qualitative researchers." That's the sort of talk that should set off your alarms. Someone is trying to sell you something that is too good to be true. And they do that when the actual nature of the good in question—here, simple comparison—doesn't justify the price.

Am I saying that techniques like QCA should never be used? Not at all! They should just never be trusted. As we saw above, inductive pattern search methods are always going to generate false positives. That's why we have things like statistics. And if QCA doesn't want to use the statistics, it can't be trusted.

But still, like LCA and its kin, QCAcan be used to generate hypotheses that can be tested by other methods. When it comes to samples too small for conventional statistical confirmations (as done by Vaisey 2007), there are other possibilities. For example, when it comes to historical questions, the best way is probably to assemble a team of historical sociologists, each of whom knows a few cases in sufficiently close detail that they will prove to us, using the same standards that a single-case researcher would use, that the aggregate pattern is as claimed. And if you *can't* get that agreement across 20 researchers for your 25 cases, then I think it's best to consider that you haven't managed to reject the null hypothesis at the $p = .05$ cutoff.

Further, I should note that there is a lot of work right now on trying to deal with the problems of QCA, and some of it is pretty good. But I still don't think it's worth five seconds of your time. The reason is, that what people *want* in QCA is the problems. You can get involved in the effort to push RWOST closer to statistical (or mathematical) theory. But you don't need to. The statistics for the sorts of questions QCA'ers are asking were worked out a century ago. We don't need to wait twenty years while QCA goes from very, very bad to bad and from there to just worse. Of course, it can be fun to recreate the wheel—I do it myself—a bit like an amateur hobbyist wanting to brew beer in the kitchen. But that doesn't mean that the rest of us should believe your findings no matter what form of QCA they come from. And once we wise up to the fact that QCA'ers have been setting themselves a lower bar than the rest of us, they'll have no choice but either to toss it and pick things up where statistics left off, or to retreat into a little enclave, like some weak theoretical traditions have done when they couldn't sell philosophers their magic beans.

Let me leave you with one last word, in case you imagine the issue just *can't* be this simple. I've pointed out that QCA is basically using eighteenth-century mathematics, although of course, there's nothing wrong with age or simplicity if it works. You still might wonder, why hasn't it already happened—or been taken over? Usually, whenever we sociologists come up with anything pretty interesting, we have only a few years before statisticians and physicists swoop down and grab it away and spiff it up. And in this case, why wouldn't epidemiologists have hit upon this? A technique that comes up with causal findings with small numbers of cases sounds like it would save a lot of lives! Even if the epidemiological statisticians had first overlooked it, once it was out there, you'd imagine it would be seized on with fervor, especially because there's much better reason to think that the ontological conditions of QCA (necessary or sufficient causality) are better met in at least some disease processes than in social processes. Yet the only uses in epidemiology have been for more sociological questions such as those having to do with drinking or social services.

The answer is simple—even those who cry up the virtues of QCA aren't going to trust their lives to it. It's not a technique anyone uses on things that matter. End of story.

Simulations

The last way that we often think that we are going to be luckier than we really are is when we rely on simulations to give us answers that they can't give. My basic argument here is simple: for almost *all* sociological questions, simulations need to be simple if they're being used to answer questions, but complex if they're being used to check the answers from *other* techniques.

Simulations are especially attractive because, better than conventional statistics, they fit what seems to be the new consensus of sociological theory. In the bad old days, theorists argued: Is society an organism? A mind? A scam? Now theorists from different traditions seem to agree that to explain social action, we don't want to appeal to vague social facts that constrain individuals, nor do we want to just do regressions of individual attributes as if everyone else were imaginary. Rather, we understand that social patterns are not laws that force our actions, they are the outcome of our actions, but that each one of us takes into account the probable actions of others when making our own. This idea was really best explicated by the British theorists like Giddens, Barnes, and American Anglophiles, but I think that you can see systems theory arising from similar presuppositions. In those terms, social order is a self-organizing open system. This isn't a very strong model, but I think it's correct. And simulations allow us to be true to this core vision in a way that other techniques don't. But that doesn't compensate for their fundamental weakness, which is . . . their strength.

From Good to Bad by Being Better

Let's say that you are interested, as are many mathematical sociology types, in processes such as interpersonal influence, and how that might affect the distribution of belief, say. Does it make a difference whether we are talking about discrete beliefs or continuous attitudes? Whether people are influenced randomly by one person, or whether they do some sort of averaging? So you decide to do a simulation. Let's say you give people a simple belief, pro (+1) or con (–1). Further, everyone (of N people) has a position in a space, which for simplicity, you make a $G \times G$ grid of squares. You assign people randomly their initial opinions, and then you let them instead adopt the belief of someone in their Moore neighborhood (anyone in the surrounding 8 cells), and you let them keep their own initial belief by treating it as just another one that they can sample from. You do a simulation using, say, 1000 people in a 25 × 25 square.[17]

You do it once, and quickly everyone switches to one side. You do it again, and the simulation stops before there is convergence. So you increase the number of iterations. And you watch plots of your little world changing over time. I've put selected plots in Figures 9.2 and 9.3.

And you try to summarize the results with a chart like fig. 9.4. What you notice is that some sense of order is established quickly, but there isn't any guarantee that it will proceed in any particular direction. It *seems* like it will always end in convergence, but you can't be sure. You can imagine it actually going up and down forever.

Then you wonder, suppose instead of picking one person to be influenced by, each person tries to *average* the opinions of those around her? So

17. All the simulations I show can be produced by the same program that I give online; the simulation numbers will tell you which parameters to pass to create this version. This first one is "S1."

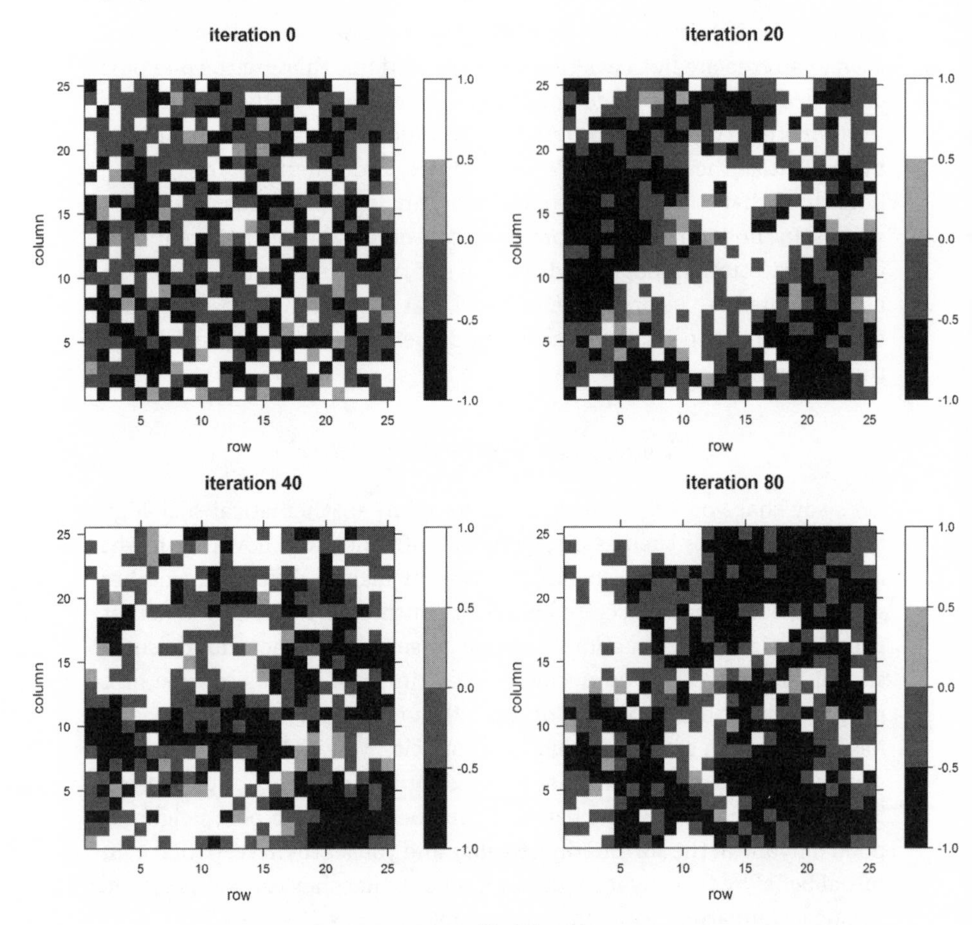

FIGURE 9.2. First Iterations

you modify the program, do it again (S2), and you see that if they always round up this average to either +1 or −1, the distribution stabilizes very quickly. That seems boring. So instead you wonder, what if folks basically *poll* those near them, and go with whichever side is more popular? Holy smokes! That (S3) also stabilizes very quickly, in the first few iterations! Fig. 9.5 below gives an example of an equilibrium achieved in iteration 7. And it doesn't lead to uniformity! You get cliques of pro and con that stay around. So you have an interesting lesson—random processes can be key for complex developments, and they make possible change, but they also tend to lead to homogeneity! You also vary the *size* of the neighborhood used. And in fact, you do this systematically, and make a graph. It shows that the larger the neighborhood from which one samples, the fewer iterations it takes before everyone has the same belief.

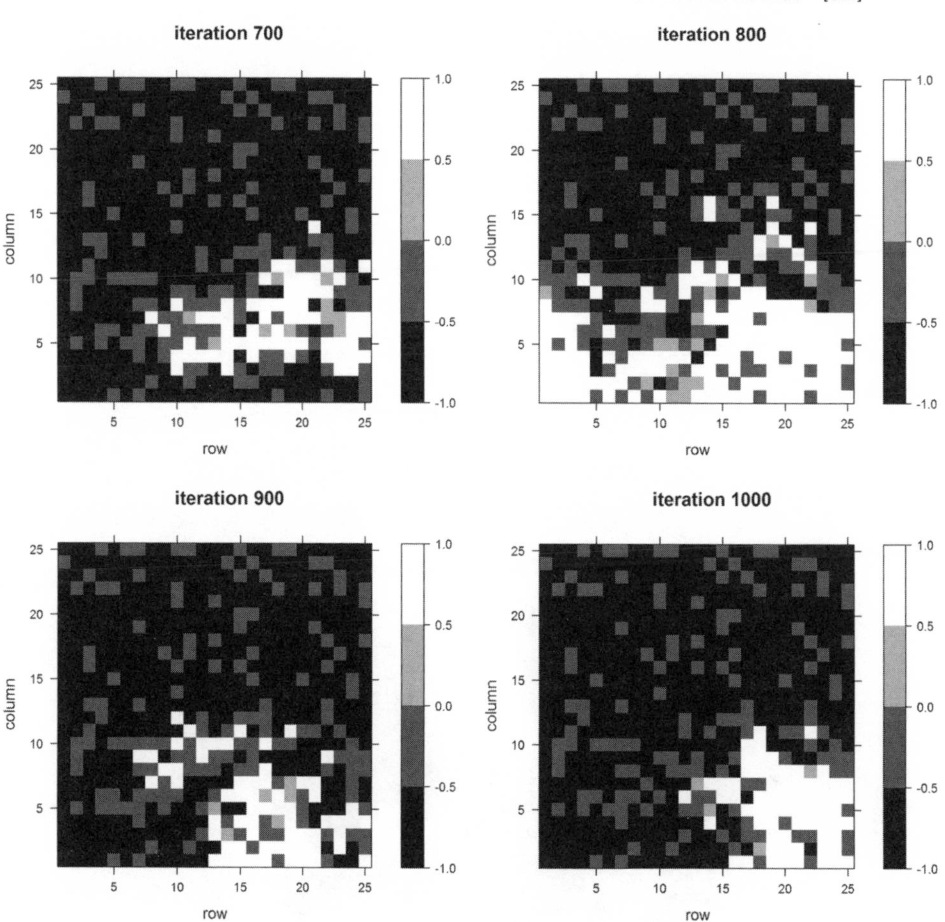

FIGURE 9.3. Final Iterations

You ask me if this is an interesting result and I say yes. So you go write a paper and present it at the "politico, culturish, and social and whatever" brownbag. And they are *so mean* to you! Someone in particular says that your simulation is *not realistic*, because you assume that people are *fixed* and simply *passively* accept beliefs, when in fact, people have *agency* and can get up and move from one place to another.

You are ashamed. You go back to your computer and stare at the code. They're right, you think. This stinks. But then, you get an idea, and you start typing away again! You make a new part in which your agents can decide to move. They investigate a neighboring square, and then compare the average belief there to their *own* belief, and, with a stochastic process, decide whether or not to move. In fact, to prevent clumping, you also make a penalty for them to move to a more densely populated square.

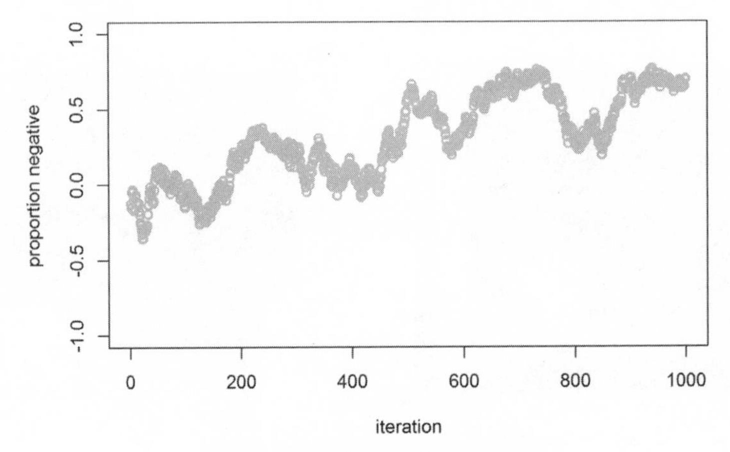

FIGURE 9.4. Evolution of Agreement

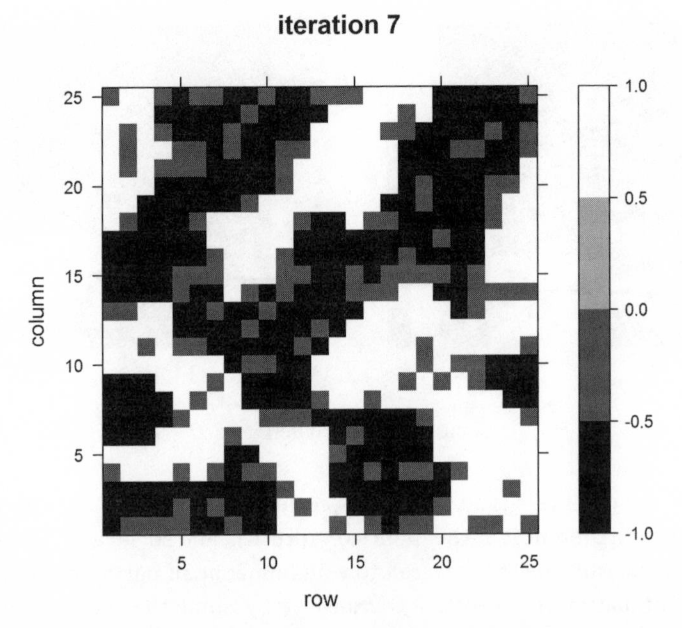

FIGURE 9.5. Distribution under a New Regime

And you even make an adjustable parameter (choose your favorite Greek letter, say, α) to express the trade-off between ideological compatibility and overpopulation.

To make it clearer, let's increase the concentration of people at any location, placing 2000 people on a 20×20 grid; first do it without the moves (S4), then with (S5). Now you are very excited! You show it to me, and don't seem to notice that my eyes are glazing over a bit. You run back to the workshop and present it again. And you are stunned: they are *so mean* to you! Someone in particular says that your simulation is *not realistic*, because you assume that people are *homogeneous* and care *exclusively* about these beliefs, when in fact, people have *cultural differences* and that this is certain to affect the process of belief change.

You are ashamed. You go back to your computer and stare at the code. They're right, you think. This stinks. But then, you get an idea, and you start typing away again! You make a new part in which each person is assigned to one of *C* different cultures. Now when people decide where to move to, they first of all, weight the number of people of *their* culture positively, while still weighting the *overall number* already there *negatively* (here you add β as a tunable parameter for the ratio of these numbers). Further, when they accept beliefs from someone else, they still sample, but they weight the influence from a member of their own culture more heavily than the influence of someone from a different culture, with the parameter γ expressing that difference.

Now you can show that the number of cultures really slows convergence, in fact, in some cases, you stop getting convergence at all! You are really excited, because now you have *explained why cultural differences persist*! You offer to show it to me, but we aren't able to work out a time. You run back to the workshop and present it again. And you are stunned: they are *so mean* to you! Someone in particular says that your simulation is *not realistic*, because you treat culture as if it were *independent* of belief, when in fact, people's *culture* is *all about* belief!

You are ashamed. And even more, you're hurt. Because, though it sounds silly, you've started to *like* your imaginary people. They have their teeny tiny culture and their teeny tiny places to live and their teeny tiny mobility decisions and their teeny tiny beliefs. They're like sea monkeys. You can imagine them wearing little crowns and teaching them tricks. It isn't nice that people were so mean to your little kingdom of sea monkeys.

You go back to your computer and stare at the code. Still, they're right, you think. This stinks. But then, you get an idea, and you start typing away again! You can make each culture have a parameter (δ_c) that indicates its distribution on the belief in question! Varying *that* can tell you how much preexisting cultural differences affect homogenization. You now, fixing α, β, and γ, vary the distribution of these δ_cs and come up with new results! You grab the 40 sheets of printout and knock on my door, because, for some reason, I haven't even returned an email from you where you announce your findings. Even though the light is on, no one opens the door. You run back to the workshop and present it again.

What will it be this time? Who knows! Maybe someone will say you need a parameter ϕ expressing individual resistance to influence. And you'll do this, each time feeling you're getting better, but you're getting worse. You've turned what was once a somewhat interesting learning experience into playing Dungeons and Dragons. You're not doing social science, you're having a computer fill out character sheets on your little dolls. Why? Because there are too many moving parts, and not a single one is really anchored to reality. There could be remarkably different models that all produce the same data. Remember, the trick is always to go from the data to a model of the world, right? It's no trick to go the other way. But that's all simulations do.

Classes of Assumption

I hope that little fictional anecdote gives you pause before assuming that you can learn a great deal from simulations. And if you think my anecdote is a priori pointless, because fictional anecdotes have no evidentiary weight, you might remember that simulations are *themselves* fictional anecdotes! What's important is less the fiction than the "faction," if I can use this word, from the Latin root *facere, to make,* to indicate the makedness of the results. The key question is what assumption we built in. There are three ways that a simulation can build on assumptions. The simplest, and most common, is that we make a set of assumptions that *obviously* have certain implications, and the simulation isn't worth doing in the first place.

But still, what is obvious to some is not always obvious to all. For example, Nancy Chodorow had a theory of gender reproduction that was basically like any storebought theory of reproduction: people grow up at time 1 in conditions of time 1, which influence their constitution at time 2; the constitution of people at time 2 influences how they raise their children at time 2. People grow up at time 2 in conditions of time 2, which influence their constitution at time 3. . . . repeat. She didn't ask for a Nobel prize for the basic idea, and if the theory is wrong (which it probably is, as it's too suspiciously like the way PLUs thought of things at exactly the time she was writing), it's wrong for the usual reasons. But Jackson (1989) argued that the problem was actually in the mechanisms of her theory, which he reproduced in more formal terms, but then said that of course we need to add error, and so then showed that after a few generations, her process produced equality, not inequality.

The way he did this was to turn it into a stochastic process. But once that's done, there's no real need to do a simulation. If you add error to a Markov process, you're not going to be able to support a conventional reproduction theory. Jackson, to his credit, emphasized just this: "No change in the assumptions of this stochastic model can make it work as Chodorow intended" (1989: 225). It's as much as to say either Chodorow is wrong, or a stochastic model isn't the right way to model it. (Or, one needs to add a new step, something like "all women [men] now select the gender attitude that is the mode in their gender.") But we need to avoid coming to conclusions that really have to do with what are supposed to be relatively innocuous assumptions about the stochastic nature of a process.

In a somewhat similar case, Elizabeth Bruch and Robert Mare (2006) claimed that the classic work by Thomas Schelling (1978; focusing on one of his earlier models) was misleading. He had shown that in a two-class system where people have preferences for the degree of integration of their neighborhood, even if the distribution of preferences is skewed toward integration, total segregation was likely to result. They argued that if one made a more stochastic model—and cranked up the error—you wouldn't see this. That in itself is quite correct—because if you crank up the error on *anything*, you wipe out the pattern. But what really got people's attention was their argument that their results held even when the error *wasn't* cranked up, which van de Rijt et al. (2009) showed was not so. There were robust findings, but they were less interesting.

That when you add noise, you get noise, we don't always need to be told. But there are other cases in which we do simulations that don't need to be done. And, indeed, most of the simulation studies I see don't do anything beyond watch their pretend world do exactly what it was set up to do. That can't be the basis for any conclusions, can it? And indeed, the more complexity that is added—the more "realistic" it becomes—the less likely a simulation is to produce counterintuitive findings, and, if they do, there are too many tunable parameters to be sure *why*. Our imaginary researcher above made a world in which the following parameters had to be fixed: {N, G, *width*, sample/average, α, β, γ, C, and C different δ_cs}. Any particular simulation is a point in a 10-dimensional space, and there is no way to systematically explore that space (Boorman 2010). And unlike conventional statistics, most simulations don't present any information on how interdependent certain parameters are for our estimates.

There's another emerging problem with much of the work in simulations. It arises because we take a well-understood *class* of problems—like the one I worked through above—and *call* it one thing. We now are really familiar with the implications of matrix multiplication to simulate Markov processes of how things would diffuse in certain kinds of networks. Calling that thing "religious commitment" in one case, and "chlamydia" in another, isn't really giving us two different pieces of social science.

When Do Simulations Work for Us?

Simulations are most likely to be useful when (1) we have a *small* number of assumptions and tunable parameters; (2) these assumptions fit *widely accepted* (and not convenient) understandings of the subject involved; (3) the simulation demonstrates something that would *not* be deduced by reasonably competent scholars on the basis of (1).

What are some examples of this? There's Schelling's demonstration that with a distribution of preferences for segregation, a community may have only one equilibrium, namely complete segregation; there's the formally very closely related demonstration by Granovetter (1978) that collective goods games among individuals with varying thresholds for contribution are incredibly sensitive to the initial conditions. Maybe there are a few others. But they're not coming to mind.

Of course, there is still the problem that Andy Clark (1997: 96) has pointed to—our simulations tend to follow our folk theory of action that ascribes too much to the actor and too little to the nature of the environment. But we don't yet have good theories of the ways in which environments differ; it won't do just to a priori construct strong environments that will guide action in the way we want to conclude.

Finally, are there times when *complex* simulations are in order? I think so, but only when three conditions are true. The first is that we have high quality data to compare our forecasts to. The second is that we have very good empirical estimates of certain parameters. And the third—and most difficult of all—is that we have a highly defensible and very restricted model of the agent involved. A good example here is HIV transmission. With a decent estimate of differential transmission, an SIR model (susceptible/infected/recovered) can tell us something; if we're not coming near to fitting the aggregate patterns, we realize that either our parameters or our model are off.[18]

Note that in these cases, we're basically interested in forecasting—applying an *accepted* theory, not building a new theory. That's why very complex simulations work very well in meteorology. And that's why they're often done—though not always quite so well—in economics (macroeconomic simulations are *not* very good predictors, though people will always work on them). Where does this work well in social life? In models of pedestrian path formation or of traffic flow, for one thing (here see the

18. A very impressive recent example of wedding real-like parameter estimates to a behaviorally defensible model of action, and then attempting to fit social science data, will be found in Bruch's (2014) later work. There are a lot of moving parts, but each one has been thought through.

work of Dirk Helbing [e.g., Helbing 1995; Helbing and Molnár 1995]). We have good measures of the distribution of reaction times for adults. In the models of single lane traffic there are only two things one can do—brake or gas—and that's pretty much the way drivers work out there in the real world.

You will probably see more and more of this sort of work emerging in the future, as physicists start moving into the social sciences; this is the new quasi-field often called "social science" (as if sociology didn't exist or wasn't scientific—and come to think of it . . .). They're going to use the methods that physicists use, which is to make models and produce data. As long as it is applied where the conditions above are met, this will be interesting. Applied to fuzzier things, it will be a waste of everyone's time.

So yes, simulations get to our notion of the nature of social emergence far better than do conventional techniques. But they're simply too sensitive to our assumptions to be a robust way that we are going to use data to learn about the world.

Stochastic Actor-Oriented Models

This general lesson, somewhat sadly, turns out to be relevant for what I thought was the most encouraging development in methods for networks, Tom Snijders's "stochastic actor-oriented models" (SAOMs) and his wonderful SIENA (Simulation Investigation for Empirical Network Analysis) program (see, e.g., Snijders, van de Bunt, and Steglich 2010). What Snijders did was to say, rather than fit structural parameters as effects, and then use these to throw out stories, let's make a plausible model of individual action, and fit the parameters to the data by "running it forward" and making it hit the data points.

Here's the basic model: each actor surveys the network as it stands (she has to have complete information). She has preferences for certain structures (for example, she might want to be friends with her friends' friends, she might want to drop ties to people who don't reciprocate). At any time, one person can make a "move." By comparing how the structure changes between two (or more) time periods (or even looking at its constitution during a single period), we can estimate the parameters that people (or at least, classes of people) have. It is a brilliant and rigorous approach.

I totally loved it, until I actually had to apply it to data. Why? Precisely *because* it's a strong behavioral model, wherever it *doesn't* apply, you're quite likely going to get bizarre results. In contrast, a correlation is a correlation is a correlation. It doesn't care if your theory is wrong. But a simulation-based technique is quite different. If your model is wrong, it's like you've followed a very precise set of instructions based on the wrong

Table 9.3. Logistic regression models of close friendship

	Model 1	*Model 2*		*Model 3*	*Model 4*
Sample	all dyads at time 1	all dyads at time 2	Sample	close at wave 1	not friends at wave 1
Upward	1.76*	2.17*	Δ Upward	0.54	1.71†
Embeddedness	[*p*<0.01]	[*p*<0.01]	Embeddedness	[*p*=0.26]	[*p*=0.09]
Sideways	0.95*	1.32*	Δ Sideways	0.05	0.34
Embeddedness	[*p*<0.01]	[*p*<0.01]	Embeddedness	[*p*=0.36]	[*p*=0.64]
Downward	-0.23	0.43	Δ Downward	0.29	-0.91†
Embeddedness	[*p*=0.19]	[*p*=0.74]	Embeddedness	[*p*=0.48]	[*p*=0.09]
Distance	-0.02	0.00	Δ Distance	0.00	-0.02
	[*p*=0.16]	[*p*=0.55]		[*p*=0.43]	[*p*=0.24]
Alter Close	2.98*	3.15*	Alter Adds	0.93	3.83*
	[*p*<0.01]	[*p*<0.01]	Close	[*p*=0.40]	[*p*<0.01]
			Alter Drops	-0.95	-1.02
			Close	[*p*=0.20]	[*p*=0.15]
Constant	-3.12	-3.67		0.01	-4.81
LL	-561.01	-484.58		-119.74	-41.01
N	2488	2909		187	755

$^{***}p < 0.001$, $^{**}p < 0.01$, $^{*}p < 0.05$, †$p < 0.1$, one-tailed QAP tests.

map. Your ass is lost but *good*. There's nothing wrong with SAOMs—there's only something wrong with thinking that they are a *general* solution to problems in network data.

Let me give you the example from the paper that Jacob Habinek, Benjamin Zablocki, and I (2015) wrote that I discussed in chapter 3 (and Jacob did the heavy lifting on the [unpublished] SAOMs we tried). We were interested in the effects of local structure on ties breaking and reforming over time, where we had two waves. In particular, we wanted to see whether if, say, Archie and Reggie had friends who implied that Reggie was more popular than Archie, then Archie would nominate Reggie but not vice versa. For example, perhaps Archie and Jughead are mutual friends, and Jughead nominates Reggie, but Reggie doesn't nominate Jughead. This would count as one indication of a relationship of "upward embeddedness" connecting Archie to Reggie, and a "downwards embeddedness" of Reggie for Archie (depending on whose tie formation processes we are predicting). Different from either of these would be the "sideways embeddedness" that could come from them both, say, having a mutual friendship with Jughead.

So we decided to study dyadic reports on friendship, and how they changed. Table 9.3 below presents cross-sectional models for close friendship, using logistic regression. The first two models are cross-sectional,

Table 9.4. Stochastic actor-oriented models of close friendship

	Model 1	Model 2	
Sample	all ties	all ties	
Function	evaluation	evaluation	creation
Transitivity	0.68+	-4.56+	4.48*
	(0.36)	(2.98)	(2.09)
Balance	-0.31	2.28*	-0.27+
	(0.67)	(0.42)	(0.15)
3-Cycles	0.14+	3.57	-2.39
	(0.08)	(3.54)	(2.32)
Distance	-0.02*	-0.01	-0.03
	(0.01)	(0.05)	(0.02)
Reciprocity	2.45*	3.66*	0.60
	(0.42)	(0.53)	(1.63)
Out-degree	-3.11*	-4.19*	-4.10*
	(0.21)	(0.50)	(0.50)
Rate	3.45*	5.08*	

and they show a tendency for dyads that are in these configurations to be friends. (These models also include the geographical distance between ego and alter, as well as alter's report about ego to tap reciprocation.) When we modeled change, we just looked separately at dyads where ego *did* name alter as a close friend at time 1 (and therefore could *drop* alter), and those where ego did *not* (and therefore could *add* alter), in models 3 and 4 respectively. They suggest that when it comes to change, upwards embeddedness might be important for adding friends, but not for not dropping them. And alter *adding* ego as a friend seems to go along with ego reciprocating, but alter *dropping* ego doesn't seem to be related to ego dropping alter.

What's wrong with these results? For one, although we do a QAP permutation test to judge the statistical significance of the parameters, the assumptions of logistic regression aren't actually met in dyadic data. So these can't be proven to be the best estimates of the parameters, and the judge of statistical significance (coming from a permutation test) is kind of a crapshoot. Let's see if we can do a better job, with defensible estimates coming from an SAOM.

The first model in table 9.4 has results from a simple SAOM model using the SAOM effects that are the closest parallels to those we included in our logistic regression. We include the outdegree parameter so that the SAOM has the proper conditional logistic function and therefore has a strong interpretation. While in our logistic model, the effect of upward paths (choosing the "locally popular") seemed very strong and was judged statistically significant by a QAP test, the SAOM model sees the corresponding effect (transitivity) as small and only perhaps marginally significant.

Why might this be? Probably because this model isn't capturing the very asymmetries in the data that we have highlighted. Fortunately, Snijders realized that this asymmetry was a problem for many applications of SAOMs, so he figured out a generalization where there are two utilities—one for *keeping* ties, and one for just *creating* them. So let's replicate (model 2), allowing for this separation. Now we find that indeed, the *creation* of these ties is highly significant—but once created, they appear to be radically *disfavored*! Holding all other effects constant, the odds that a tie yielding a transitive triplet will be retained are extremely low (= exp [−4.56] = 0.01), whereas the odds that one will be created are about even (= exp [−4.56 + 4.48] = 0.9). People apparently like making them, but they don't like keeping them.

That might indeed be a possible state of the world, but I would tend to doubt it. Practitioners know that finding large magnitude, but opposite sign, terms for the same effects in a dynamic model usually indicates a problem with the model.[19] The results of both change models above (models 3 and 4 in the table 9.3), in contrast, are more consistent with each other, with the cross-sectional models, with simple crosstabs, and with the existing literature on tie formation and retention. They certainly don't give us any reason to believe that upward paths were *more* likely to disintegrate over time.

So why does the SAOM get things so gosh dang backward? Simply because *its* assumptions aren't met. And they're a darn sight more consequential than the convenient math assumptions underlying a logistic regression. The SAOM model assumes the case in which a set of actors, all of whom are aware of the existence of other actors, and have correct knowledge of the state of the network, are in a position to send out and to reject tie formation initiatives, and that we have rather rapid waves of measurement (or continuous time). One of the reasons that the model doesn't work so well in this example is that we have only two time periods, and there isn't that much change (the Jacard index—the measure of stability in ties—between the two periods is quite high, 0.816). But most important, the changes that we *do* see don't come from the process

19. At the time of writing, Habinek and I could find only two published papers that employ creation and evaluation parameters, and both report estimates with similar large magnitudes but opposite signs. They come from the same team: see Cheadle, Stevens, Williams, and Goosby (2013); Cheadle and Williams (2013).

that the SAOM assumes (an actor surveys his or her local network and makes changes). How do we know this? Because our people have been spread out across the country for years. Instead, what's happening is probably that ego reaches out to alter because a third party moves closer, *tells ego* that alter lives nearby, and then ego reaches out. We learned all this with models whose *statistical* assumptions weren't always defensible, but whose *behavioral* assumptions were negligible.

So what is the SAOM doing here? It's giving us very nice estimates of very wrong parameters. And it does so in a way that makes it rather difficult to know that the model is wrong. Remember, its job is just to tell you how, if the world *did* work the way it assumes (and it doesn't), what numerical values those parameters (which don't correspond to anything meaningful) *would* have. It's totally incorrect—indeed, borderline insane—for us to reject the results from the logistic regression because its *formal* assumptions aren't met, but accept the results of the SAOM whose *substantive* assumptions aren't met, even though these latter turn out to be hugely consequential!

Now here's the awkward thing—there's basically *no* critique that one can make of this approach that can't be handled. Take the assumption that each actor has complete information. It's true that all implementations so far do make this assumption, and removing it would complicate things considerably. But it isn't necessary. So why not just work on relaxing the assumptions? Indeed, that's basically what Snijders has been doing. And in fact, if you have appropriate data, and you can get a version made for you, I'd be quite interested in the results you get and would trust them over most other modeling approaches. But this still isn't going to be a cure-all for network analysis. Aside from the practical problem of only those people who can get Snijders to modify his program actually successfully using it, there's a bigger problem: as we move away from a *single* basic model and approach, and start adding in more and more complications, we can fit any set of data, but only by doing that thing we saw undermining the use of simulations—namely, putting in too many parameters. Sure, we fit the data, but we aren't sure *why*.

In other words, the success of the model is to use substantive assumptions to help identify parameters that other, less behaviorally oriented models, would struggle with. Where the assumptions are wrong, one can adapt them, but then one moves away from using assumptions to fit a model comparable to what other methods would fit, and toward doing one of those "miniworld simulations." Yes, you are relaxing assumptions. But this sort of relaxation is totally different from adding more parameters in a conventional model. There, our parameters are usually the slopes of covariates tied to observations. Each one is an anchor to the world. In contrast, when we start to *also* estimate the "knowledge" that people have of the network, in addition to their utility in making, breaking, and keeping relations, we are adding flexibility, and *cutting* the ropes of data that bind us to the world.

Of course, *if* you know that this *is* the way the world works, then hooray! You've got the best estimates. But chances are, you *don't* know how the world works. You just aren't that lucky. These approaches take the problem that we've seen lying at the base of our methods—that they are guaranteed to work only if we already know the answer—and increases their sensitivity to that problem.

Coda: When We Must *Do Simulations*

Remember, I'm not arguing against doing simulations. This book has been full of them. We want to do simulations *whenever* we have a complicated model, especially a new one we've devised ourselves. We want to simulate different kinds of random data and see what happens when we feed it into our model. We want to simulate data that fit the conception of the world that we are most earnestly arguing *against*, and see what the model does with it. Especially for TGTBT methods, we want to see what happens if we scramble or reverse the data. More often then we'd like, we'll learn that the model that seems to support us would give equally positive results if we were definitely wrong. That's good to know.

But I think that just as we've been making our simulations-to-learn-from too complex, we've been making our simulations-to-check too simple. That is, when we compare one method to another, we often start by assuming a "true" version of some linear model, with "true" coefficients. Then we see if our method reproduces them. This might sound like some easy picking—like those commercial fishing "holes" where they dump in 200 trout a day. (Recall that Baumgartner and Thiem thought that anything else was unfair.) Even at this sort of generous standard, our methods often break down, as we saw with sequence clustering.

We need to go further, however. We need to test our methods not on true *models*, but on mocked-up worlds that involve our best guess as to the actual *processes* involved. These can produce data that doesn't look like our regression-based imaginary version of the same basic claims. An example is how I tested the ERGMs in chapter 8—not by assuming a true *parametric model* but making my best shot of a *plausible world*. Because that's what we're trying to get from our statistics—a plausible inference *outside* of the pretend world.

Conclusion

"My opponent says there are no easy answers.
I say, he's not looking hard enough!"

BART SIMPSON, campaign speech

Things that sound too good to be true usually are. I admit that I'm attracted to such techniques just like you are. I want to squeeze a bit more out the same data matrix than I do now, and if someone says there's a way, I'm all ears. And we should understand that there *are* breakthroughs. For example, LED lights give the same illumination as the incandescent at a fraction of the heat output. Your rules of thumb that come from the old-lightbulb-world don't apply any more. Similarly, mixed models can get large sample approximations that are pretty darn robust, and they squeeze more out of data than a lot of us thought was possible.

But—and here's the central thing—they use statistics. And usually, the reason no one developed them earlier is that they require *more* (and not *less*) advanced statistics. It's one thing for advances in solid state physics to make possible an LED light. It's another to say that by adding a simple household substance to your gas tank, you can double your mileage. Sorry, fellow anarcho-syndicalists—in science, real advances obey the laws. And that means that they take the uncertainty of small samples or unknown models into account. And *that* means that when you ask them to fit a model, they often tell you, "I have no idea."

You *want* techniques that say that to you. The problem with the methods that we looked at in this chapter is that they basically never give you that bad news. And so they are like the shysters who promise you an investment that is guaranteed. That's a con job, even if you get it from someone who is 100% convinced.

And the real problem for us is that bad statistics, if allowed to flourish, will chase out good ones as sure as your lawn will turn to dandelions. Start weeding, my friends.

Conclusion

"How'm I gonna find out what I gotta find out
if he no find out what I gotta find out?"

CHICO MARX to Groucho Marx, *Duck Soup*

Who Is Your Friend?

I am. I'm not saying that statisticians are your enemy, but their job is not to help you. It is to help someone else, someone who has a much easier job. That person knows how the world works, pretty much, and just needs to put the finishing touches on her knowledge, by getting the best estimates of the parameters from a true model. But your problems have to do with the fact that you don't know what is going on. That you don't have all the relevant variables, and the omitted *are* correlated with the included. Take that for granted. What are you going to do about this? *Not* come up with "bestimates" for your model, but figure out how to guard against the *worst* types of error that will interfere with inference.

You're in Chico's position—and, like him, you probably have a Harpo as a partner. Or advisor.

Further, I'm your friend because I haven't shied away from working through failures in our research, even when the research was conducted by decent people who are still alive. We usually avoid doing that. But there's a version of Gresham's law we need in the social sciences: "bad statistics chase out good statistics." By this I mean that techniques that are worse tend to replace those that are better. Why? Because the essence of a bad technique is that it favors false positives, and makes it hard to notice that this has happened. Often it's so complicated, either in its guts or in its output, and has such an indirect relation to the data, that you can be wrong and never know it. That means that unless we hack this stuff back regularly, like you do with weeds and poison ivy, it's going to totally take over. And, when we do hack away, sometimes we'll find that part of it is worth keeping—a really important methodological innovation, something that you can use. But as for the rest, don't shed a tear for its loss.

Here I want to do two things in conclusion. First, I want to consider some ethical issues that I think are never treated. And then I want to leave you with a few watchwords to help guide later practice. But before that, I need to really re-emphasize my main point here, because it's so incredibly simple, but so completely avoided by our current approach to methodology, that it's easy to ignore.

Having the Wrong Model

I've noted that we basically *always* have the "wrong" model. You might—especially if you lean toward the statistician-side of life—think that this has all been a waste of time. Of course, we all know that if we have the wrong model, we get wrong results. Why work through this again and again? But what I think few have appreciated is that there are classes of questions in which we are predisposed to have *massive* misspecification which leads to *predictable* forms of bias that often are taken as strong findings. And in fact, we've been spending disproportionate attention on those sorts of questions (having to do with time, for example, or aggregated cases) precisely *because* we can count on misspecification to hand us false positives.

The issue of having the wrong model doesn't only have to do with the problem of omitted predictors, although that's usually our biggest worry. I mean that even if you have the right predictors, the chance that you have the correct *specification* is near zero. There's nothing about the relation between our predictors that means that they have linear, or log-linear, or even quadratic relations. So you might be fitting a model that regresses y on $b_1x_1 + b_2x_1^2 + b_3\ln(x_2) + \varepsilon$, and feel like you're being really smart to take into account the possible quadratic effect of x_1 and the fact that x_2 is a zero-bounded variable that probably scales logarithmically with y.

But let's say that the "real" model is $b_1x_1 + b_2x_1^{1.4} + b_4x_1^{2.2} + b_3\log_5(x_2) + \varepsilon$? "Oh please, why split hairs?" you justifiably respond. "My model will do well enough." *Precisely*. But most of what you learn in statistics *is* splitting hairs. Ask your statistics teacher whether, for a dichotomous variable, if the model assumptions aren't exactly right, you have reason to suspect that your parameter estimates are closer to "right" than those of an OLS. She or he will admit that no, there's no guarantee. There's no reason to think that a Poisson model gives better predictions than an OLS if the specification is wrong. And so on. Most of what we teach you just doesn't matter for the real world. And even more important, complex models can make it harder for you to get a "feel" for your data; even if you plan to go on to more complex models, you should start with ones that are more transparent.

So look, I'm definitely not saying that you should never go beyond OLS! First, you should be using various non-linear models to check whether certain assumptions could be driving your results (plus, frankly, those models are really beautiful). Second, there are times when our simple techniques have *known* biases. Not as in "this might be wrong," but "given your own assumptions, your answers are certainly too big." (An example is the deflation of standard errors in data from clustered sampling frames; another is the negative predictions for count models that OLS can give you.) In no way am I suggesting that you ignore what statisticians say here, though if the only "fix" is one that involves serious assumptions, you can't rest with a single "state of the art" analysis. Third, there are techniques that turn out to be really good for helping us use data to answer our questions—ways of dealing with plausible noise that we need to separate from the signal we're looking for. Sometimes, those actually require some pretty complex math. But what we end up with is robust tools—and not "the correct" ritual to perform. It's with robust tools that we can, really, do social science. I swear it. I've seen it happen, and I've lived to tell.

Toward a Serious Social Science

The best way I have of describing this sort of social science is that it is *serious*. By *serious*, I mean the term how Aristotle used it (*spoudaios*)—to mean a person who is really living up fully to the implicit potential in what she is doing. A lot of sociologists I know *don't* take their job seriously in this sense. Rather, they think that social science is like a game: it has rules, and so long as you work within the rules, what you do is okay. For example, if the maximum likelihood estimate of a parameter from a model including controls agrees with your theory, you get to publish it. You don't worry if there is actually more support overall for the three different theories you claim to be rejecting.

This is a great way to have a decent career. But it's a pretty crappy life, so far as I can see. You have some limited time on planet Earth. Social science is poorly remunerated, receives no respect from anyone, but it offers you the chance to spend your days learning about the world. If you're not going to do *that*, I don't know what you're doing here.

I think that social statistics is something that can be done in a meaningful way, by serious people who are living lives that are not barren of all worth. And that means just as in other fields of endeavor that involve hard work by serious people, we have to confront ethical issues.

Ethics in Statistics

Cherry-Picking

Unfortunately, almost all discussions of ethics in sociology have to do with human subjects. There's often an assumption that there no ethical issues for those who only use already created data. But practitioners know that isn't true. Maybe the IRB doesn't have to worry about them, but you do. Let's start with the issue of when and what to "'fess up." There are two main versions of this dilemma. One is when you made a mistake, and the other is about contrary evidence. Now the first thing I have to say is that the bulk of sociologists really need a wake-up call to hold themselves to higher standards. But I've definitely seen some students who are so rigorous in their application of moral strictures to themselves, that they need a little calming down.

For example, we all know that it's wrong to "cherry-pick"—to selectively present the findings that work for your claim. For this reason, the best analysts often feel a need to unburden themselves of *all* the analyses they've done, and all the different ways it can work out. No one wants to hear all that. We want *you* to determine what we need to see and what we don't. We want to trust you, which means you should be trustworthy. Is the problem that you don't trust yourself? Become the person we *should* trust . . . and then present your results selectively, but honestly.

Some kinds of selective marshaling of data are completely nonproblematic. For example, if you are fitting a model in which something in the past affects the present, it's conventional for analysts to let the data tell them how long that lag should be. We try a whole bunch, and go with the one that has the strongest relation. Is this cherry-picking? No, it's learning from the data. We just tell people what we did (for example, that we searched through 20 different models). It's great if you can turn this into an estimable parameter, but most readers will have an intuitive understanding of how much to dock a claim for searching through different lag times (not much, unless the length of time is substantively implausible, or the data series short).

What if you think that "education" might be related to some outcome, but you're not sure if it's about *years* of education, presence of a *degree*, or your position in a *percentile* distribution, and you try them all? Is that cheating? No—again, you're learning from the data. Does this mean your statistical tests (e.g., *p*-values) are wrong? Of course, but as I've said in chapter 4, take that for granted. If you've really pursued your interpretation—you've tried to make it go away, by testing implications in your data—you've done as good a job as anyone can. You shouldn't be confusing readers by walking through the many things you did that didn't turn out right. If you're *not* confident that your result is robust—that it would be found by a similar study but using a radically different sample—don't go ahead and still put your claim forward, only peppering it with one qualification or complication after another. Go back to the data. But if you *are* confident that your claim holds, that's fine.[1]

As I've emphasized before, we need to learn how to be okay with multiple tests, with trying to learn from our data, pursuing the implications of ideas. And that's why I'm against the notion of "registration," that we all call out our hypothesis before we collect the data. But I make an exception for the "game players," those for whom anything done by the rules is okay to publish, even if they know it's wrong. I'm not simply saying that registration would be a good idea for the interaction-mongers who like the approaches that tend to produce one false positive after another. I'm saying that they should have nonremovable GPS-tagging ankle bracelets attached to their registered hypotheses.

Which Test Values to Report?

And here's another awkward situation which I'm sure many of us have actually faced. Let's say you are doing an analysis that relies on simulation or permutation to make significance tests, and you get a result of $p = .0502$ for your favorite coefficient. Do you look at the ceiling and let your pinky accidentally-on-purpose hit "enter" and run it again, writing your paper and submitting it if now $p = .0498$? It's a pretty hard temptation to resist. For this reason, when I was writing a permutation-based QAP program (Martin 1999), what I did was to automate a routine where, if the p-value was close to a critical value (.1, .05, .01), the program would continue to do more simulations to be really sure of which side of the line it was on. I didn't want to tempt myself to have discretion, because I didn't think I would resist the temptation to find a reason to redo the ones that didn't work the way I wanted them to. . . . If we are going to fetishize which side of the line we're on—and we do—we'd better go to any lengths necessary to make sure we aren't fudging it. Do you think it's silly to make such big deal about a small difference? Note to IRS: Audit this reader.

1. When would you have a false sense of confidence? Usually *not* when you've run the bizillion analyses, but have just tested your "theory" and it was right. Or your claim relies on an *interaction coefficient*, and only appears when two highly correlated (or inversely correlated) variables are in the same model.

What about 'fessing up when you are wrong? Sometimes you will actually present a paper, or publish one, where the findings aren't right. On the first one (presentation): I've seen some students frantically thinking that they have to email every person who was in the Webster Ballroom, East, at 8:10 a.m. Saturday, August 23rd (which is where and when they presented their paper), because they just realized that they computed the quadratic effect wrong and so the coefficient for age isn't actually .018 but is in fact .030. Presentation findings aren't interpreted as set in stone. In fact, chances are good no one is going to even remember what they were—or who *you* were!—let alone need a retraction a week later. It would be nice if it were otherwise, but that's about the size of things. We're lucky if the audiences remembers where *they* were that morning.

But what if you actually *publish* something that might be wrong? First, what if someone wants to destructively challenge your work? There is only one right response, which is to make your data and your procedures accessible to the critic. And *that* means that you should preserve these in an interpretable form. (If the data itself can't be shared, you share, say, a correlation matrix that will allow your results to be reproduced.)[2] If your critic is going to publish a critical comment on your work, don't try to defend your original finding if you think the critic hasn't made a fundamental mistake. You thank the person, move on.[3]

What if *you* find that something is wrong? First, people are *always* finding that their things are wrong, even when they did everything right. The data you used will turn out to have some errors in it, even if *you* didn't make a mistake. The rule of thumb is that *if* correcting the error changes your *main* conclusions, you print a retraction. If someone else brings your error to your attention, you thank her and offer to coauthor a comment pointing out the error in your own work, if she doesn't want to go to the trouble of replicating your work to show you were wrong.

Do you think you'll look like an idiot for making a mistake? Spoiler alert: you already looked like an idiot to those who knew what was going on. Now you look like someone who has a spine and who can learn. That isn't bad.

2. We older folks take for granted that one can, in good faith, lose all sorts of things. But the more digital younger generations won't believe that data or control files were "lost." They'll take that as an admission of guilt.

3. I'm not even talking about how to retract wrong things you said/posted on blogs. Because you should not be there in the first place. Leave the blogosphere for the loudmouths and fools who don't mind being wrong. It's the drunk dialing of the academic world. If you post your findings on blogs before they've been picked over by experts, *you will live to regret it*. I personally guarantee it.

FINAL REMINDERS

1) *Don't treat ignorance as knowledge.*
   But that's what we do when we *assume our measures are perfect*, and thus assume that if we haven't seen it, it isn't there.
2) *Remember where you are.*
   That is, are you near the floor, or near the ceiling? For some models, such as marital homogamy, chances are, you can only come up with a floor estimate—because your measures aren't perfect, and there's reason to think that actors are in fact oriented to information you don't have, real class homogamy is unlikely to be *lower* than your estimate, but it may be higher. For other models, such as influence, chances are, you can only come up with a ceiling estimate—because you can't perfectly capture selectivity, and there's reason to think that actors are in fact oriented to information you don't have, real influence is unlikely to be *higher* than your estimate, but it may be lower.
3) *Know your data.*
   That means, where they came from, who made them, and how, but it also means their numerical characteristics. How many are missing? What are the patterns of missingness? How is this related to the interview that produced the data? Where is the variation? The covariation?
4) *You can be too smart.*
   Don't trust the results from techniques that are so complicated that you aren't really sure if they make sense. If you do use a complicated technique to form patterns, and then you can interpret these patterns, there's a good chance that your claim has implications that you can test in a more straightforward manner. Do.

References

Abrutyn, Seth, and Anna S. Mueller. 2014. "Are Suicidal Behaviors Contagious in Adolescence? Using Longitudinal Data to Examine Suicide Suggestion." *American Sociological Review* 79: 211–227.

Achen, Christopher H. 1975. "Mass Political Attitudes and the Survey Response." *American Political Science Review* 69: 1218–1231.

Achen, Christopher H., and W. Phillips Shively. 1995. *Cross-Level Inference*. Chicago: University of Chicago Press.

Aisenbrey, Silke, and Anette Fasang. 2017. "The Interplay of Work and Family Trajectories over the Life Course: Germany and the United States in Comparison." *American Journal of Sociology* 122: 1448–1484.

Alba, Richard, Noura E. Insolera, and Scarlett Lindemann. 2016. "Is Race Really So Fluid? Revisiting Saperstein and Penner's Empirical Claims." *American Journal of Sociology* 247–262.

Allison, Paul D. 1982. "Discrete-Time Methods for the Analysis of Event Histories." *Sociological Methodology* 13: 61–98.

Allison, Paul D. 1987. "Introducing a Disturbance into Logic and Probit Regression Models." *Sociological Methods and Research* 15: 355–374.

Allison, Paul D. 1999. "Comparing Logic and Probit Coefficients across Groups." *Sociological Methods and Research* 28: 186–208.

Almquist, Zack W., and Carter T. Butts. 2014. "Logistic Network Regression for Scalable Analysis of Networks with Joint Edge/Vertex Dynamics." *Sociological Methodology* 44: 273–321.

Anderson, David R. 2012. *Model Based Inference in the Life Sciences*. New York: Springer.

Angrist, Joshua D., and Jörn-Steffen Pischke. 2009. *Mostly Harmless Econometrics*. Princeton: Princeton University Press.

Anselin, Luc. 1988. *Spatial Econometrics: Methods and Models*. Dordrecht: Kluwer.

Anselin, Luc. 2002. "Under the Hood: Issues in the Specification and Interpretation of Spatial Regression Models." *Agricultural Economics* 27: 247–267.

Aronow, Peter M., and Cyrus Samil. 2015. "Does Regression Produce Representative Estimates of Causal Effects?" *American Journal of Political Science* 60: 250–267.

Arum, Richard, and Josipa Roksa. 2011. *Academically Adrift*. Chicago: University of Chicago Press.

Austin, S. Bryn, Najat Ziyadeh, Laurie B. Fisher, Jessica A. Kahn, Graham A. Colditz, and A. Lindsay Frazier. 2004. "Sexual Orientation and Tobacco Use in a Cohort Study of US Adolescent Girls and Boys." *Archives of Pediatric Adolescent Medicine* 158: 317–322.

Bacon, Francis. 1901 [1620]. *Novum Organum*. Edited by Joseph Devey. New York: P. F. Collier and Son.

Baker, Frank B., and Lawrence J. Hubert. 1981. "The Analysis of Social Interaction Data: A Nonparametric Technique." *Sociological Methods and Research* 9: 339–361.

Baldassarri, Delia, and Amir Goldberg. 2014. "Neither Ideologues nor Agnostics: Alternative Voters in an Age of Partisan Politics." *American Journal of Sociology* 120: 45–95.

Baraffa, Aaron J., Tyler H. McCormick, and Adrian E. Raftery. 2016. "Estimating Uncertainty in Respondent-Driven Sampling Using a Tree Bootstrap Method." *PNAS* 113: 14668–14673.

Barton, Allen H. 1968. "Bringing Society Back In: Survey Research and Macromethodology." *American Behavioral Scientist* 12: 1–9.

Baumgartner, Michael, and Alrik Thiem. 2017. "Often Trusted but Never (Properly) Tested: Evaluating Qualitative Comparative Analysis." *Sociological Methods and Research* 46: 345–357.

Bearman, Peter S., and Paolo Parigi. 2004. "Cloning Headless Frogs and Other Important Matters: Conversation Topics and Network Structure." *Social Forces* 83: 535–557.

Beck, Nathaniel, and Jonathan N. Katz. 2001. "Throwing Out the Baby with the Bath Water: A Comment on Green, Kim, and Yoon." *International Organization* 55: 487–495.

Bennett, Claudette. 2000. "Racial Categories Used in the Decennial Censuses, 1790 to the Present." *Government Information Quarterly* 17: 161–180.

Berg-Schlosser, Dirk, Gisèle de Meur, Benoît Rihoux, and Charles C. Ragin. 2008. "Qualitative Comparative Analysis (QCA) as an Approach." In *Configurational Comparative Methods: Qualitative Comparative Analysis (QCA) and Related Techniques*, edited by Benoît Rihoux and Charles C. Ragin, 1–18. Thousand Oaks, CA: Sage.

Berkson, Joseph. 1950. "Are There Two Regressions?" *Journal of the American Statistical Association* 45: 164–180.

Bernard, H. Russell, Eugene C. Johnsen, Peter D. Killworth, and Scott Robinson. 1991. "Estimating the Size of an Average Personal Network and of an Event Subpopulation: Some Empirical Results." *Social Science Research* 20: 109–121.

Berry, William D., Matt Golder, and Daniel Milton. 2012. "Improving Tests of Theories Positing Interaction." *Journal of Politics* 74: 653–671.

Besag, Julian. 1974. "Spatial Interaction and the Statistical Analysis of Lattice Systems." *Journal of the Royal Statistical Society, Series B* 36: 192–236.

Blau, Peter M. 1977. *Inequality and Heterogeneity: A Primitive Theory of Social Structure*. New York: Free Press.

Bloome, Deirdre. 2014. "Racial Inequality Trends and the Intergenerational Persistence of Income and Family Structure." *American Sociological Review* 79: 1196–1225.

Boltanski, Luc, and Laurent Thénevot. 2006 [1991]. *On Justification: Economies of Worth*. Translated by Catherine Porter. Princeton: Princeton University Press.

Bonacich, Phillip. 1987. "Power and Centrality: A Family of Measures." *American Journal of Sociology* 92: 1170–1182.

Bonikowski, Bart, and Paul DiMaggio. 2016. "Varieties of American Popular Nationalism." *American Sociological Review* 81: 949–980.

Boorman, Scott A. 2010. "A Larger Model-Building Context for Visual Models." Paper presented at European University Institute workshop on "Visualization and History," Florence, Italy, March 18, 2010.

Borsboom, Denny, Gideon J. Mellenbergh, and Jaap Van Heerden. 2003. "The Theoretical Status of Latent Variables." *Psychological Review* 110: 203–219.

Boutyline, Andrei. 2017. "Improving the Measurement of Shared Cultural Schemas with Correlational Class Analysis." *Sociological Science*. doi:10.15195/v4.a15.

Bowers, Jake. 2014. "Comment: Method Games—And a Proposal for Assessing and Learning about Methods." *Sociological Methodology* 44: 112–117.

Bradley, R. A., and M. B. Terry. 1952. "Rank Analysis of Incomplete Block Designs, I: The Method of Paired Comparisons." *Biometrika* 39: 324–345.

Brambor, Thomas, William Roberts Clark, and Matt Golder. 2006. "Understanding Interaction Models: Improving Empirical Analyses." *Political Analysis* 14: 63–82.

Brand, Jennie E., Fabian T. Pfeffer, and Sara Goldbrick-Rab. 2014. "The Community College Effect Revisited: The Importance of Attending to Heterogeneity and Complex Counterfactuals." *Sociological Science* 1: 448–464.

Braumoeller, Bear F. 2015. "Guarding against False Positives in Qualitative Comparative Analysis." *Political Analysis* 23: 471–487.

Breault, Kevin D. 1989a. "New Evidence on Religious Pluralism, Urbanism, and Religious Participation." *American Sociological Review* 54: 1048–1053.

Breault, Kevin D. 1989b. "A Reexamination of the Relationship between Religious Diversity and Religious Adherents." *American Sociological Review* 54: 1057–1059.

Breen, Richard, Seongsoo Choi, and Anders Holm. 2015. "Heterogeneous Causal Effects and Sample Selection Bias." *Sociological Science* 2: 351–369.

Breen, Richard, Anders Holm, and Kristian Bernt Karlson. 2014. "Correlations and Non-linear Probability Models." *Sociological Methods & Research* 43: 571–605.

Breen, Richard, and Kristian Bernt Karlson. 2013. "Counterfactual Causal Analysis and Nonlinear Probability Models." In *Handbook of Causal Analysis for Social Research*, edited by Stephen L. Morgan, 167–187. Dordrecht: Springer.

Breiger, Ronald L. 2000. "A Tool Kit for Practice Theory." *Poetics* 27: 91–115.

Breiger, Ronald L. 2009. "On the Duality of Cases and Variables: Correspondence Analysis (CA) and Qualitative Comparative Analysis (QCA)." In *The SAGE Handbook of Case-Based Methods*, edited by David Byrne and Charles C. Ragin, 243–259. London: Sage.

Breiger, Ronald L., and David Melamed. 2014. "The Duality of Organizations and Their Attributes: Turning Regression Modeling 'Inside Out.'" *Research in the Sociology of Organizations* 40: 261–274.

Browne, W. J., S. V. Subramanian, K. Jones, and H. Goldstein. 2005. "Variance Partitioning in Multilevel Logistic Models that Exhibit Overdispersion." *Journal of the Royal Statistical Association Series A* 168: 599–613.

Bruch, Elizabeth. 2014. "How Population Structure Shapes Neighborhood Segregation." *American Journal of Sociology* 119: 1221–1278.

Bruch, Elizabeth, and Robert Mare. 2006. "Neighborhood Chance and Neighborhood Change." *American Journal of Sociology* 112: 667–709.

Brunsdon, Chris, Stewart Fotheringham, and Martin Charlton. 1998. "Geographically Weighted Regression—Modelling [*sic*] Spatial Non-stationarity." *Statistician* 47: 431–443.

Burnham, Kenneth P., and David A. Anderson. 2004. "Multimodel Inference: Understanding AIC and BIC in Model Selection." *Sociological Methods and Research* 33: 261–304.

Cameron, A. Colin, Jonah B. Gelbach, and Douglas L. Miller. 2011. "Robust Inference with Multiway Clustering." *Journal of Business and Economic Statistics* 29: 238–249.

Carbonaro, William J. 1998. "A Little Help from My Friend's Parents: Intergenerational Closure and Educational Outcomes." *Sociology of Education* 71: 295–313.

Carbonaro, William J. 1999. "Opening the Debate: On Closure and Schooling Outcomes." *American Sociological Review* 64: 682–686.

Chamberlain, T. C. 1965 [1890]. "The Method of Multiple Working Hypotheses." *Science*, New Series 148: 754–759.

Cheadle, Jacob E., Michael Stevens, Deadric T. Williams, and Bridget J. Goosby. 2013. "The Differential Contributions of Teen Drinking Homophily to New and Existing Friendships." *Social Science Research* 42: 1297–1310.

Cheadle, Jacob E., and Deadric Williams. 2013. "The Role of Drinking in New and Existing Friendships across High School Settings." *Health* 5: 18–25.

Cheng, Simon, and Brian Powell. 2015. "Measurement, Methods, and Divergent Patterns: Reassessing the Effects of Same-Sex Parents." *Social Science Research* 52: 615–626.

Christakis, Nicholas A., and James H. Fowler. 2007. "The Spread of Obesity in a Large Social Network over 32 Years." *New England Journal of Medicine* 357: 370–379.

Christakis, Nicholas A., and James H. Fowler. 2013. "Social Contagion Theory: Examining Dynamic Social Networks and Human Behavior." *Statistics in Medicine* 32: 556–577.

Clark, Andy. 1997. *Being There: Putting Brian, Body, and World Together Again*. Cambridge, MA: MIT Press.

Coleman, James S. 1988. "Social Capital in the Creation of Human Capital." *American Journal of Sociology* 94: S95–S120.

Congdon, Peter. 2006. "A Model for Non-parametric Spatially Varying Regression Effects." *Computational Statistics and Data Analysis* 50: 422–445.

Converse, Philip E. 1964. "The Nature of Belief Systems in Mass Publics." In *Ideology and Discontent*, edited by David E. Apter, 206–261. International Yearbook of Political Research, vol. 5. New York: Free Press.

Cramer, Jan S. 2005. "Omitted Variables and Mis-speficied Disturbances in the Logit Model." Amsterdam: Tinbergen Institute Discussion Paper 05–084/4.

Crawford, Forrest W. 2016. "The Graphical Structure of Respondent-Driven Sampling." *Sociological Methodology* 46: 187–211.

Davenport, Lauren D. 2016. "The Role of Gender, Class, and Religion in Biracial Americans' Racial Labeling Decisions." *American Sociological Review* 81: 57–84.

Davis, James A., and Samuel Leinhardt. 1972. "The Structure of Positive Interpersonal Relations in Small Groups." In *Sociological Theories In Progress*, vol. 2, edited by Joseph Berger, Morris Zelditch Jr., and Bo Anderson, 218–251. Boston: Houghton Mifflin.

Dekker, David, David Krackhardt, and Tom A. B. Snijders. 2007. "Sensitivity of MRQAP Tests to Collinearity and Autocorrelation Conditions." *Psychometrika* 72: 563–581.

DeMaris, Alfred. 2002. "Explained Variance in Logistic Regression: A Monte Carlo Study of Proposed Measures." *Sociological Methods and Research* 31: 27–74.

Desmond, Matthew. 2012. "Disposable Ties and the Urban Poor." *American Journal of Sociology* 117: 1295–1335.

Diaz, Christina J., and Jeremy E. Fiel. 2016. "The Effect(s) of Teen Pregnancy: Reconciling Theory, Methods, and Findings." *Demography* 53: 85–116.

DiPrete, Thomas A., and Whitman T. Soule. 1986. "The Organization of Career Lines: Equal Employment Opportunity and Status Advancement in a Federal Bureaucracy." *American Sociological Review* 51: 295–309.

Donohue, John, and Steven Levitt. 2001. "The Impact of Legalized Abortion on Crime." *Quarterly Journal of Economics* 116: 379–420.

Duncan, Otis Dudley. 1966. "Methodological Issues in the Analysis of Social Mobility." In *Social Structure and Mobility in Economic Development*, edited by Neil J. Smelser and Seymour Martin Lipset, 51–97. Chicago: Aldine.

Duncan, Otis Dudley. 1984a. *Notes on Social Measurement, Historical and Critical.* New York: Russell Sage Foundation.

Duncan, Otis Dudley. 1984b. "Measurement and Structure." In *Surveying Subjective Phenomena*, vol. 1, edited by Charles F. Turner and Elizabeth Martin, 179–229. New York: Russell Sage Foundation.

Duncan, Otis Dudley. 1984c. "Rasch Measurement in Survey Research: Further Examples and Discussion." In *Surveying Subjective Phenomena*, vol. 2, edited by Charles F. Turner and Elizabeth Martin, 367–404. New York: Russell Sage Foundation.

Durkheim, Emile. 1951 [1897]. *Suicide*. Translated by John A. Spaulding and George Simpson. New York: Free Press.

Elhorst, J. Paul. 2010. "Applied Spatial Econometrics: Raising the Bar." *Spatial Economic Analysis* 5: 9–28.

Eliason, Scott R., and Robin Stryker. 2009. "Goodness-of-Fit Tests and Descriptive Measures in Fuzzy-Set Analysis." *Sociological Methods and Research* 38: 102–146.

Evans-Pritchard, E. E. 1974 [1956]. *Nuer Religion*. New York: Oxford University Press.

Fearon, James D., and David Laitin. 2003. "Ethnicity, Insurgency, and Civil War." *American Political Science Review* 97: 75–90.

Feld, Scott L. 1991. "Why Your Friends Have More Friends than You Do." *American Journal of Sociology* 96: 1464–1477.

Ferraro, Kenneth F., Markus H. Schafer, and Lindsay R. Wilkinson. 2016. "Childhood Disadvantage and Health Problems in Middle and Later Life: Early Imprints on Physical Health?" *American Sociological Review* 81: 107–133.

Finke, Roger, and Rodney Stark. 1988. "Religious Economies and Sacred Canopies: Religious Mobilization in American Cities, 1906." *American Sociological Review* 53: 41–49.

Finke, Roger, and Rodney Stark. 1989. "Evaluating the Evidence: Religious Economies and Sacred Canopies." *American Sociological Review* 53: 203–218.

Firebaugh, Glenn, and Jack P. Gibbs. 1985. "User's Guide to Ratio Variables." *American Sociological Review* 50: 713–722.

Fisher, Sir R. A. F. 1956. *Statistical Methods and Scientific Inference*. Edinburgh: Oliver and Boyd.

Foote, Chris, and Christopher Goetz. 2005. "Testing Economic Hypotheses with State-Level Data: A Comment on Donohue and Levitt (2001)." Federal Reserve Bank of Boston Working Paper 05–15, November 22, 2005.

Frank, Kenneth A., and Kyung-Seok Min. 2007. "Indices of Robustness for Sample Representation." *Sociological Methodology* 37: 349–392.

Frank, Kenneth A., Chandra Muller, and Anna S. Mueller. 2013. "The Embeddedness of Adolescent Friendship Nominations: The Formation of Social Capital in Emergent Network Structures." *American Journal of Sociology* 119: 216–253.

Friedkin, Noah E. 1998. *A Structural Theory of Social Influence*. New York: Cambridge University Press.

Fuguitt, Glenn V., and Stanley Lieberson. 1974. "Correlation of Ratios or Difference Scores Having Common Terms." *Sociological Methodology* 1973/1974: 128–144.

Gamoran, Adam, Sarah Barfels, and Ana Cristina Collares. 2016. "Does Racial Isolation in School Lead to Long-Term Disadvantages? Labor Market Consequences of High School Racial Composition." *American Journal of Sociology* 121: 1116–1167.

Gauchat, Gordon. 2012. "Politicization of Science in the Public Sphere: A Study of Public Trust in the United States, 1974 to 2010." *American Sociological Review* 77: 167–187.

Gelman, Andrew. 2007. "Struggles with Survey Weighting and Regression Modeling." *Statistical Science* 22: 153–164.

Gelman, Andrew, Aleks Jakulin, Maria Grazia Pittaue, and Yu-Sing Su. 2008. "A Weakly Informative Default Prior Distribution for Logistic and Other Regression Models." *Annals of Applied Statistics* 2: 1360–1383.

Gigerenzer, Gerd. 1991. "From Tools to Theories: A Heuristic of Discovery in Cognitive Psychology." *Psychological Review* 98: 254–267.

Glynn, Adam N., and Jon Wakefield. 2014. "Alleviating Ecological Bias in Poisson Models Using Optimal Subsampling: The Effects of Jim Crow on Black Illiteracy in the Robinson Data." *Sociological Methodology* 44: 159–184.

Goel, S., and M. J. Salganik. 2009. "Respondent-Driven Sampling as Markov Chain Monte Carlo." *Statistics in Medicine* 28: 2202–2229.

Goel, S., and Salganik, M.J. 2010. "Assessing Respondent-Driven Sampling." *Proceedings of the National Academy of Sciences* 107: 6743–6747.

Goldberg, Amir, Sameer B. Srivastava, V. Govind Manian, William Monroe, and Christopher Potts. 2016. "Fitting In or Standing Out? The Tradeoffs of Structural and Cultural Embeddedness." *American Sociological Review* 81: 1190–1222.

Goodman, Leo A. 1964. "Mathematical Methods for the Study of Systems of Groups." *American Journal of Sociology* 70: 170–192.

Goodman, Leo A. 1969. "How to Ransack Social Mobility Tables and Other Kinds of Cross-Classification Tables." *American Journal of Sociology* 75: 1–40.

Goodman, Leo A. 1972. "A Modified Multiple Regression Approach to the Analysis of Dichotomous Variables." *American Sociological Review* 37: 28–46.

Goodman, Leo A. 1974. "The Analysis of Systems of Qualitative Variables When Some of the Variables Are Unobservable, 1: A Modified Latent Structure Approach." *American Journal of Sociology* 79, reprinted in *Analyzing Qualitative/Categorical Data*, edited by Jay Magdison. Cambridge: Abt, 1978.

Granovetter, Mark. 1978. "Threshold Models of Collective Behavior." *American Journal of Sociology* 83: 1420–1443.

Green, Donald P., Dara Z. Strolovitch, and Janelle S. Wong. 1998. "Defended Neighborhoods, Integration, and Racially Motivated Crime." *American Journal of Sociology* 104: 372–403.

Greenland, Sander. 2000. "Principles of Multilevel Modeling." *International Journal of Epidemiology* 20: 158–167.

Greenland, Sander. 2001. "Ecologic versus Individual-Level Sources of Bias in Ecologic Estimates of Contextual Health Effects." *International Journal of Epidemiology* 30: 1343–1359.

Griffith, Daniel A. 2008. "Spatial-Filtering-Based Contributions to a Critique of Geographically Weighted Regression." *Environment and Planning A* 40: 2751–2769.

Gross, Neil. 2013. *Why Are Professors Liberal and Why Do Conservatives Care?* Cambridge, MA: Harvard University Press.

Guillot, Michel. 2005. "The Momentum of Mortality Change." *Population Studies* 59: 283–294.

Guo, Guang, Michael E. Roettger, and Tianji Cai. 2008. "The Integration of Genetic Propensities into Social-Control Models of Delinquency and Violence among Male Youths." *American Sociological Review* 73: 543–568.

Guo, Guang, Xiao-Ming Ou, Michael E. Roettger, and Jean C. Shih. 2008. "The VNTR 2 Repeat in *MAOA* and Delinquent Behavior in Adolescence and Young Adulthood: Associations and *MAOA* Promoter Activity." *European Journal of Human Genetics* 16: 626–634.

Guo, Guang, Yi Li, Tianji Cai, Hongyu Wang, and Greg Duncan. 2015. "Peer Influence, Genetic Propensity, and Binge Drinking: A Natural Experiment and a Replication." *American Journal of Sociology* 121: 914–954.

Guo, Guang, and Yuying Tong. 2006. "Age at First Sexual Intercourse, Genes, and Social Context: Evidence from Twins and the Dopamine D4 Receptor Gene." *Demography* 43: 747–769.

Haberstick, B. C., J. M. Lessem, C. J. Hopfer, A. Smolen, M. A. Ehringer, D. Timberlake, and J. K. Hewitt. 2005. "Monoamine Oxidase A (MAOA) and Antisocial Behaviors in the Presence of Childhood and Adolescent Maltreatment." *American Journal of Medical Genetics B—Neuropsychiatric Genetics* 135B: 59–64.

Habinek, Jacob, John Levi Martin, and Benjamin Zablocki. 2015. "Double-Embeddedness: Spatial and Relational Contexts of Tie Persistence and Re-formation." *Social Networks* 42: 27–41.

Hagen, Ryan, Kinga Makovi, and Peter Bearman. 2013. "The Influence of Political Dynamics on Southern Lynch Mob Formation and Lethality." *Social Forces* 92: 757–787.

Halaby, Charles N. 2004. "Panel Models in Sociological Research: Theory into Practice." *Annual Review of Sociology* 30: 507–544.

Halle, David. 1984. *America's Working Man*. Chicago: University of Chicago Press.

Hallinan, Maureen T., and Warren N. Kubitschek. 1999. "Conceptualizing and Measuring School Social Networks." *American Sociological Review* 64: 687–693.

Hällsten, Martin and Fabian T. Pfeffer. 2017. "Grand Advantage: Family Wealth and Grandchildren's Educational Achievement in Sweden." *American Sociological Review* 82: 328–360.

Handcock, Mark S. 2003. "Statistical Models for Social Networks: Inference and Degeneracy." In *Dynamic Social Network Modeling and Analysis*, edited by Ronald Breiger, Kathleen Carley, and Philippa Pattison, 229–240. Washington, DC: National Academies Press.

Harding, David J. 2003. "Counterfactual Models of Neighborhood Effects: The Effect of Neighborhood Poverty on Dropping Out and Teenage Pregnancy." *American Journal of Sociology* 109: 676–719.

Heckathorn, Douglas D. 1997. "Respondent-Driven Sampling: A New Approach to the Study of Hidden Populations." *Social Problems* 44: 174–199.

Heckathorn, D. D., Broadhead, R. S., Anthony, D. L., and Weakliem, D. L. 1998. "AIDS and Social Networks: HIV Prevention through Network Mobilization." *Sociological Focus* 32: 159–179.

Helbing, Dirk. 1995. "Improved Fluid-Dynamic Model for Vehicular Traffic." *Physical Review E* 51: 3164–3169.

Helbing, Dirk, and Péter Molnár. 1995. "Social Force Model for Pedestrian Dynamics." *Physical Review E* 51: 4282–4286.

Herndon, Thomas, Michael Ash, and Robert Pollin. 2014. "Does High Public Debt Consistently Stifle Economic Growth? A Critique of Reinhart and Rogoff." *Cambridge Journal of Economics* 38: 257–279.

Hicks, Alexander, Joya Misra, and Tang Nah Ng. 1995. "The Programmatic Emergence of the Social Security State." *American Sociological Review* 60: 329–349.

Hochschild, Jennifer L., and Brenna Marea Powell. 2008. "Racial Reorganization and the United States Census 1850–1930: Mulattoes, Half-Breeds, Mixed Parentage, Hindoos, and the Mexican Race." *Studies in American Political Development* 22: 59–96.

Hoff, Peter D. 2003. "Random Effects Models for Network Data." In *Dynamic Social Network Modeling and Analysis: Workshop Summary and Papers*, edited by Ronald Breiger, Kathleen Carley, and Philippa Pattison, 303–312. Washington, DC: National Academies Press.

Holland, Paul W., and Samuel Leinhardt. 1970. "A Method for Detecting Structure in Sociometric Data." *American Journal of Sociology* 70: 492–513.

Holland, Paul W., and Samuel Leinhardt. 1971. "Transitivity in Structural Models of Small Groups." *Comparative Group Studies* 2: 107–124.

Holland, Paul W., and Samuel Leinhardt. 1976. "Local Structure in Social Networks." In *Sociological Methodology*, edited by David Heise, 1–45. San Fransisco: Jossey-Bass.

Holland, Paul W., and Samuel Leinhardt. 1981a. "An Exponential Family of Probability Distributions for Directed Graphs." *Journal of the American Statistical Association* 76: 33–50.

Holland, Paul W., and Samuel Leinhardt. 1981b. "Rejoinder." *Journal of the American Statistical Association* 76: 62–65.

Hout, Michael, and Claude S. Fischer 2014. "Explaining Why More Americans Have No Religious Preference: Political Backlash and Generational Succession, 1987–2012." *Sociological Science* 1: 423–447.

Hubert, Lawrence. 1985. "Combinatorial Data Analysis: Association and Partial Association." *Psychometrika* 50: 449–467.

Hubert, Lawrence, and James Schultz. 1976. "Quadratic Assignment as a General Data Analysis Strategy." *British Journal of Mathematical and Statistical Psychology* 29: 190–241.

Huckfeldt, Robert. 1986. "Social Contexts, Social Networks, and Urban Neighborhoods: Environmental Constraints on Friendship Choice." *American Journal of Sociology* 89: 651–669.

Hug, Simon. 2013. "Qualitative Comparative Analysis: How Inductive Use and Measurement Error Lead to Problematic Inference." *Political Analysis* 21: 252–265.

Hyman, Herbert H. 1954. *Interviewing in Social Research*. With William J. Cobb, Jacob J. Feldman, Clyde W. Hart, and Charles Herbert Stember. Chicago: University of Chicago Press.

Imai, Kosuke, Luke Keele, Dustin Tingley, and Teppei Yamamoto. 2011. "Unpacking the Black Box of Causality: Learning about Causal Mechanisms from Experimental and Observational Studies." *American Political Science Review* 105: 765–789.

Institute on Aging, University of Wisconsin. 2004. "Documentation of Scales." Madison: University of Wisconsin. www.midus.wisc.edu/midusl/documenta tionofscales.pdf.

Jackson, Robert Max. 1989. "The Reproduction of Parenting." *American Sociological Review* 54: 215–232.

Jacobs, David, and Lindsey Myers. 2014. "Union Strength, Neoliberalism, and Inequality: Contingent Political Analyses of U.S. Income Differences since 1950." *American Sociological Review* 79: 752–774.

Johnston, Ron, Kelvyn Jones, Simon Burgess, Carol Propper, Rebecca Sarker, and Anne Bolster. 2004. "Scale, Factor Analyses, and Neighborhood Effects." *Geographical Analysis* 36: 350–368.

Kasarda, John D., and Patrick D. Nolan. 1979. "Ratio Measurements and Theoretical Inference in Social Research." *Social Forces* 58: 212–227.

Koehly, Laura M., Steven M. Goodreau, and Martina Morris. 2004. "Exponential Family Models for Sampled and Census Network Data." *Sociological Methodology* 34: 241–270.

Krackhardt, David. 1987. "QAP Partialling as a Test of Spuriousness." *Social Networks* 9: 171–186.

Krackhardt, David. 1988. "Predicting with Networks: Nonparametric Multiple Regression Analysis of Dyadic Data." *Social Networks* 10: 359–381.

Krackhardt, David. 1992. "A Caveat on the Use of the Quadratic Assignment Procedure." *Journal of Quantitative Anthropology* 3: 279–296.

Kramer, Rory, Robert DeFina, and Lance Hannon. 2016. "Racial Rigidity in the United States: Comment on Saperstein and Penner." *American Journal of Sociology* 122: 233–246.

Krogslund, Chris, Donghyun Danny Choi, and Mathias Poertner. 2014. "Fuzzy Sets on Shaky Ground: Parameter Sensitivity and Confirmation Bias in fsQCA." *Political Analysis* 24: 1–21.

Kurzman, Charles. 2014. "World Values Lost in Translation." Monkey Cage, *Washington Post*, September 2, 2014. https://www.washingtonpost.com/news/monkey-cage/wp/2014/09/02/world-values-lost-in-translation/?utm_term=.d6c124d95194

Latour, Bruno. 1986. "Visualization and Cognition: Thinking with Eyes and Hands." *Knowledge and Society* 6: 1–40.

Laumann, Edward O., John H. Gagnon, Robert T. Michael, and Stuart Michaels. 1994. *The Social Organization of Sexuality: Sexual Practices in the United States*. Chicago: University of Chicago Press.

Lazarsfeld, Paul F., and Neil W. Henry. 1968. *Latent Structure Analysis*. Boston: Houghton-Mifflin.

Leach, William. 1993. *Land of Desire*. New York: Pantheon.

Lee, Byungkyu, and Peter Bearman. 2017. "Important Matters in Political Context." *Sociological Science* 4: 1–30.

Lee, Monica, and John Levi Martin. 2015. "Coding, Counting, and Cultural Cartography." *American Journal of Cultural Sociology* 3: 1–33.

Leifeld, Philip, Skyler J. Cranmer and Bruce A. Desmarais. 2017. "Temporal Exponential Random Graph Models with btergm: Estimation and Bootstrap Confidence Intervals." *Journal of Statistical Software*.

Lesage, James P., and Manfred M. Fischer. 2008. "Spatial Growth Regressions: Model Specification, Estimation and Interpretation." *Spatial Economic Analysis* 3: 275–304.

Lesage, James P., and R. Kelley Pace. 2009. *Introduction to Spatial Econometrics*. Boca Raton: CRC.

Lesage, James P., and R. Kelley Pace. 2010. "Omitted Variable Biases of OLS and Spatial Lag Models." In *Progress in Spatial Analysis*, edited by Antonio Páez, Ron N. Buliung, Sandy Dall'erba, and Julie Gallo, 17–28. Berlin: Springer-Verlag.

Levin, Maurice, Harold G. Moulton, and Clark Warburton. 1934. *America's Capacity to Consume*. Washington, DC: Brookings Institution.

Lewis, Kevin, Marco Gonzalez, and Jason Kaufman. 2012. "Social Selection and Peer Influence in an Online Social Network." *Proceedings of the National Academy of Sciences of the United States of America* 109: 68–72.

Lewontin, R. C. 1995. "Sex, Lies, and Social Science." *New York Review of Books*, April 20, 24–29.

Lieberson, Stanley. 1985. *Making It Count*. Berkeley: University of California Press.

Lieberson, Stanley. 2001. "Review of *Fuzzy Set Social Science*, by Charles C. Ragin." *Contemporary Sociology* 30: 331–334.

Lucas, Samuel R., and Alisa Szatrowski. 2014. "Qualitative Comparative Analysis in Critical Perspective." *Sociological Methodology* 44: 1–79.

Lui, Yujia, and David B. Grusky. 2013. "The Payoff to Skill in the Third Industrial Revolution." *American Journal of Sociology* 118: 1330–1374.

Lyons, Russell. 2011. "The Spread of Evidence-Poor Medicine via Flawed Social-Network Analysis." *Statistics, Politics, and Policy* 2(1), Article 2.

MacMillan, Alexander, and Richard L. Daft. 1980. "Relationships among Ratio Variables with Common Components: Fact or Artifact?" *Social Forces* 58: 1109–1128.

Mahoney, James. 2008. "Toward a Unified Theory of Causality." *Comparative Political Studies* 41: 412–436.

Mahoney, James. 2003. "Long-Run Development and the Legacy of Colonialism in Spanish America." *American Journal of Sociology* 109: 50–106.

Martin, John Levi. 1999a. "A General Permutation-Based QAP Analysis for Dyadic Data from Multiple Groups." *Connections* 22: 50–60.

Martin, John Levi. 1999b. "Entropic Measures of Belief System Constraint." *Social Science Research* 28: 111–134.

Martin, John Levi. 2000. "The Relation of Aggregate Statistics on Belief to Culture and Cognition." *Poetics* 28: 5–20.

Martin, John Levi. 2001. "*The Authoritarian Personality*, 50 Years Later: What Lessons Are There for Political Psychology?" *Political Psychology* 22: 1–26.

Martin, John Levi. 2002. "Some Algebraic Structures for Diffusion in Social Networks." *Journal of Mathematical Sociology* 26: 123–146.

Martin, John Levi. 2006. "Jointness and Duality in Algebraic Approaches to Dichotomous Data." *Sociological Methods and Research* 35: 159–192.

Martin, John Levi. 2009. *Social Structures*. Princeton: Princeton University Press.

Martin, John Levi. 2011. *The Explanation of Social Action*. New York: Oxford.

Martin, John Levi. 2014. "Spatial Processes and Galois/Concept Lattices." *Quality and Quantity* 48: 961–981.

Martin, John Levi. 2015. *Thinking Through Theory*. New York: Norton.

Martin, John Levi. 2016a. "The Dimensionality of Discrete Factor Analyses." *Quality and Quantity* 50: 2451–2467.

Martin, John Levi. 2016b. "Comment on Guo, Li, Wang, Cai and Duncan." *SocArXiv* osf.io/99jsp/.

Martin, John Levi. 2017. *Thinking Through Methods*. Chicago: University of Chicago Press.

Martin, John Levi, and James Wiley. 2000. "Algebraic Representations of Beliefs and Attitudes, 2: Microbelief Models for Dichotomous Belief Data." *Sociological Methodology* 30: 123–164.

Martin, John Levi, Adam Slez, and Chad Borkenhagen. 2016. "Some Provisional Techniques for Quantifying the Degree of Field Effect in Social Data." *Socius* 2: 1–18.

Martin, John Levi, James Wiley, and Dennis Osborn. 2003. "Social Networks and

Unobserved Heterogeneity for Risk of AIDS." *Population Research and Policy Review* 22: 65–90.

Marx, Axel, and Adrian Dusa. 2011. "Crisp-Set Qualitative Comparative Analysis (csQCA), Contradictions and Consistency Benchmarks for Model Specification." *Methodological Innovations Online* 6: 103–148.

Massey, Douglas S., Jorge Durand, and Karen A. Pren. 2016. "Why Border Enforcement Backfired." *American Journal of Sociology* 121: 1557–1600.

McPherson, Miller, Lynn Smith-Lovin, and Matthew E. Brashears. 2006. "Social Isolation in America: Changes in Core Discussion Networks over Two Decades." *American Sociological Review* 71: 353–375.

Micceri, Theodore, Pradnya Parasher, Gordon W. Waugh, and Charlene Herreid. 2009. "The Gulliver Effect: The Impact of Error in an Elephantine Subpopulation on Estimates for Lilliputian Subpopulations." Paper presented at the Florida Association for Institutional Research annual conference, Cocoa Beach, Florida, February 25–27.

Monk, Ellis P. 2015. "The Cost of Color: Skin Color, Discrimination, and Health among African-Americans." *American Journal of Sociology* 121: 396–444.

Montgomery, James D. 2000. "The Self as a Fuzzy Set of Roles, Role Theory as a Fuzzy System." *Sociological Methodology* 30: 261–314.

Mood, Carina. 2010. "Logistic Regression: Why We Cannot Do What We Think We Can Do, and What We Can Do about It." *European Sociological Review* 26: 67–82.

Morgan, Stephen, and Aage Sorensen. 1999. "Parental Networks, Social Closure, and Mathematics Learning: A Test of Coleman's Social Capital Explanation of School Effects." *American Sociological Review* 64: 661–681.

Morgan, Stephen L., and Jennifer J. Todd. 2009. "Intergenerational Closure and Academic Achievement in High School: A New Evaluation of Coleman's Conjecture." *Sociology of Education* 82: 267–286.

Morgan, Stephen L., and Christopher Winship. 2007. *Counterfactuals and Causal Inference*. Cambridge: Cambridge University Press.

Mouw, Ted, and Ashton M. Verdery. 2012. "Network Sampling with Memory: A Proposal for More Efficient Sampling from Social Networks." *Sociological Methodology* 42: 206–256.

Munsch, Christin L. 2015. "Her Support, His Support: Money, Masculinity, and Marital Infidelity." *American Sociological Review* 80: 469–495.

Nederhof, Esther, Jay Belsky, Johan Ormel, and Albertine J. Oldehinkel. 2012. "Effects of Divorce on Dutch Boys' and Girls' Externalizing Behavior in Gene-Environment Perspective." *Development and Psychopathology* 24: 929–939.

Nee, Sean, Nick Colegrave, Stuart A. West, and Alan Grafen. 2005. "The Illusion of Invariant Quantities in Life Histories." *Science* 309: 1236–1239.

Newman, Mark E. J., Steven H. Strogatz, and Duncan J. Watts. 2001. "Random Graphs with Arbitrary Degree Distributions and Their Applications." *Physical Review E* 64: 026118.

Noel, Hans, and Brendhan Nyhan. 2011. "The 'Unfriending' Problems: The Consequences of Homophily in Friendship Retention for Causal Estimates of Social Influence." *Social Networks* 33: 211–218.

Olson, Daniel V. A. 1998. "Religious Pluralism in Contemporary U.S. Counties." *American Sociological Review* 63: 759–761.

Opportunity. 1925. Editorial: "The Vanishing Mulatto." *Opportunity* 3: 291.

Orsini, Chiara, et al. (11 coauthors). 2015. "Quantifying Randomness in Real Networks." *Nature Communications* 6: 8627. doi:10.1038/ncomms9627.

Páez, Antonio, Steven Farber, and David Wheeler. 2011. "A Simulation-Based Study of Geographically Weighted Regression as a Method for Investigating Spatially Varying Relationships." *Environment and Planning A* 43: 2992–3010.

Paik, Anthony, and Kenneth Sanchagrin. 2013. "Social Isolation in America: An Artifact." *American Sociological Review* 78: 339–360.

Pattison, Philippa, and Tom Snijders. 2013. "Modeling Social Networks: Next Steps." In *Exponential Random Graph Models for Social Networks*, edited by Dean Lusher, Johan Koskinen, and Garry Robins, 287–301. Cambridge: Cambridge University Press.

Pattison, Philippa, and Stanley Wasserman. 1999. "Logit Models and Logistic Regressions for Social Networks, 2: Multivariate Relations." *British Journal of Mathematical and Statistical Psychology* 52: 169–193.

Pearson, Karl. 1978 [1921–1933]. *The History of Statistics in the 17th and 18th Centuries against the Changing Background of Intellectual, Scientific and Religious Thought*, edited by E. S. Pearson. London: C. Griffin.

Peirce, Charles S. 1985 [1865–1866]. "The Logic Notebook." In *Writings of Charles S. Peirce*, vol. 1, edited by Nathan Houser et al., 337–350. Bloomington: Indiana University Press.

Peirce, Charles S. 1985 [1866]. "The Logic of Science; Or, Induction and Hypothesis, Lecture III, Lowell Lectures of 1866." In *Writings of Charles S. Peirce*, vol. 2, edited by Nathan Houser et al., 357–504. Bloomington: Indiana University Press.

Peirce, Charles S. 1986 [1870]. "On the Theory of Errors of Observations, Coast Survey Report 1870, 200–224." In *Writings of Charles S. Peirce*, vol. 3, edited by Nathan Houser et al., 114–160. Bloomington: Indiana University Press.

Peterson, Richard R. 1996. "A Re-evaluation of the Economic Consequences of Divorce." *American Sociological Review* 61: 528–536.

Pinderhughes, Howard. 1997. *Race in the Hood: Conflict and Violence among Urban Youth*. Minneapolis: University of Minnesota Press.

Popper, Karl. 1959 [1934]. *The Logic of Scientific Discovery*. New York: Basic.

Qian, Zhenchao. 1998. "Changes in Assortative Mating: The Impact of Age and Education, 1970–1990." *Demography* 35: 279–292.

Raftery, Adrian. 1985. "A Note on Bayes Factors for Log-Linear Contingency Table Models with Vague Prior Information." *Journal of the Royal Statistical Society, Series B* 48: 249–250.

Ragin, Charles C. 2008. *Redesigning Social Inquiry: Fuzzy Sets and Beyond*. Chicago: University of Chicago Press.

Rasch, Georg. 1960. *Probabilistic Models for Some Intelligence and Attainment Tests*. Copenhagen: Danish Institute of Educational Research.

Regnerus, Mark. 2012. "How Different Are the Adult Children of Parents Who Have Same-Sex Relationships? Findings from the New Family Structures Study." *Social Science Research* 41: 752–770.

Regnerus, Mark, and Christian Smith. 1998. "Selective Deprivatization among American Religious Traditions: The Reversal of the Great Reversal." *Social Forces* 76: 1347–1372.

Reinhart, Carmen M., and Kenneth S. Rogoff. 2010. "Growth in a Time of Debt." *American Economic Review* 100: 573–578.

Roberts, J. M., Jr., and D. D. Brewer. 2001. "Measures and Tests of Heaping in Discrete Quantitative Distributions." *Journal of Applied Statistics* 28: 887–896.

Rogin, Michael Paul. 1969. *The Intellectuals and McCarthy: The Radical Specter*. Cambridge, MA: MIT Press.

Rosa, Eugene A., and Thomas Dietz. 2012. "Human Drivers of National Greenhouse-Gas Emissions." *Nature Climate Change* 2: 581–586.

Rosenbaum, Paul R., and Donald B. Rubin. 1983. "The Central Role of the Propensity Score in Observational Studies for Causal Effects." *Biometrika* 70: 41–55.

Salganik, Matthew J. 2006. "Variance Estimation, Design Effects, and Sample Size Calculation for Respondent-Driven Sampling." *Journal of Urban Health* 83: 98–111.

Salganik, Matthew J. 2012. "Commentary: Respondent-Driven Sampling in the Real World." *Epidemiology* 23: 148–150.

Salganik, Matthew, and Douglas Heckathorn. 2004. "Sampling and Estimation in Hidden Populations Using Respondent-Driven Sampling." *Sociological Methodology* 34: 193–239.

Saperstein, Aliya, and Andrew Penner. 2012. "Racial Fluidity and Inequality in the United States." *American Journal of Sociology* 118: 676–727.

Savin-Williams, Ritch C., and Kara Joyner. 2014. "The Dubious Assessment of Gay, Lesbian, and Bisexual Adolescents of Add Health." *Archives of Sexual Behavior* 43: 413–422.

Schelling, Thomas C. 1978. *Micromotives and Macrobehavior*. New York: Norton.

Schneider, Daniel. 2012. "Gender Deviance and Household Work: The Role of Occupation." *American Journal of Sociology* 117: 1029–1072.

Seawright, Jason. 2014. "Comment: Limited Diversity and the Unreliability of QCA." *Sociological Methodology* 44: 118–121.

Sharkey, Patrick, and Felix Elwert. 2011. "The Legacy of Disadvantage: Multigenerational Neighborhood Effects on Cognitive Ability." *American Journal of Sociology* 116: 1934–1981.

Skocpol, Theda. 1984. "Emerging Agendas and Recurring Strategies in Historical Sociology." In *Vision and Method in Historical Sociology*, edited by Theda Skocpol, 356–391. Cambridge: Cambridge University Press.

Skvoretz, John. 2013. "Diversity, Integration, and Social Ties: Attraction versus Repulsion as Drivers of Intra- and Intergroup Relations." *American Journal of Sociology* 119: 486–517.

Slez, Adam, Heather A. O'Connell, and Katherine J. Curtis. 2017. "A Note on the Identification of Constant Geographies." *Sociological Methods and Research* 46: 288–299.

Smith, Anna, Catherine A. Calder, and Christopher R. Browning. 2016. "Empirical Reference Distributions for Networks of Different Sizes." *Social Networks* 47: 24–37.

Smith, Jeffrey A., and James Moody. 2013. "Structural Effects of Network Sampling Coverage, I: Nodes Missing at Random." *Social Networks* 35: 652–668.

Smith, Sandra S. 2005. "'Don't Put My Name on It': (Dis)Trust and Job-Finding Assistance among the Black Urban Poor." *American Journal of Sociology* 111: 1–57.

Smith, Sanne, Frank van Tubergen, Ineke Maas, and Daniel A. McFarland. 2016. "Ethnic Composition and Friendship Segregation: Differential Effects for Adolescent Natives and Immigrants." *American Journal of Sociology* 121: 1223–1272.

Snijders, Tom A. B. 2002. "Markov Chain Monte Carlo Estimation of Exponential Random Graph Models." *Journal of Social Structure* 3: 2.

Snijders, Tom A. B., Gerhard G. van de Bunt, and Christian E. G. Steglich. 2010. "Introduction to Stochastic Actor-Based Models for Network Dynamics." *Social Networks* 32: 44–60.

Song, Xi, Cameron D. Campbell, and James Z. Lee. 2015. "Ancestry Matters: Patrilineage Growth and Extinction." *American Sociological Review* 80: 574–602.

Sorokin, Pitrim. 1927. *Social Mobility*. New York: Harper and Brothers.

Stigler, Stephen M. 1986. *The History of Statistics: The Measurement of Uncertainty before 1900*. Cambridge, MA: Harvard University Press.

Stinchcombe, Arthur. 1983–1984. "Linearity in Loglinear Analysis." *Sociological Methodology* 1983–4: 104–125.

Strully, Kate. 2014. "Racially and Ethnically Diverse Schools and Adolescent Romantic Relationships." *American Journal of Sociology* 120: 750–797.

Sullins, D. Paul. 1999. "Catholic/Protestant Trends on Abortion: Convergences and Polarity." *Journal for the Scientific Study of Religion* 38: 354–369.

Sullins, D. Paul. 2017. "Sample Errors Call into Question Conclusions Regarding Same-Sex Married Parents." *Demography* doi:10.1007/s13524-016-0501-y.

Swait, Joffre. 2001. "Choice Set Generation within the Generalized Extreme Value Family of Discrete Choice Models." *Transportation Research Part B* 35: 643–666.

Thiem, Alrik, and Michael Baumgartner. 2016. "Modeling Causal Irrelevance in Evaluations of Configural Comparative Methods." *Sociological Methodology* 46: 345–357.

Tippins, Nancy T., and Margaret Hilton. 2010. A Database for a Changing Economy: Review of the Occupational Information Network (O*NET). Washington, DC: National Academies Press.

Tobler, Waldo. 1970. "A Computer Movie Simulating Urban Growth in the Detroit Region." *Economic Geography* 46: 234–240.

Turney, Kristin, and Christopher Wildeman. 2015. "Detrimental for Some? The Heterogeneous Effects of Maternal Incarceration on Child Wellbeing." *Criminology & Public Policy* 14: 125–156.

Vaisey, Stephen. 2007. "Structure, Culture, and Community: The Search for Belonging in 50 Urban Communes." *American Sociological Review* 72: 851–873 .

Vaisey, Stephen. 2014. "Comment: QCA Works—When Used with Care." *Sociological Methodology* 44: 108–112.

Vaisey, Stephen, and Andrew Miles. 2014. "What You Can—and Can't—Do with Three-Wave Panel Data." *Sociological Methods and Research* 46: 44–67.

Van de Rijt, Arnout, David Siegel, and Michael Macy. 2009. "Neighborhood

Chance and Neighborhood Change: A Comment on Bruch and Mare." *American Journal of Sociology* 114: 1166–1180.

Voas, David, Daniel V. A. Olson, and Alasdair Crockett. 2002. "Religious Pluralism and Participation: Why Previous Research Is Wrong." *American Sociological Review* 67: 212–230.

Voas, David, and Mark Chaves. 2016. "Is the United States a Counterexample to the Secularization Thesis?" *American Journal of Sociology* 121: 1517–1556.

Wang, Cheng, Carter T. Butts, John R. Hipp, Rupa Jose, and Cynthia M. Lakon. 2016. "Multiple Imputation for Missing Edge Data: A Predictive Evaluation Method with Application to Add Health." *Social Networks* 45: 89–98.

Wang, Dan J., Xiolin Shi, Daniel A. McFarland, and Jure Leskovec. 2012. "Measurement Error in Network Data: A Re-classification." *Social Networks* 34: 396–409.

Warren, John Robert, Liying Luo, Andrew Halpern-Manners, James M. Raymo, and Alberto Palloni. 2015. "Do Different Methods for Modeling Age-Graded Trajectories Yield Consistent and Valid Results?" *American Journal of Sociology* 120: 1809–1856.

Washington, Scott Leon. 2011. *Hypodescent: A History of the Crystallization of the One-Drop Rule in the United States, 1880–1940*. Unpublished PhD diss., Princeton University.

Wasserman, Stanley, Carolyn J. Anderson, and Bradley Crouch. 1999. "A *p** Primer: Logit Models for Social Networks." *Social Networks* 21: 37–66.

Wasserman, Stanley, and Philippa Pattison. 1996. "Logit Models and Logistic Regressions for Social Networks, 1: An Introduction to Markov Graphs and *p**." *Psychometrika* 61: 401–425.

Wasserstein, Ronald L., and Nicole A. Lazar. 2016. "The ASA's Statement on P-values: Context, Process and Purpose." *American Statistician* doi:10.1080/00031305.2016.1154108.

Weitzman, Lenore J. 1985. *The Divorce Revolution: The Unexpected Social and Economic Consequences for Women and Children in America*. New York: Free Press.

Weitzman, Lenore J. 1996. "The Economic Consequences of Divorce Are Still Unequal: Comment on Peterson." *American Sociological Review* 61: 537–538.

Western, Bruce. 1996. "Vague Theory and Model Uncertainty in Macrosociology." *Sociological Methodology* 26: 165–192.

Wheeler, David, and Michael Tiefelsdorf. 2005. "Multicollinearity and Correlation among Local Regression Coefficients in Geographically Weighted Regression." *Journal of Geographic Systems* 7: 161–187.

White, Harrison. 1962. "Chance Models of Systems of Casual Groups." *Sociometry* 25: 153–172.

Whyte, William Foote. 1981 [1943]. *Street Corner Society*. 3rd ed. Chicago: University of Chicago Press.

Wierzbicka, Anna. 1996. *Semantics : Primes and Universals*. Oxford: Oxford University Press.

Wiley, David E., and James A. Wiley. 1970. "The Estimation of Measurement Error in Panel Data." *American Sociological Review* 35: 112–17.

Wiley, James A., and John Levi Martin. 1999. "Algebraic Representations of Beliefs and Attitudes: Partial Order Models for Item Responses." *Sociological Methodology* 29: 113–146.

Wiley, James A., John Levi Martin, Stephen Herschkorn, and Jason Bond. 2015. "A New Extension of the Binomial Error Model for Responses to Items of Varying Difficulty in Educational Testing and Attitude Surveys." *PLOS ONE* 10(11): e0141981. doi:10.1371/journal.pone.0141981.

Winship, C., and R. D. Mare. 1983. "Structural Equations and Path Analysis for Discrete Data." *American Journal of Sociology* 88: 54–110.

Woodwell, Douglas. 2014. *Research Foundations*. Thousand Oaks, CA: Sage.

Yeung, King-To, and John Levi Martin. 2003. "The Looking Glass Self: An Empirical Test and Elaboration." *Social Forces* 81: 843–879.

York, Richard, and Ryan Light. 2017. "Directional Asymmetry in Sociological Analyses." *Socius* 3:1-13. doi:10.1177/2378023117697180.

Zerubavel, Eviatar. 1997. *Social Mindscapes: An Invitation to Cognitive Sociology*. Cambridge, MA: Harvard University Press.

Zipf, George K. 1949. *Human Behavior and the Principle of Least Effort*. Cambridge, MA: Addison-Wesley.

Index